Advances in Intelligent Systems and Computing

Volume 1235

The series "Advances in Intelligent Systems and Computing" contains publications on theory, applications, and design methods of Intelligent Systems and Intelligent Computing. Virtually all disciplines such as engineering, natural sciences, computer and information science, ICT, economics, business, e-commerce, environment, healthcare, life science are covered. The list of topics spans all the areas of modern intelligent systems and computing such as: computational intelligence, soft computing including neural networks, fuzzy systems, evolutionary computing and the fusion of these paradigms, social intelligence, ambient intelligence, computational neuroscience, artificial life, virtual worlds and society, cognitive science and systems, Perception and Vision, DNA and immune based systems, self-organizing and adaptive systems, e-Learning and teaching, human-centered and human-centric computing, recommender systems, intelligent control, robotics and mechatronics including human-machine teaming, knowledge-based paradigms, learning paradigms, machine ethics, intelligent data analysis, knowledge management, intelligent agents, intelligent decision making and support, intelligent network security, trust management, interactive entertainment, Web intelligence and multimedia.

The publications within "Advances in Intelligent Systems and Computing" are primarily proceedings of important conferences, symposia and congresses. They cover significant recent developments in the field, both of a foundational and applicable character. An important characteristic feature of the series is the short publication time and world-wide distribution. This permits a rapid and broad dissemination of research results.

Indexed by DBLP, INSPEC, WTI Frankfurt eG, zbMATH, Japanese Science and Technology Agency (JST).

All books published in the series are submitted for consideration in Web of Science.

More information about this series at http://www.springer.com/series/11156

Ashish Kumar Luhach · Ramesh Chandra Poonia ·
Xiao-Zhi Gao · Dharm Singh Jat
Editors

Second International Conference on Sustainable Technologies for Computational Intelligence

Proceedings of ICTSCI 2021

Springer

Editors
Ashish Kumar Luhach
The PNG University of Technology
Lae, Papua New Guinea

Xiao-Zhi Gao
University of Eastern Finland
Kuopio, Finland

Ramesh Chandra Poonia
CHRIST (Deemed to be University)
Bangalore, India

Dharm Singh Jat
Namibia University of Science
and Technology
Windhoek, Namibia

ISSN 2194-5357 ISSN 2194-5365 (electronic)
Advances in Intelligent Systems and Computing
ISBN 978-981-16-4640-9 ISBN 978-981-16-4641-6 (eBook)
https://doi.org/10.1007/978-981-16-4641-6

This Springer imprint is published by the registered company Springer Nature Singapore Pte Ltd.
The registered company address is: 152 Beach Road, #21-01/04 Gateway East, Singapore 189721, Singapore

Preface

Second International Conference on Sustainable Technologies for Computational Intelligence (ICTSCI-2021) targeted state-of-the-art as well as emerging topics pertaining to Sustainable Technologies for Computational Intelligence and their implementation for engineering applications. The objective of this international conference is to provide opportunities for researchers, academicians, industry persons, and students to interact and exchange ideas, experience, and expertise in the current trend and strategies for information and communication technologies. Besides this, participants will also be enlightened about vast avenues and current and emerging technological developments in the field of advanced informatics, and their applications will be thoroughly explored and discussed.

Second International Conference on Sustainable Technologies for Computational Intelligence (ICTSCI-2021) was held at Graphic Era Deemed to be University, Dehradun, India, in association with Graphic Era Hill University, Dehradun, India; Namibia University of Science and Technology, Namibia; and MRK Institute of Engineering and Technology, Haryana, India, on May 22–23, 2021.

We are highly thankful to our valuable authors for their contribution and our technical program committee for their immense support and motivation for making the first edition of ICTSCI—2021 a success. We are also grateful to our keynote speakers for sharing their precious work and enlightening the delegates of the conference. We express our sincere gratitude to our publication partner, Springer AISC Series, for believing in us.

Lae, Papua New Guinea
Bangalore, India
Kuopio, Finland
Windhoek, Namibia
June 2021

Ashish Kumar Luhach
Ramesh Chandra Poonia
Xiao-Zhi Gao
Dharm Singh Jat

Contents

About the Editors

Dr. Ashish Kumar Luhach received PhD degree in department of computer science from Banasthali University, India. Dr. Luhach is working as Senior lecturer at The Papua New Guinea University of Technology, Papua New Guinea. He has more than a decade of teaching and research experience. Dr. Luhach also worked with various reputed universities and also holds administrate experience as well. Dr. Luhach has published more 100 research paper in reputed journals and conferences, which are indexed in various international databases. He is also edited various special issues in reputed journals and he is Editor/Conference Co-chair for various conferences. Dr. Luhach is also editorial board members of various reputed journals. He is member of IEEE, CSI, ACM and IACSIT.

Dr. Ramesh Chandra Poonia is working as associate professor at CHRIST (Deemed to be University) Bangalore, India; Postdoctoral Fellow at Cyber-Physical Systems Laboratory (CPS Lab), Department of ICT and Natural Sciences, Norwegian University of Science and Technology (NTNU), Alesund, Norway. He is having a rich experience of research and conducted a lot of special issues in various reputed journals having international repute. He conducted various conferences in India as conference chair. He has published more than 100 research paper in various journals and conferences. He is an active member of IEEE, CSI and ACM.

Prof. Xiao-Zhi Gao is working at the University of Eastern Finland, Finland. He has published than 350 publications in reputed journals and conferences. He also serves in the editorial boards of many journals published by Springer and Elsevier.

Prof. Dharm Singh Jat is a Professor of Computer Science at Namibia University of Science and Technology (NUST). He has guided about 8 PhD and 24 master research scholars. He is the author of more than 146 peer-reviewed articles and the author or editor of more than 16 books. His interests span the areas of multimedia communications, wireless technologies, mobile communication systems, edge, roof computing, Software Defined Networks, Network security, and the Internet of things. He has given several Guest Lecturer/Invited talks at various prestigious conferences. He has been the recipient of more than 19 prestigious awards. He is a Fellow of The

Institution of Engineers (I), Computer Society of India and Chartered Engineer (I), Senior Member IEEE, Distinguished ACM Speaker, and IEEE CS DVP Speaker.

Alleviating the Issues of Recommendation System Through Deep Learning Techniques

Bhupesh Rawat, Ankur Singh Bist, Purushottam Das, Jitendra Kumar Samriya, Suresh Chandra Wariyal, and Nitin Pandey

Abstract Social media and other online sources generate abundance of data which leads to the problem of information overload. Recommender systems have emerged as a solution to deal with this problem which suggest items to users based on their information need. Despite the success of these systems, they also suffer from few issues such as cold start problem, sparsity, scalability, and overspecialization, among others. Although several attempts were made in the past to deal with these issues, several current recommender systems continue to face these issues. Deep learning algorithms can also be used to deal with recommendation system issues. *In this paper, we discuss the issues of recommendation systems and how their impact can be alleviated through deep learning algorithms.*

Keywords Convolutional neural network (CNN) · Recurrent neural networks · Stacked auto encoder · Natural language processing (NLP)

1 Introduction

A plethora of online systems including social networking sites, e-learning systems have emerged in the recent years particularly in the last six months due to covid-19 pandemic which have generated huge amount of data for researchers to explore [1]. In particular many e-learning systems providing online education to learners have come up. However, these systems make it difficult for learners to choose a learning resource based on their preferences [2]. To mitigate this issue, recommendation system has developed which suggest learning items to learners based on their profile. The profile is varied from learner to learner, and it consists of user preferences such as what item

B. Rawat · A. S. Bist (✉) · P. Das
Graphic Era Hill University, Bhimtal, Uttarakhand 263136, India

J. K. Samriya
Dr B R Ambedkar National Institute of Technology, Jalandhar 144 011, India

S. C. Wariyal · N. Pandey
Amrapali Institute of Technology and Sciences, Haldwani, Nainital, Uttarakhand, India

A. K. Luhach et al. (eds.), *Second International Conference on Sustainable Technologies for Computational Intelligence*, Advances in Intelligent Systems and Computing 1235,
https://doi.org/10.1007/978-981-16-4641-6_1

was like by learner in the past, what is the level of knowledge of a learner, the learning style among others [3]. In this direction machine learning algorithms have also contributed significantly. Although recommender systems have been employed in several domains such as e-commerce, education among others our focus in this paper is on improving the performance of e-learning recommendation system using deep learning algorithms. Recommendation system offers a set of items to users based on their profile for consideration. For example, YouTube presents list of videos to users based on the types of videos a user has already watched in the past. Another example, Facebook suggests list of friends to users based on their social connections. For example, if a user clicks on a book on Amazon, then the site also shows other pair of books which is frequently bought by other customers. According to a report, 35% of Amazon.com profit comes from recommendation algorithms. Despite the huge success of recommendation system in all domains they are suffering from several issues including cold start, representing usage data among others [4]. This paper first discusses these issues and how they negatively impact the performance of a recommendation algorithm then we elaborate on how deep learning algorithms can help resolve these issues and, in the end, we provide an insight in the future research direction.

2 Research Question

2.1 RQ1 What are the critical issues to be addressed for leveraging the benefits of a recommender system?

2.2 RQ2 How challenges of recommendation system can be better deal with new techniques?

3 Research Methodology

Table 1 list various links of reputed publisher whose papers have been studied to conduct this comprehensive survey. The details of the sources are given in Table 1.

Table 1 Data sources

Sources	URL
ACM	https://www.elsevier.com
SPRINGER	http://link.springer.com/
IEEE	http://ieeexplore.ieee.org/
Elsevier	http://dl.acm.org/

4 Recommendation System

Recommendation system (RS) started way back in the year 2005 when the first paper was published on collaborative filtering [5]. They suggest item to user from vast repository based on their profile. They are used in e-commerce, e-learning and e-governance, among others.

The rest of the article is organized as follows: Firstly, we provide a brief overview of recommendation system which is followed by a presenting various issue in recommendation system. We also present deep learning algorithms. Then we discuss how the impact of these issues can be alleviated using deep learning algorithms. Then we provide future research direction for developing intelligent recommender system followed by conclusion section.

5 Issues in Recommendation System

5.1 Cold Start Problem

It refers to sparse ratings of learners or items for making recommendations which is commonly known as cold start problem and is of three types. The new user problem occurs when a user has not consumed the item from the repository of items. The new item problem means an item is just added to the database and not ratings have been given on the items by user. One of the challenges of recommendation system is to suggest the most relevant item to new user or to recommend a new item to user. From the study of existing **literature,** it is found that several solutions were proposed to deal with this issue [6]. Moreover, missing ratings which are caused by cold start problem can be filled with their default values such as middle values of average user or item.Furthermore, the content filtering approach [6] and autonomous agents have also been used to deal with missing ratings.

5.2 Sparsity

It means insufficient ratings in user item rating matrix [7]. Due to sparsity the recommendation algorithms are unable to find similar user based on their profile. Collaborative algorithm works by estimating the similarity between different users based on their preferences which performs poorly due to the sparsity issue. Furthermore, a rule-based recommender system also needs rating matrix to make list of suggested items which also does not work due to the same issue. Among other factors, the main reason of sparsity is that many users do not rate several items hence make it difficult to find similar items or users.

5.3 *Overspecialization Problem*

A recommendation algorithm is heavily dependent on ratings in overspecialization [8]. For example, if a user prefers action movie, then the algorithm will not recommend the movie which is a mix of action and comedy. In other words, a user may not get recommendation other than his/her profile. Genetic algorithm have come up to deal with this problem though there is a need to discover new techniques which takes data apart from user profile.

6 Deep Learning Techniques

It is a subfield of machine learning which refers to neural network which mimics human brain. It has been hugely successful in performing machine learning tasks particularly supervised and unsupervised tasks [9]. In this section we present several deep learning techniques:

6.1 *Convolutional Neural Network (CNN)*

CNN is a deep learning algorithm which takes input data such as image and assigns weight to various objects for classification. CNN requires relatively less pre-processing as compared to other classification algorithms. The design of the CNN is similar to neurons in human brain, and it was inspired by the organization of the visual cortex [10].

6.2 *Autoencoder (AE)*

Autoencoder is a kind of feedforward NN where input and output are same. They compress the input into lower dimensional code and subsequently reconstruct the output image from this compact representation. This compact code is also called latent space representation. An autoencoder has three components namely encoder, code and decoder (Fig. 1).

6.3 *Multilayer Perceptron*

MP [11] is the feed forward neural network with multiple hidden layers between input and output layer. Each layer has large number of perceptron so this whole

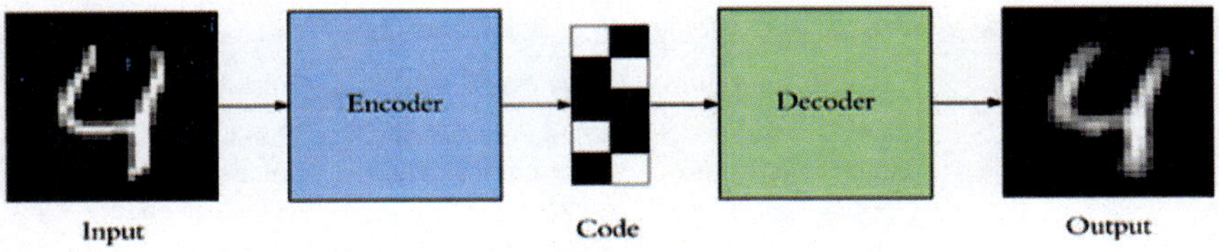

Fig. 1 Working of autoencoder

system can become quite complex. MLP utilizes a supervised learning algorithm for training. It also utilizes nonlinear activation function that differentiate MLP from linear perceptron. A perceptron refers to neural network unit which is used to detect features in the input data.

6.4 Recurrent Neural Network

RNN is a type of ANN utilizing sequential data or temporal data [12]. In the traditional neural network, the input and output are independent. however, RNN has loops and memories to remember previous computations or when the next word of a sentence needs to be predicted using the previous words. The unique feature of recurrent neural network is its hidden state which remember some information about a state. Some of the variation of RNN are long short-term memory, gated recurrent unit which are frequently used to deal with vanishing gradient problem.

6.5 Adversial Networks

AN [13] are algorithm architecture that consists of two neural networks using once against the other hence the name Adversial for generating new, synthetic instances of data that can pass for real data. Such networks find their application in the field of image generation, video generation and voice generation.

6.6 Attentional Models

Attentional models [14] are input processing techniques that allows a network to focus on a specific aspect of an image. This method is like how a human brain works when it comes across a new complex problem and divides it into simpler task, solve each task and finally delivers the output by integrating these smaller tasks. This is also called divide and conquer strategy technique. The goal is not to translate the

Table 2 Deep learning techniques for solving cold start issues

Ideas	Typical algorithms	Advantages	Disadvantages
item-to-item deep learning matcher	Doc2vecwith contextual features [16]	Allows document embedding and matching	Sub linear relationships are not clearly defined
Proposes a model based on latent representation of input data	Dropout net algorithms [17]	Able to handle both cold and warm start problem	Unable to include preference information directly into the model
The proposed solution works by identifying a community of similar users and then applying deep learning model	Matrix factorization algorithm [18]	Allows grouping users into communities	Unable to use information from cross communities
The proposed framework is based on attribute graph neural network	Attribute graph neural network [19]	Aggregate various information in neighbourhood which improve model capacity	Relies on the interaction graph
The proposed model consists of GCN encoder and decoders combined with supervised learning	GCN based algorithm [20]	The proposed framework is generic and can be used in other applications	Performance is bound to the number of interactions

sentence word for word, but rather pay attention to the general, "high level" overall sentiment. AM is widely used in the domain of image processing and NLP.

6.7 Deep Reinforcement Learning (DRL)

DRL [15] works on trial-and-error paradigm. A framework based on deep reinforcement learning has following components: agents, environments, state, actions and rewards. e combination between deep neural networks and reinforcement learning (Tables 2 and 3).

7 Future Research Direction

Although many deep learning algorithms/techniques have been used for improving the results of recommendation system, a lot needs to be explored. Deep learning has lot of potential for recommender systems. Deep learning can advance the recommendation system field in many ways. First, user and item representation can be better modelled through various data sources and the models using these representations.

Table 3 Deep learning approaches for solving recommendation system issues

Ideas	Approach description	Results
DNNRec: the approach uses hybrid model [21] Uses multiple criteria collaborative approach [22]	Side information and deep network combination is used Model uses user's item features and deep learning framework to meet out the goal	DNNRec provides state-of-the-art techniques on the whole and in cold start case Proposed process gives state of art results using deep learning on multi-criteria in recommendation systems
MoodleREC: The approach utilizes moodle[23]	Process provides learning Objects ranking and identifying using system recommendations	Proposed process gives state of art results by creating hybrid recommender system
A hybrid recommendation system with many-objective evolutionary algorithm [24]	Process includes mixing several recommendation techniques to advance recommendation performance	Proposed techniques give clustering techniques are used to decrease recommended consumption
Exploiting implicit social relationships [25]	Model is proposed using three sources 1. a user-item matrix 2. explicit 3. implicit relationships	Proposed pipeline is outperforming while compared with eight standard procedures like singular value decomposition

Dynamic behavior modelling is another approach which can enhance the modelling of user and item. Context aware recommendation utilizes the contextual information of user and thus helps to enrich a user profile. Special recommendation methods using deep learning techniques for incorporating the content itself into the model (text, image, audio, video). Another important area is architectural advancement of deep learning techniques required for recommendation system problems.

8 Conclusion

This paper focused on the key issues of recommendation algorithms in the current scenario to enhance the precision of Recommendation systems. We also discussed the recommendation approaches and provide a detailed discussion on how these approaches could be improved. We also provided an overview of the new technologies for mitigating the issues. Although deep learning techniques have been used in several domain such as e-commerce and movie among others more unexplored area need to be addressed such as e-learning which has shown its significance in the time of COVID-19 by continued support for teaching learning process and examination process.

References

1. S. Fitzek, "BOOK REVIEW, Cace Sorin, Fitzek Sebastian (Eds.): ICCV 2020 social report. Covid19 in Romania, data, analysis, evolutions and statistics, Bucharest: Romanian Academy, 2020, p. 273." *Jurnalul Practicilor Comunitare Pozitive* 20.3 (2020): 77–80
2. H. Wang, W. Fu, Personalized learning resource recommendation method based on dynamic collaborative filtering. Mob. Netw. Appl. 1–15 (2020)
3. B.A. Rogowsky, B.M. Calhoun, P. Tallal, Providing instruction based on students' learning style preferences does not improve learning. Front. Psychol. **11**, 164 (2020)
4. X. Wang et al., Optimizing data usage via differentiable rewards. in *International Conference on Machine Learning*. PMLR, (2020)
5. J.F. Chartier, P. Mongeau, J. Saint-Charles, Predicting semantic preferences in a socio-semantic system with collaborative filtering: A case study. Int. J. Inf. Manage. **51**, 102020 (2020)
6. S. Natarajan et al., Resolving data sparsity and cold start problem in collaborative filtering recommender system using linked open data. Expert Syst. Appl. **149**, 113248 (2020)
7. M.C. Brouwers et al., Development and validation of a tool to assess the quality of clinical practice guideline recommendations. JAMA Netw. open **3**(5), e205535-e205535 (2020)
8. L. Wang et al., Diversified service recommendation with high accuracy and efficiency. Knowl.-Based Syst. **204**, 106196 (2020)
9. M. Caron et al., Unsupervised learning of visual features by contrasting cluster assignments. arXiv preprint arXiv:2006.09882 (2020)
10. S.Y. Wang et al. CNN-generated images are surprisingly easy to spot... for now. in *Proceedings of the IEEE/CVF Conference on Computer Vision and Pattern Recognition*, (2020)
11. A.A. Ewees et al., Improving multilayer perceptron neural network using chaotic grasshopper optimization algorithm to forecast iron ore price volatility. Resour. Policy **65**, 101555 (2020)
12. Y. Luo, Z. Chen, T. Yoshioka, Dual-path rnn: efficient long sequence modeling for time-domain single-channel speech separation. in *ICASSP 2020–2020 IEEE International Conference on Acoustics, Speech and Signal Processing (ICASSP)*. (IEEE, 2020)
13. J. Ma et al., DDcGAN: A dual-discriminator conditional generative adversarial network for multi-resolution image fusion. IEEE Trans. Image Process. **29**, 4980–4995 (2020)
14. C. Zheng et al., Gman: A graph multi-attention network for traffic prediction. in *Proceedings of the AAAI Conference on Artificial Intelligence*, vol. 34. No. 01 (2020)
15. R. Portelas et al., Automatic curriculum learning for deep rl: A short survey. arXiv preprint arXiv:2003.04664 (2020)
16. A. Onan, Sentiment analysis on product reviews based on weighted word embeddings and deep neural networks. Concurrency Comput. Pract. Experience e5909 (2020)
17. S.I. Mirzadeh, M. Farajtabar, H. Ghasemzadeh, Dropout as an implicit gating mechanism for continual learning. in *Proceedings of the IEEE/CVF Conference on Computer Vision and Pattern Recognition Workshops*. (2020)
18. X. Lin, P.C. Boutros, Optimization and expansion of non-negative matrix factorization. BMC Bioinformatics **21**(1), 1–10 (2020)
19. Z. Guo, H. Wang, A deep graph neural network-based mechanism for social recommendations. IEEE Trans. Industr. Inf. **17**(4), 2776–2783 (2020)
20. M. Dai, W. Guo, X. Feng, Over-smoothing algorithm and its application to GCN semi-supervised classification. in *International Conference of Pioneering Computer Scientists, Engineers and Educators*. (Springer, Singapore, 2020)
21. R. Kiran, P. Kumar, B. Bhasker, DNNRec: A novel deep learning based hybrid recommender system. Expert Syst Appl **144**, 113054 (2020)
22. N. Nassar, A. Jafar, Y. Rahhal, A novel deep multi-criteria collaborative filtering model for recommendation system. Knowl.-Based Syst. **187**, 104811 (2020)
23. C. De Medio et al., MoodleREC: A recommendation system for creating courses using the moodle e-learning platform. Comput. Hum. Behav. **104**, 106168 (2020)

24. X. Cai et al., A hybrid recommendation system with many-objective evolutionary algorithm. Expert Syst. Appl. **159**, 113648 (2020)
25. A.M.A. Al-Sabaawi, H. Karacan, Y.E. Yenice, Exploiting implicit social relationships via dimension reduction to improve recommendation system performance. PloS One **154**, e0231457 (2020)

Communication Assistant Using IoT-Based Device for People with Vision, Speech, and Hearing Disability

Chirag Umraniya, Mayank Timbal, Karmishth Tandel, Dhiraj Prajapati, and Pradip Patel

Abstract According to the census of 2011, about 10.6 million people are visually impaired, 1.2 million people have hearing disability, and 1.6 million people are speech impaired in India. In this paper, we propose an Internet of Thing (IoT)-based integrated device for these people in order to provide them a user-friendly and interactive environment through which they can easily communicate with others and interact with their surroundings. The device provides advanced features like facial recognition, emotion recognition, object recognition, sign language recognition, cloth pattern and color recognition, distance from an object, and optical character recognition. By giving input in the form of voice, text, and sign, one can use different functionalities of the device. Algorithms like image-to-text, text-to-speech, and speech-to-text conversion are implemented to provide intended functionality. Convolutional Neural Network (CNN) is used for feature extraction and classification. The proposed device considers various communications between speech, vision, and hearing impaired people with others.

Keywords Facial recognition · Object detection and recognition · Emotion recognition · Optical character recognition · Sign language conversion

1 Introduction

A large number of people, all over the world, suffer from speech, hearing, and vision disabilities. Recently, there is a huge advancement in the field of Human–Computer Interaction (HCI) in order to develop an automatic system to support these people. Many systems have been proposed in this area where machines are trained on the basis of different algorithms and can easily interact with different environments. Visual computing plays a vital role in HCI were images are captured

C. Umraniya (✉) · M. Timbal · K. Tandel · D. Prajapati · P. Patel
LD College of Engineering, Gujarat Technological University, Ahmedabad, Gujarat, India

P. Patel
e-mail: pradippatel@ldce.ac.in

A. K. Luhach et al. (eds.), *Second International Conference on Sustainable Technologies for Computational Intelligence*, Advances in Intelligent Systems and Computing 1235,
https://doi.org/10.1007/978-981-16-4641-6_2

through camera and used to extract various features with the help of different techniques. In this paper, we propose an integrated device that supports such application of HCI. The device takes different inputs in form of image, text or speech, which is then given to various algorithms, and an output as per the disability of the person is generated. The data is first preprocessed and then it is utilized by algorithms to generate desired outcomes. Different optimizers and libraries were used, and with the help of that, the entire system is developed. The system is aimed to enhance the life of people with disabilities by making their day-to-day communication with others in the society barrier free. While developing the various algorithms of the device, many important parameters have been considered such as time complexity, space complexity, efficiency, accuracy, and overall performance. This paper briefly describes the algorithms used and the techniques that were synthesized for the development of the device. Intended features like facial recognition, emotion recognition, object recognition, sign language recognition, cloth pattern and color recognition, distance from an object, and optical character recognition are provided by the device. One can use this device through inputs in form of voice, text or by capturing images.

2 Literature Review

Many algorithms were developed with the use of generic methodologies to implement facial recognition, emotion recognition, object recognition, cloth pattern and color recognition as well as optical character recognition. Many such techniques were presented by Kavita and Manjeet Kaur [1] in their survey paper which used methods like Support Vector Machine (SVM), Principal Component Analysis (PCM), and different transformation techniques like Discrete Cosine Transform (DCT) and Fast Fourier Transform (FFT) for feature extraction to provide face detection and recognition. Technique like FFT in inception model and deep neural network along with Long Short Term Memory (LSTM) is used by Nithya [2] for emotion recognition from facial expression. In addition to that, Monika and Lokesh [3] used a Naive Bayesian Classifier to classify six different emotions. Moreover, they have also presented outcomes of different algorithms based on hybrid approach and SVM with AdaBoost algorithm. Dilbag [4] used the action unit and Hidden Markov Model (HMM) to extract features for emotion recognition and came up with high-performance system. The object recognition and detection techniques, mainly imposed on YOLO. Jiwoong [5], describe the RCNN and RFCN methods with an increase in efficiency of YOLOv3 as compared to YOLO and YOLOv2. Vidyavani [6] used Darknet-53 and utilized COCO dataset to increase efficiency of YOLOv3 using ground truth. Sandeep Kumar [7] also contributed with his work on object detection and recognition using EASYNET model, RCNN, and FRCNN for feature extraction. Juan [8] used Hough transform techniques and then applied various models for object recognition. A noteworthy work for Indian sign language can be reflected by Joyeeta and Karen [9]. They used Eigen value weighted Euclidean distance-based classification technique to classify 26 alphabets with an accuracy rate of 94%. Pradip Patel and Narendra Patel [10] used HOG for feature extraction to detect hand gestures to identify sign language. Various steps like skin color detection, transforming into HSV color space, and finally detecting hand using BLOB was described in his work. Debesh Chaudhary and Sunitha [11] tried skin detection using YCbCr and HSV color space with different lighting conditions. They also used SVM for sign language classification. Furthermore, Sachin [12] used flex sensors for sign language recognition and converted them into speech. Apart from that, very little work related to cloth pattern recognition is done till date. Mrs. Anuradha and Thogaricheti [13] proposed pattern recognition using canny edge detection and HOG transformation approaches in their work. SVM was used for feature extraction by them and implemented using the raspberry pi module. Yannis et al. [14] implemented a cloth recognition system using a prior probability map which was followed by segmentation and clustering. Multiprobe LSH index was used for classification. In addition to that, measurement of the distance of object was also calculated. Huda and Husain [15] used the camera to calculate the dimensions and distance of an object using laser spots. Shrivastava et al. [16] have highlighted the use of ultrasonic sensor for calculating the distance of obstacle/object using P89C51RD2. Moreover, research in the field of paper currency recognition is also noteworthy. Manjunath [17] used the image processing method to extract different features from 20 different currencies of different countries and recognized them on the basis of specific symbols and denominations of note. Kiran [18] used similar method to identify currency notes. She used image processing and edge detection for extracting different features to recognize different currencies and their authenticity. Muhammad [19] recognized Saudi real on the basis of Radial Basis Neural Network (RBNN) and had an accuracy of 91% with a dataset of 110 images. An IoT-based device was also developed by Karmel [20] and team in which they used Google vision API and were mainly focused on OCR. In this paper, we propose an integrated device that provides various features like facial recognition, emotion recognition, object recognition, sign language recognition, cloth pattern and color recognition distance from an object, and optical character recognition.

3 Proposed System

3.1 Data Acquisition

The first step while dealing with Data Science in carrying out research work is to collect appropriate data which is also known as data acquisition. We have gathered data from different sources as input in various algorithms. The dataset for facial recognition was generated by us, where 100 sample images of 4 different persons were captured and stored. This data was then split into two different categories: test and train data. These images were converted to 89×89 pixel resolution and were provided as input to Local Binary Pattern Histogram (LBPH) algorithm. Some of the sample images are illustrated as shown in Fig. 1. The dataset for Indian Sign

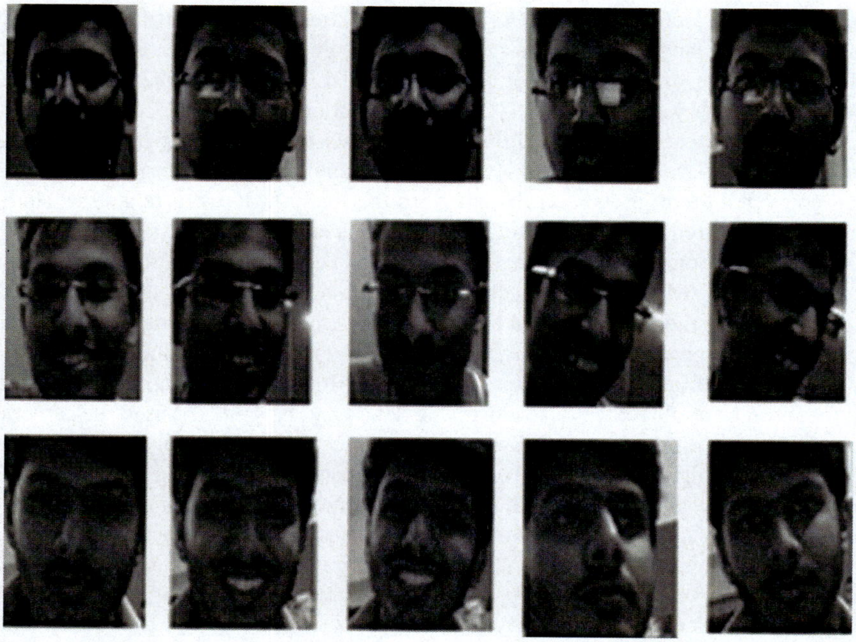

Fig. 1 Sample images for facial recognition

Language Recognition was generated collectively by us, where 190 images of size 110×110 are used as input for the algorithm. Some of the sample images are shown in Fig. 2. The figure depicts preprocessed images of 26 alphabets and 10 digits in binary form. Additionally, we have also created a dataset for cloth pattern recognition as shown in Fig. 3.

25 sample images of 250×250 resolution each of 4 different classes (plain, checks, graphics or printed and lines) were taken into consideration. TThese sample images were furthur divided into two classes for training and testing purpose. These pictures were downloaded from the internet and some of them were captured by us so that different parameters like color intensity, noise, and orientation of cloth can be considered. For Indian Currency Recognition, we used different paper currency of denominations 5, 10, 20, 50, 100, 200, 500, and 2000 both old and new as shown in Fig. 4 on the previous page to extract the desired feature and recognize it. Twenty-four images including both sides were considered for implementing this algorithm.

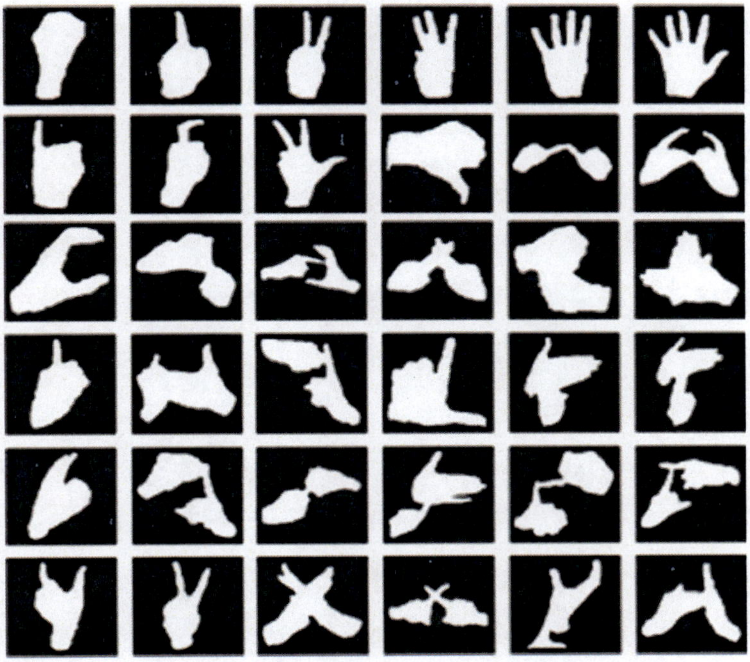

Fig. 2 Sample images of digits (0–9) and alphabets (A-Z) in Indian Sign Language

Fig. 3 Sample image consisting of four different categories (Checks, Lines, Plain, and Printed)

3.2 Data Pre-processing

The next step after data acquisition is data preprocessing. This step helps us to obtain useful data out of the data collected. Many times dataset may contain noise or missing values. This may reduce the efficiency of the algorithm and hence may generate undesired output. In our case, we have preprocessed all the images by converting them to

Fig. 4 Images of Indian Paper Currency including both sides

equal resolution so that the input size does not vary. Also, for sign language recognition, the data was converted from RGB color space to binary, so that the algorithm takes less execution time in order to identify the gesture. Whereas in case of cloth pattern recognition, the data was cleaned and its brightness and contrast were adjusted to increase the efficiency of the algorithm. Furthermore, while running the algorithms like facial recognition, emotion recognition, and sign language recognition, images of the dataset were converted from RGB color space to YCbCr or HSV color space to extract useful hidden information from the images.

3.3 Feature Extraction

The goal or objective of the algorithms highly depends on the features extracted from the dataset during feature extraction. In case of sign language recognition, the features are extracted using a convolution neural network that used 2 fully connected hidden layers. Adaptive Moment Estimation (Adam) optimizer was used to optimize given inputs and categorical cross entropy was used as loss function while training the model. The algorithm detected the skin pixels using YCbCr. For this, RGB image is first converted to YCbCr image using the following equation.

$$\begin{aligned} Y &= 0.299 * R + 0.587 * G + 0.114 * B \\ Cr &= 128 + 0.5 * R - 0.418 * G - 0.081 * B \\ Cb &= 128 - 0.168 * R - 0.331 * G + 0.5 * B \end{aligned} \tag{1}$$

Threshold values were used as shown in Eq. 2 to find out skin pixels. This results in binary image.

$$75 < \text{Cb} < 135 \text{ and } 130 < \text{Cr} < 180 \text{ and } \text{Y} > 80 \tag{2}$$

Facial recognition used LBPH algorithm of OpenCV library, where histograms of images are generated using sampling method and features are extracted accordingly. Convex hull and polygon are used to identify the area of interest and then face detection and recognition is performed. Emotion recognition algorithm also used similar methodologies and approaches to extract the desired feature. The optical character recognizer algorithm of our system implemented the tesseract library of python. For the case of cloth recognition, features are extracted using CNN with three hidden layers: optimizer, softmax, and ReLU activation function. The trained CNN categorizes input into one of the four different classes. The entire CNN architecture can be visualized as shown in Fig. 5.

In object detection and recognition, the COCO dataset is used and the YOLOv3 model is implemented. The use of ultrasonic sensor AJ-SR04M added the feature of distance measurement from the person to that object. The following equation is used to measure the distance accurately:

$$\text{Distance} = (\text{Time Elapsed} * 34300)/2 \tag{3}$$

Here, 34,300 is the sonic speed of the wave of the sensor in cm/s. Various readings were taken into consideration and analyzed. The distance from 20 to 100 cm has been measured with an accuracy of about 96%. The algorithm for recognition of Indian paper currency is implemented using the Brute-Force matcher of OpenCV. Various

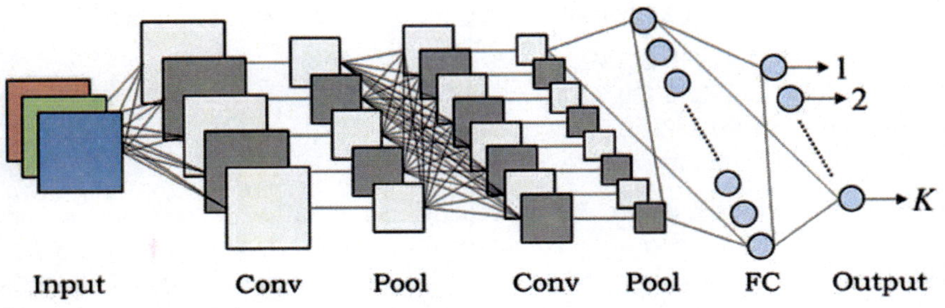

Fig. 5 CNN model used for cloth pattern recognition

Fig. 6 Device consisting of different components

templates based on ROI were used for feature matching and hence the currency is recognized. The accuracy obtained by the implementation of this algorithm is near 90%. Also, by using the backside of currency note that uses different architectural structures, feature extraction was easy, and due to that, the accuracy obtained was 98%. Hence, on the basis of both sides, the currency could be recognized.

3.4 System Design and Working

As the system is intended to build an integrated device to assist various impaired people, we have implemented these algorithms on Raspberry pi 4 B model along with the use of various components. Figure 6 shows the circuit of the entire device with connected components. The entire working of the device can be well understood from the flow diagram shown in Fig. 7.

4 Experimental Results

The developed system is implemented using the following hardware systems:

- Raspberry Pi 4B (2 GB RAM).
- Lenovo Ideapad 310 (8 GB RAM, i5 Intel processor).
- Ultrasonic Sensor (AJ-SR04M).
- Logitech C310 webcam.

The results achieved during various experiments on different algorithms along with inputs, outputs, and accuracies are shown in Table 1. All algorithms as mentioned in the table were tested on different environments to ensure high accuracy. Different data were fetched as inputs to different algorithms to test the nature of the output obtained. The accuracy achieved for various features is shown in Table 2.

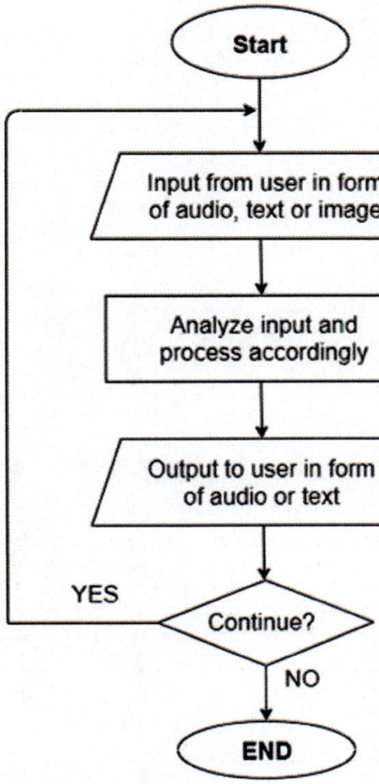

Fig. 7 Flow diagram of the system

We have also experimented with different optimizers to find the accuracy of our CNN model for cloth pattern recognition. We used the ADAM optimizer to train our model with 20 epochs. Figure 8 shows the accuracy and loss of our trained model. In addition, the accuracy and loss of sign recognition model is also illustrated in Fig. 9 using the ADAM optimizer.

Furthermore, the readings from the ultrasonic sensor were also noted at an interval of 5 cm and the following observations were obtained. Thus, from the observation table, it can be concluded that the distance between 20 and100 cm can be efficiently measured with minimum error.

5 Conclusion

An Internet of Things (IoT)-based integrated device for speech, vision, and hearing-impaired people is presented in this paper. This device is intend to provide user-friendly environment for various disabled people to interact with each other and

Table 1 Experimental outcome of various algorithms

S. no	Feature	Algorithm used	Input	Output
1	Facial Recognition	LBPH		The person is Chirag (in text and speech)
				The person is Mayank
2	Sign Language Recognition	CNN		C
				T
3	Image-to-text-speech	Tesseract OCR	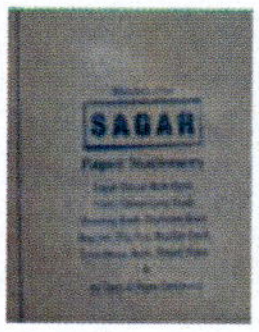	SAGAR Paper Stationay Super elute Note book, Super deluns Note Book, Drowing Book, Duplicate Book
4	Object Recognition	YOLOv3		The object detected is person and cell phone
5	Cloth Pattern Recognition	CNN		The cloth contains checks

(continued)

Table 1 (continued)

S. no	Feature	Algorithm used	Input	Output
				The cloth is plain
				The cloth contains lines
				The cloth contains graphics or is printed
6	Indian Currency Recognition	Brute-Force Matcher		It is 10 Rupees note
				It is 20 Rupees note
				It is 100 Rupees note
				It is 200 Rupees note

Table 2 Accuracy achieved for various features

S. No	Feature	Accuracy (%)
1	Facial recognition	66
2 5	Sign language recognition	74
3	Image-to-text-speech	83
4	Object recognition	87
5	Cloth pattern recognition	88
6	Indian currency recognition	94

Table 3 Observations of ultrasonic sensor (AJ-SR04M)

Actual Distance (in cm)	Measured Distance (in cm)	Error (in %)
5	17.80	256.00
10	18.60	86.00
15	17.90	19.33
20	18.25	8.75
25	23.60	5.60
30	27.70	7.66
35	32.65	6.71
40	37.30	6.75
45	43.23	3.93
50	47.10	5.80
55	51.80	5.81
60	58.80	2.00
65	61.75	5.00
70	67.33	5.00
75	71.10	5.20
80	73.40	8.25
85	80.00	6.25
90	85.75	4.72
95	92.35	2.78
100	96.70	3.30

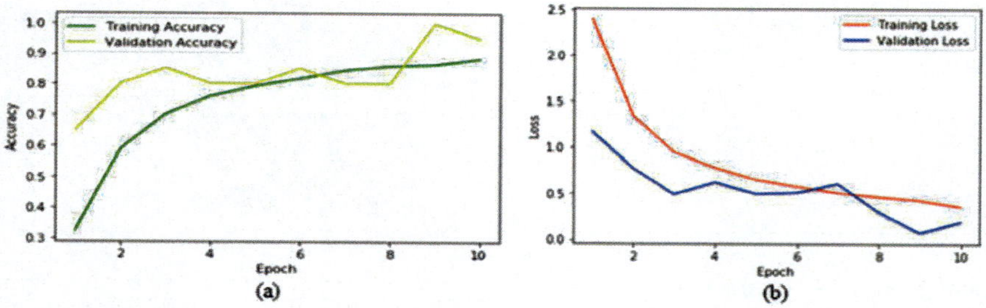

Fig. 8 Cloth Pattern Recognition. **a** Accuracy using ADAM **b** Loss using ADAM

to be aware of their surrounding. The device provides, with high accuracy, various features useful for this purpose such as facial recognition, emotion recognition, object recognition, sign language recognition, cloth pattern and color recognition distance from an object, and optical character recognition. The device is able to take the input in the form of voice, text, and sign, depending on the type of algorithm used. Convolutional Neural Network (CNN) is used in various parts of the algorithm for

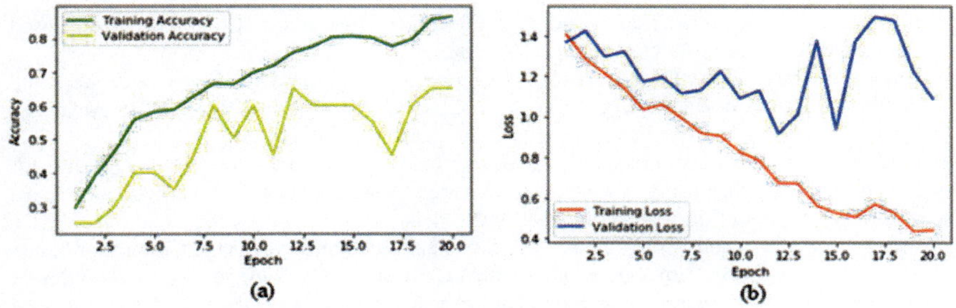

Fig. 9 Sign Language Recognition. **a** Accuracy using ADAM **b** Loss using ADAM

feature extraction and classification. During the experiment, it is observed that most of the algorithms work efficiently under different situations and circumstances. The proposed device is able to serve disabled persons by providing basic requirements.

6 Future Scope

As this device works properly with the algorithms mentioned, we can also add more modules and functionalities like object color recognition, emotion recognition, navigating visually impaired people indoors and outdoors, as well as biometric authentication of the user of this device, to ensure confidentiality of their data stored like facial data of various peoples.

References

1. Kavita and Manjeet Kaur, A survey paper for face recognition technologies. International journal for scientific and research publications **6**(7), 441–445 (2016)
2. N. Roopa, Emotion recognition from facial expression using deep learning. International journal of engineering and advanced technology **8**(6), 91–95 (2019)
3. Monika Dubey and Prof. Lokesh Singh. Automatic emotion recognition using facial expression: A review. International research journal of engineering and technology, Volume 3, issue 2, pp 488–492 (2016).
4. Dilbag Singh. Human Emotion recognition system. I.J. Image, Graphics and Signal Processing, 8, 50–56, 14, pp 50–56 (2012).
5. Jiwoong Choi, Dayoung Chun Hyun Kim and Hyuk-Jae Lee. Gaussian YOLOv3: An Accurate and Fast Object Detector Using Localization Uncertainty for Autonomous Driving. IEEE, pp 502–511.
6. A. Vidyavani , K. Dheeraj, M. Rama Mohan Reddy, KH. Naveen Kumar. Object Detection Method Based on YOLOv3 using Deep Learning Networks.International Journal of Innovative Technology and Exploring Engineering (IJITEE) ISSN: 2278–3075, Volume-9 Issue- 1, pp 1414–1417 (2019).
7. S. Kumar, A. Balyan, M. Chawla, Object Detection and Recognition in Images. International Journal of Engineering Development and Research **5**(4), 1029–1034 (2017)

8. Juan Wu1, Bo Peng1, Zhenxiang Huang1, and Jietao Xie2Research on Computer Vision-Based Object Detection and Classification, pp 183–188.
9. Joyeeta Singha and Karen Das Indian Sign Language Recognition Using Eigen Value Weighted Euclidean Distance Based Classification Technique. International Journal of Advanced Computer Science and Applications, Vol. 4, No. 2, pp 188–195 (2013).
10. Pradip Patel And Narendra Patel, Vision Based Real-time Recognition of Hand Gestures for Indian Sign Language using Histogram of Oriented Gradients Features. International Journal of Next-Generation Computing **10**(2), 92–102 (2019)
11. Debesh Chaudhary and Sunitha Beevi K.Spotting and Recognition of Hand Gesture for Indian Sign Language using Skin Segmentation with YCbCr and HSV Color Models under different Lighting Conditions. International Journal of Innovations & Advancement in Computer Science Volume 6, Issue 9, pp 426–435 (2017).
12. Sachin Bhat, Amruthesh M, Ashik, Chidanand Das and Sujith. Translating Indian Sign Language to text andvoice messages using flex sensors. International Journal of Advanced Research in Computer and Communication Engineering Vol. 4, Issue 5, pp 430–434 (2015).
13. Mrs. Anuradha.S.G, Thogaricheti Ashwini.Clothing Color and Pattern Recognition for Impaired people. International Journal Of Engineering And Computer Science ISSN: 2319–7242 Volume 5, Issue 4, pp 16317–16324 (2016).
14. Yannis Kalantidis, Lyndon Kennedy and Li-Jia Li. Getting the Look: Clothing Recognition and Segmentation for Automatic Product Suggestions in Everyday Photos. Conference Paper (2013).
15. H.M. Jawad, T.A. Husain, Measuring Object Dimensions and its Distances Based on Image Processing Technique by Analysis the Image Using Sony Camera. Eurasian Journal of Science & Engineering **3**(2), 100–110 (2017)
16. A.K. Shrivastava, A. Verma, S.P. Singh, Distance Measurement of an Object or Obstacle by Ultrasound Sensors using P89C51RD2. International Journal of Computer Theory and Engineering **2**(1), 64–68 (2010)
17. Vedasamhitha Abburu, Saumya Gupta, S. R. Rimitha, Manjunath Mulimani, Shashidhar G. Koolagudi.Currency recognition system using image processing. Proceedings of 2017 Tenth International Conference on Contemporary Computing (IC3), Noida, India (2017).
18. Kiran Swami, Kshitija Murumkar, Dr. Premanand Ghadekar, Yash Solanki.Currency Recognition and fake currency identification using image processing.International Journal of Advance Research, Ideas and Innovation in Technology.Volume 5, issue 3, pp 659–661 (2019).
19. Muhammad Sarfraz.An intelligent paper currency recognition system. International Conference on Communication, Management and Information Technology, pp 538-545 (2015).
20. Karmel A, Anushka Sharma, Muktak Pandya, Diksha Garg. IoT based assistive device for deaf, dumb and blind people. International conference on recent trends in advanced computing, pp 259–269 (2019)

An Optimized Object Detection System for Visually Impaired People

Meenal Vardar and Prashant Sharma

Abstract Vision is one of the main senses that enables us to interact with the natural environment. Globally, there are almost two hundred million blind people and visually impaired people who hinder multiple daily activities. It is, therefore, really important for blind persons to understand their surroundings and know what things they are communicating with. This article discusses many of the camera-based software methods and tools that help the blind person read text patterns written on handheld items. This is the method to help people with visual impairments understand and convert text patterns into audio output. The framework initially proposes the solution to taking a picture from the camera and the goal region to extract the item from the background and draw a text pattern from the object. Different algorithms are evaluated in different scenes. The observed text is connected to the blueprint and converted through voice output. Text patterns located and binaries using Optical Character Recognition (OCR). The text is converted into an audio output. The result shows that the proposed study reaches the precision of the object description in terms of 98.30.

Keywords OCR · Visually blind person · Machine learning · Deep learning · SVM · Accuracy

1 Introduction

Computer science has taken a fundamental role in the development of the daily activities of human being, presenting tools that provide solutions to problems in different areas. There are lots of research focuses on artificial intelligence; emphasizes in his machine learning course how applications based on bio-inspired algorithms,

M. Vardar (✉) · P. Sharma
Department of Computer Science, Pacific University, Udaipur, India

P. Sharma
e-mail: prashant.sharma@pacific-it.ac.in

A. K. Luhach et al. (eds.), *Second International Conference on Sustainable Technologies for Computational Intelligence*, Advances in Intelligent Systems and Computing 1235,
https://doi.org/10.1007/978-981-16-4641-6_3

machine learning and evolutionary techniques, allow for example to have information about traffic, make weather predictions, generate security with biometric recognition, control crops, obtain location thanks to automatic mapping or even allow us to interact on social networks. The challenge of computer science is to extract useful information from the environment in which humans interact, in order to create mathematical, statistical or quantitative models that can represent these natural processes of man [1].

The development of computers was then limited to the advancement of new technologies, the machines that were created were the size of rooms. Still, downsizing these machines wasn't the only important issue; scientists were looking for a way to make these machines increasingly intelligent [2]. That is why the investigations of relevant people in history began to appear, characters like Alan Turing, considered as the father of computer science, who began to make an abstraction of the human brain to represent it in the world of computers, understanding that, in this way, the machines could be made better not only in the hardware part, but also in the software part.

It was then Messrs. Waltter Pitts and Warren McCulloch, a neurophysiologist and a mathematician, respectively, who conceived the foundations of neural computing, modeled in 1943, a simple neural network using electrical circuits [3]. From this moment on, it became practically a new line of research to represent the functioning of the brain in mathematical techniques and models applied to computer computers.

In 1980, the decade considered to be the age of illustration in computer science, Ray Solomonoff, inventor of algorithmic probability, built the foundations for the beginning of a really important and fundamental computational technique in computer applications currently used, machine learning "Machine Learning", an area whose purpose is to create programs capable of generalizing behaviors from unstructured information supplied in the form of examples. This technique was complemented over the years with topics that were developed in parallel, including all the knowledge acquired from neural networks to optimize computational processes. In summary, these machines that are provided as already mentioned, a series of examples with their respective outputs, Companies such as Facebook, Google or YouTube today use learning algorithms to make the interaction of their platforms and their users increasingly intelligent and personalized, taking as reference tastes, customs, and recurring activities [4].

In 2006, Geoffrey Hinton, a specialist in cognitive psychology and neural networks established the concept of deep learning "Deep Learning", which can be considered an evolution of machine learning, with a similar idea and through more robust algorithms. This process work on texts, audios, videos, and imges for problem-solving [5], making the machines look more and more like a person. In the same way, as "machine learning", with the aim of creating more complete applications, it was important to link with more computer science techniques [6, 7], including artificial vision, data mining, bio-inspired systems, robotics, and artificial intelligence.

Image Recognition Based on Deep Learning [8]. In this work author apply Deep Learning, convolutional artificial neural networks and deep belief networks, in classification problems of images, with this they manage to determine the high efficiency of these two models. Result are compare with their performance, this work leads them to the conclusion that convolutional neural networks have a superior performance than their competition. Hybrid Deep Learning for Face Verification [9]. In this research, developed in 2016, it was sought to make a hybrid between a convolutional neural network algorithm and the Boltzmann machine algorithm. The main objective was to carry out a study for the verification of faces, starting from the extraction of local visual characteristics of the face compared in two images, these data are processed through multiple layers.

In this work [10], the researchers focus their efforts on a highly complex problem such as the recognition of easily deformable objects, in this case, garments were hung from a single point, the system was designed and implemented in a robotic platform in charge of manipulating the garment and data extraction from it, achieving a high degree of performance in the classification task. This research shows the versatility and high efficiency of convolutional networks applied to digital images.

Andrew Ng [11] took advances and new techniques of artificial vision to determine depth in images, with the aim of classifying objects in RGB images with a model based on the combination of convolutional neural networks and recursive neural networks. To carry out the tests, a database with 51 kinds of household objects was obtained. With 300 examples for each class, each one of them was seen from three different angles to make the training with a more robust base.

The previous articles show how the solution to problems of detection and classification of objects in videos or images are oriented towards the use of convolutional neural networks, thanks to the high performance rates shown and the use or splicing that occurs from probabilistic techniques and artificial vision techniques.

A. **Background**

This work focused on developing a technological tool based on computer science, it is really important to contextualize its birth, some of its history and development in the service of human beings, and then deepen the techniques that They are planned to be implemented in the development Computer science has been historically recorded since the construction of the first useful devices to keep accounts and solve mathematical problems. Over the years, there were important contributions from researchers such as Leibniz, Pascal, and Babbage [12] to achieve an approximation to the first computer and the first algorithms.

B. **Justification**

Technology seeks to transform the environment to meet the needs of man, sometimes. The needs arise due to the existence of barriers that limit their capabilities, these barriers can be cultural, regional, intellectual or even physical. This degree project seeks to eliminate some of the limitations created by these barriers by developing a tool that, by applying computational intelligence techniques, is capable of giving those who use it a better understanding of their environment and in certain cases improve their quality of life. On the other hand, autonomous learning techniques are one of the most relevant topics today. Hundreds of researchers work daily for the improvement of these techniques, which is why it has been decided to apply this recent and growing technique to implement a useful tool for humans.

2 Literature Review

Accurately extracting the DCNNs needs significant hardware and connectivity resources. We have proposed a swift and low-energy detection system that enables the impaired to explore their surroundings without hassle [13]. We have created a highly refined 8-bit DCNN algorithm that evaluates data to the 32nd decimal place by using only 5-bit indexing, allowing us to save hardware resources. Some hardware accelerators use customizable process engines that dramatically lower or exclude the transition off-to-chip process data. A convolution search table is used to find all the multiplicands that feed into the final calculation and all of all the multipliers are used to reduce capacity. The architecture is composed of SMIC 55-nm and 1.2 V/1.1 GB/s with a maximum of 2.2 Tb performance that has been calculated for this case after the simulation is complete.

This article illustrates the usefulness of an electromagnetic sensor [14] to autonomously move visually impaired and blind individuals. It is known that people with vision disorders appear to be helped by the most common white cane. Our design is a traditional white cane microwave radar that allows customers to be aware of an obstacle in a broader, more safe range. The technology suggested is better than the existing electronic means of transport, noise tolerance, and lower measurements. The following are recent advances with an emphasis on the miniaturization of circuit boards and antennas in this research activity. A laboratory prototype has been developed and conducted and the first barrier detection test results prove the effectiveness of the unit.

A number of obstruction identification (OD) methods [15] for monocular vision were created to help visually disabled individuals. Using the vector differences or the object size of two consecutive frames, most traditional OD methods detect obstacles. However, short-term OD efficiency is greatly impacted by the tracking mistake, which results in unreliable OD results. This paper proposes a new OD approach focused on a new framework named Deformable Grid to tackle this issue (DG). The DG is initially a standard grid format, but may be increasingly deformed, based on the object's movement in the scene. The suggested approach detects the collision danger entity depending on the deformation degree in the DG. Experimental findings reveal that the OD system proposed, beats the traditional approach in terms of processing time and precision.

Echolocation helps individuals who are blind and who have sensory impairments to detect their surroundings by echoes. However, one must have good attention to detail to follow this method precisely. In addition, it is often essential to produce sound and analyze the collected data. This paper discusses and tries out the LAR (Latent Audio Range) method to handle these shortcomings by equipping a LID (Lass) sensor with a stereo Pit Sensor. This sense of stereophonic sound decreases the isolation, and therefore, often enhances the spatial abilities of the hearing impaired ones. In Phase I, the LASS computer's hardware and software are mounted, and in Phase II, the device is tested to make sure it is stable.

The Penn State Administrative Review Board approved 18 volunteers from the Department of Psychology in Penn. This paper shows that individuals with blindfolded LASS systems can quantitatively classify exterior barriers, discern their relative distances and differentiate the angular orientation of different objects at low levels. Because of the exponential growth of mobile technologies, a technology [16] is evolving which can identify banknotes and coins to help visually disabled citizens utilizing smartphone embedded cameras. In previous research, robust features have often been handmade, such as scale-invariant characteristics turn or accelerate, and cannot yield robust identification results for banknotes or corners recorded in complex environments and contexts. With recent developments in deep learning technologies, some banknote and coin recognition experiments have been performed utilizing a profound neural network (CNN). However, these experiments demonstrated degraded efficiency according to context and climate shifts.

This paper offers a 3-stage identification technology with quick regional CNN, geometric constraints, and residual networking for new banknotes and coins (ResNet). The experiment carried out by Jordanian dinar and 6,400 images of 8 types of Korean winning banknotes and coins on our smartphones produced more results than the state-of-the-art methods based on manufactured features and profound functionality. The experiment was carried out with a Jordanian Dinar (JOD).

Navigation support system for visually disabled persons (NAVI) applied [17] to devices that may aid or direct vision deficient individuals with sound signals, varying from partly sighted to fully blind. In this article, a new NAVI method focused on visual and range knowledge is introduced. We choose a single unit, a consumer RGB-D camera, instead of using several sensors, and gain from both spectrum and visual knowledge. The key contribution is, in particular, the combination of the depth details and the picture strength, contributing to a rigorous extension of the segmentation of the surface. On the one side, the accurate yet restricted profundity information is improved by the long-range visual information. It encourages and enhances the difficulty and vulnerability to error picture processing with depth knowledge. The framework proposed detects and classifies the key structural components of the scene, which enable the user to move in an unknown manner. The device has been evaluated in a broad spectrum of conditions and data sets to prove performance and that the framework is stable and operates in demanding indoor conditions.

The creation of electronic sensing devices for visually disabled persons [18] involves awareness of their needs and ability. This paper provides a rough study that can be used to describe the parameters for the construction of these devices properly. The emphasis would be on clear-cut metrics, stressing their position in orientation and mobility activities. A new computer is presented that belongs to this class. The detector is designed on a multisensory technique and uses intelligent signal processing to warn the consumer of the presence of artifacts hindering its trajectory. Experimental result show the device's effectiveness in real time.

Many assistance mechanisms for the identification of items (or obstacles) of significance for visually disabled persons (VIPs) have long been studied. However, the functional frameworks in the modern world are also very demanding because of standardized entity types in incredibly humiliating scenes (or complicated background) [19]. In this essay, we suggest a novel framework to detect and approximate the complete model of the typical artifacts in the everyday life of the VIP. Proposed method include relevant facts such as size and protection instructions for catching on a flat surface, but also solve the issue of where the item is questioned. The pipelines use a rigorous estimator to incorporate a sequence of point cloud representation, table plane identification, entity detection, and the whole model estimation. In this paradigm work, we discuss the new benefits of profound learning (e.g., RCNN, YOLO), which may be an effective means of calculating the mission when geometry-based methods approximate a complete 3D model. This scheme would not involve the isolation (or segmentation) of the items involved from the context of the scenes. The method suggested is contrasted to other methods and tested on the existing datasets gathered in typical scenes such as the kitchen or cafeteria room. The suggested structures fulfill the criteria of high precision, cycle time, and suitability for VIPs in these assessments. The assessment data sets are written.

Deep neural network recognition (DCNN) is an effective solution to visual perceptions that demands tremendous computational and coordination costs [20]. We have proposed a fast and low-power object recognition processor that lets visually disabled people recognize their environment. A DCNN automatic measuring algorithm has been developed that successfully quantizes data with 32 values at 8-bit fixed points and uses 5-bit indexes for display, reducing the hardware costs to marginal precision relative to the 16-bit DCNN. The design of a certain hardware accelerator uses reconfigurable process engines for multi-layer pipelines to minimize or significantly eliminate the temporary transfer of off-chip files. To significantly minimize power, a search table is used to force all multiplication in convolutions. The design of proposed method is constructed as SMIC 55-nm and only 68-mw with a 1,1-v voltage of 155 GB/s with a maximum output of 2.2 Top/w after a structure simulation has been done. In this article, we established an indoor image processing method for the color recognition of related objects [21]. Our device utilizes color recognition technologies to assess the position of a consumer with very high precision for real-time applications. We used our method to filter a picture for a particular color and to collect pixel coordinates for the image.

The position of the consumer is then calculated by contrasting the matrix with the pre-created matrix of the training images with these values. We have successfully performed indoor tests and have obtained really positive results. After reviewing the findings, we suggest that our localization systems be combined with indoor navigation systems, where precision is the most essential aspect for blind persons. We have also created an Android-based framework to ease the navigation phase.

3 Theoretical Framework

For the development of this research, it has been necessary to conceptualize several terms, that will help to better understand the purpose of this research. Background of Artificial Network explain concepts that must be needed for the realization of this paper. Framework used in this work focused and applied mainly to the study of the storage, transformation, and transfer of information in computers [22, 23].

4 Proposed Work

Artificial vision is a discipline that aims to simulate the different processes and elements that give vision to a machine. These include geometric and other properties such as color, lighting, texture, and composition. Vision, both for a man and for a computer. This discipline consists mainly of two phases: acquiring an image and interpreting it. The image acquisition process is carried out by means of a camera. In this way, it only remains to implement tools to interpret the images, distinguish the objects in the scene, extract information from them, and solve more particular aspects according to the needs to be satisfied. Equipment can detect static and dynamic objects in a video stream. The implementation of the system on the smartphone constitutes a great help for the mobility of the visually impaired population, since smartphones are increasingly portable and with greater processing capacity [9, 10]. However, the detection of objects in poorly lit places and those that are in constant motion cannot be detected efficiently. Artificial vision has become a widely used tool for the development of assistive technologies for the visually impaired population, the full potential it can offer has not yet been reached, taking advantage of current developments in hardware, with the appearance of computer processor's great capacity to implement algorithms in software [24]. A common problem in the development of new technologies is that they are not designed based on the specific needs of people in this condition, for this reason, the true requirements of users are unknown, causing the developed devices not to obtain the desired results. Another significant problem is that the devices are large and difficult to use [25]. For the development of this project, it has been necessary to conceptualize several terms that will help to better understand the purpose of this research. These being explained from the most general aspects to the most specific concepts that must be needed for the realization of this Paper. Sciences focused and applied mainly to the study of the storage, transformation, and transfer of information in computers. Computer science can be divided into two perspectives: a theoretical part, where the design of algorithms is understood following mathematical techniques such as optimization, probabilistic techniques. The practical part will then be the implementation of these algorithms that together form a software that will operate on a specific hardware.

Machine learning, also called machine learning or machine learning, is a technique derived from computer science where the main objective is to make computer equipment capable of learning. In this context, learning refers to identifying patterns in millions of data and through them predicting future behaviors in an environment or situation [26] using statistical algorithms and probability theories.

Two main subareas can be identified in machine learning: supervised learning and unsupervised learning. Through the first method, it is sought by means of the meaningful collection of a set of examples of which the answer is known, to generate a descriptive equation of the system (hypothesis) to give a possible solution to a new entry. Two types of problems can be recognized from this same method:

Regression problems: Its purpose is to generate a continuous value according to a series of examples with their respective answers, for example, the value of a house taking into account its area, or its location, on the other hand, unsupervised learning commonly works with random inputs to the system and the output represents the degree of familiarity or similarity between the information that is being presented to the input and the information that has been shown until then.

Artificial neural networks: Neural networks have become relevant in machine learning applications due to their ability to solve non-linear equations and their cognitive learning ability [14]. Its purpose is to emulate the behavior of the nervous system and the way it can process information through neurons in conjunction with the brain.

Thus, in this paradigm of artificial intelligence, there is a unit analogous to the biological neuron; the perceptron [27, 28]. This being an element with several inputs that emulate the dendrites of the neuron and an output variable representing the axon, the inputs that are multiplied by a vector of weights, are combined with a basic sum to later be compared or analyzed by a function that determines the output. An artificial neural network will then be a set of elementary units (Perceptron) connected in a concrete way [29]. The architecture of these neural networks is basically made up of groups of perceptron's grouped by levels or layers; an input layer where the data is entered into the network, a hidden layer, and a final layer where the network outputs are presented. The layers that lie between the two mentioned above are commonly known as hidden layers.

5 Proposed Algorithm

1. Call the function random (0 to 20,10) = x
2: Call the function random (0 to 20,10) = y
3: Call hsv method to call function convert RGB to HSV
4: Assign the value of hp = h and ($H = 2 - 10$: $H = 2 + 10$; $W = 2 < -10$: $W = 2 + 10$)
5: Assign the value of $sp < -s$ ($H = 2 - 10$: $H = 2 + 10$; $W = 2 < -10$: $W = 2 + 10$)
6: $vp < -v$ ($H = 2 - 10$: $H = 2 + 10$; $W = 2 < -10$: $W = 2 + 10$)
7: rct; gct; bct; yct; $blkct$; $wct < -0$
8: $t < -0$

(continued)

(continued)

1. Call the function random (0 to 20,10) = x
9: *for* $t < -0$: 10 *do*
10: if $sp(x[t]; y[t]) \geq 0$: 2 then
11: if $hp(x[t]; y[t]) \geq 330$ then
12: $rct < -r + 1$
13: else if $(hp(x[t]; y[t]) \geq 140)$ and$(hp(x[t]; y[t]) \leq 180)$ then
14: $gct < -gct + 1$
15: else if $(hp(x[t]; y[t]) \geq 210)$ and$(hp(x[t]; y[t]) \leq 250)$ then
16: $bct < -bct + 1$
17: else if $(hp(x[t]; y[t]) \geq 5)$ and $(hp(x[t]; y[t]) \leq 40)$ then
18: $yct < -yct + 1$
19: else
20: color value are not define by classifiers
21: end if
22: else
23: if $(vp(x[t]; y[t]) \geq 0$: 7) then
24: $wct < -wct + 1$
25: else if $(vp(x[t]; y[t]) \leq 0$: 2) then
26: $blkct < -blkct + 1$
27: else
28: gray2bin()
29: end if
30: end if
31: end for
32: index argmax ([*rct*; *gcr*; *bct*; *yct*; *blkct*; *wct*])
33: Label color dictionary(*index*)

Mode selection is a very helpful device mechanism enabling the consumer to adjust system features on demand that only requires user color or object detection, since blind or visually disabled audience members depend on auditory capability. The mode selection block output enters into one image processing block, such as color and object recognition. The results of the chosen block are analyzed and the user may produce feedback with a speech synthesizer. Each sub-block is specified in the following subdivisions [30] (Table 1).

Table 1 Color threshold value

Color	Hue	Saturation
Red	00.94–1.0	00.4–1.01
Green	00.35–00.46	00.4–1.01
Blue	00.56–00.66	00.67–1.01
Yellow	00.16–00.21	00.55–1.01
Cyan	00.56–00.66	00.5–00.66
Purple	00.67–00.76	00.4–00.62
Brown1	00.89–00.98	00.0–00.42
Brown2	00.0–00.15	00.4–00.56
Brown 3	00.0–00.15	00.55–00.66

6 Result and Analysis

The Color Recognition module has been deemed significant, in particular, for people with impaired vision and for those who have damaged their eyes in an accident. We use the color space HSI to describe color (hue, saturation, and intensity) as color space. For these purposes, color space RGB (Red, Green, Blue) [31] was encoded into the color space of HSI and further processed. For the color of an item next to the sensor, each channel is first divided into an HSI color space, and for each channel, a small 20×20 window is chosen around the middle pixel of the picture. In this sub-picture, we sample pixels randomly and the computer determines the color of an object on a cardinal basis using the observed threshold, producing a colored label for production. We've tested some of our cameras in our laboratory [32] for Hue, Saturation, and Intensity, like Logitech C110, Logitech C120, and iball CHD20.

The camera is used to take pictures from the actual world, it is 1.3MP Low power, CMOS, USB monitor. It can be wired to one of the two on-board USB2.0 ports. The implementation is performed using the library and the programming language of OpenCV image processing.

The type of classification (classification output) is shown in the figure. In the case of group box, the chances for a box are also 98.98—which is obviously right, since the box is first for data selection, the chances for the correct grouping of second and third groups are (keyboard, cell phone) (Figs. 1, 2 and 3).

7 Conclusion

As a classification system, support for vector type machines and Yolo was extremely accurate because it offered enormous versatility when combining SVM with each of the classes. Automatic grid search is important when each class is trained in the system. The optimal value of the parameters SVM and Yolo used differs between classes. This allows the system to be fully autonomous and continuously improved

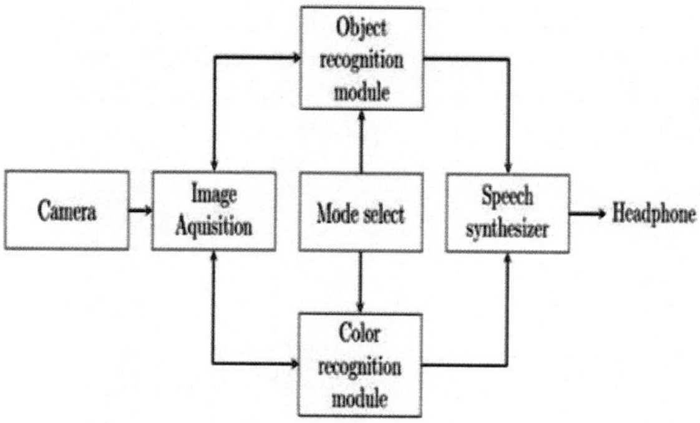

Fig. 1 Proposed system block diagram

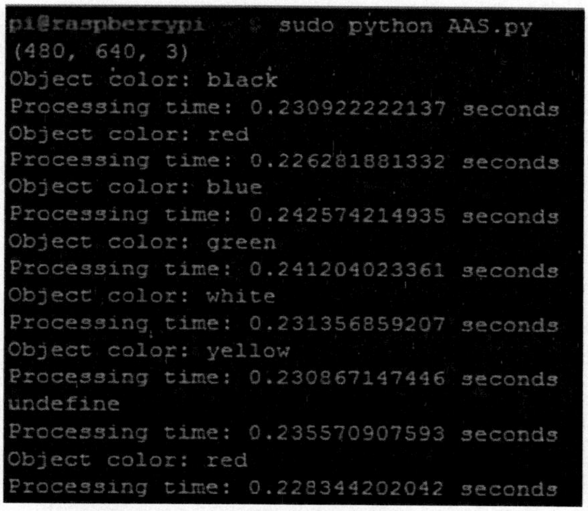

Fig. 2 Proposed system computation

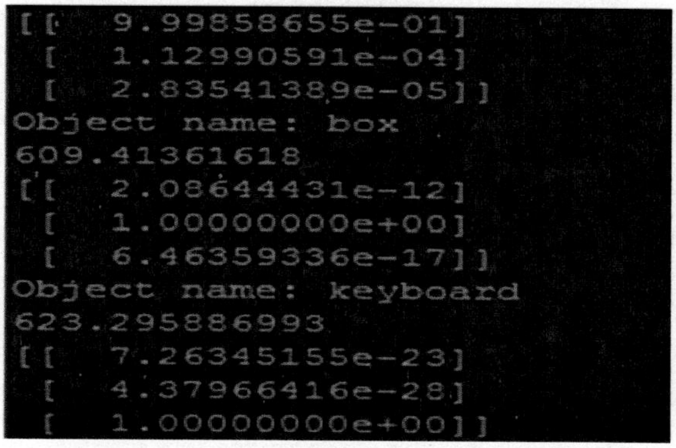

Fig. 3 Proposed system classification

by user data in addition to designing multiple SVM (one per class). This continuous learning extends not only to the increasing accuracy of current classes as the training data is improved, but also to the ability to learn from new students. At first, a grouping technique was identified for the similarity and significance of the artifacts.

References

1. C. Park, S. W. Cho, N. R. Baek, J. Choi, K. R. Park, "DeepFeature-Based Three- Stage Detection of Banknotes and Coinsfor Assisting Visually Impaired People," IEEE Access, vol. 8,pp. 184 598–184 613,2020
2. S. Bhatlawande, M. Mahadevappa, J. Mukherjee, M. Biswas, D. Das, S. Gupta, Design, Development, and Clinical Evaluation of the Electronic Mobility Canefor Vision Rehabilitation. IEEE Trans. Neural Syst. Rehabil. Eng. **22**(6), 1148–1159 (2014)
3. Aladren, G. Lopez-Nicolas, L. Puig, and J. J. Guerrero, "Navigation Assistance for the Visually Impaired Using RGB-D Sensor with Range Expansion," IEEE Systems Journal, vol. 10, no. 3, pp. 922–932, 2016
4. D. Dakopoulos, N.G. Bourbakis, "Wearable Obstacle Avoidance Electronic Travel Aids for Blind: A Survey",IEEE Transactions on Systems, Man, and Cybernetics. Part C (Applications and Reviews) **40**(1), 25–35 (2010)
5. Ton, LIDAR Assist Spatial Sensing for the Visually Impaired and Performance Analysis," IEEE Transactionson Neural Systems and Rehabilitation Engineering, vol.26, pp. 1727–1734 (2018)
6. M.-C. Kang, S.-H. Chae, J.-Y. Sun, J.-W. Yoo, S.-J. Ko, A novel obstacle detection method based on deformable grid for the visually impaired. IEEE Trans. Consum. Electron. **61**(3), 376–383 (2015)
7. Cardillo, "An Electromagnetic Sensor Prototype toAssist Visually Impaired and Blind People in Autonomous Walking," IEEE Sensors Journal, vol. 18, no. 6, pp. 2568– 2576, 2018
8. J. Chen, Z. Xu, and Yu, "A 68-mw 2.2 Tops/w Low Bit Width and Multiplierless DCNN ObjectDetection Processor for Visually Impaired People," IEEE Transactions on Circuits and Systems for Video Technol- ogy, vol. 29, pp. 3444–3453, 2019
9. B. Ando, S. Graziani, Multisensor Strategies toAssistBlind People: A Clear- Path Indicator. IEEETransactions on Instrumentation and Measurement **58**(8), 2488–2494 (2009)
10. S. Pehlivan, M. Unay, A. Akan, "Designing an Obstacle Detection and Alerting System for Visually Impaired People on Sidewalks", in. Medical Technologies Congress (TIPTEKNO) **2019**, 1–4 (2019)
11. F. Al-Muqbali, N. Al-Tourshi, K. Al-Kiyumi, and F.Hajmohideen, "Smart Technologies for Visually Impaired:Assisting and conquering infirmity of blind people using AI Technologies," in 2020 12th Annual Undergraduate Research Conference on Applied Computing (URC), 2020, pp. 1–4
12. A. N. Zereen and S. Corraya, "Detecting real time object along with the moving direction for visually impaired people," in 2016 2nd Inter- national Conference on Electrical, 2016, pp. 1–4
13. W. Lin, M. Su, W. Cheng, and W. Cheng, "An Assist System for Visually Impaired at Indoor Residential Environment using Faster-RCNN," in 2019 8th International Congress on Advanced Applied Informatics (IIAI- AAI), 2019, pp. 1071–1072
14. S. Deshpande and R. Shriram, "Real time text detection and recognition on hand held objects to assist blind people," in 2016 International Conference on Automatic Control and Dynamic Optimization Techniques (ICACDOT), 2016, pp. 1020–1024. [15.] N. Khaled, S. Mohsen, K. E. El-Din, S. Akram, H. Metawie, and A. Mohamed, "In-Door Assistant Mobile Application Using CNN and TensorFlow," 2020International Conference on Electrical

15. S. Shah, J. Bandariya, G. Jain, M. Ghevariya, and S. Dastoor, "CNN based Auto- Assistance System as a Boon for Directing Visually Im- paired Person," in 2019 3rd International Conference on Trends in Electronics andInformatics (ICOEI), 2019, pp. 235–240
16. T. Saeteng, T. Srionuan, C. Choksuchat, and N. Trakulmaykee, "Reform- ing Warning and Obstacle Detection Assisting Visually Impaired People on mHealth," in 2019 IEEE International Conference on Consumer Electronics - Asia (ICCE-Asia), 2019, pp. 176–179
17. K. Shishir, S. R. Fahim, F. M. Habib, and T. Farah, "Eye Assistant: Using mobile application to help the visually impaired," in 2019 1st International Conference on Advances in Science, Engineering and Robotics Technology (ICASERT), 2019, pp. 1–4
18. K. Lakde and P. S. Prasad, "Navigation system forvisually impaired people," in 2015 International Conference on Computation of Power, Energy, Information and Communication (ICCPEIC), 2015, pp. 93–0098
19. A. Adishesha, B. Desai, "3D Imprinting of the Environment for the Visually Impaired", in. IEEE European Modelling Symposium (EMS) **2015**, 148–153 (2015)
20. S. Yadav, R. C. Joshi, M. K. Dutta, M. Kiac, and P. Sikora, "Fusion of Object Recognition and Obstacle Detection approach for Assisting Visually Challenged Person," in 2020 43rd International Conference on Telecommunications and Signal Processing, 2020, pp. 537–540
21. S. Alghamdi, R. V. Schyndel, and I. Khalil, "Safe trajectory estimation at a pedestrian crossing to assist visuallyimpairedpeople,"in2012Annual International Conference of the IEEE Engineering in Medicine and Biology Society, 2012, pp. 5114–5117
22. T. Patel, V.J. Mistry, L.S. Desai, Y.K. Meghrajani, "Multisensor - Based Object Detection in Indoor Environment for Visually Impaired People", in. Second International Conference on Intelligent Com- puting and Control Systems (ICICCS) **2018**, 1–4 (2018)
23. M. Kamal, A. I. Bayazid, M. S. Sadi, M. M. Islam, and N. Hasan, "To- wards developing walking assistants for the visually impaired people," in 2017 IEEE Region 10 Humanitarian Technology Conference (R10- HTC),2017, pp. 238–241.
24. H. Dahiya, M. K. Gupta, and Dutta, "A Deep Learning based Real Time Assistive Framework for Visually Impaired," in 2020 International Conference on Contemporary Computing and Applications (IC3A), 2020, pp. 106–109.
25. Adishesha and B. Desai, "3D Imprinting of the Environment for the Visually Impaired," IEEE European Modelling Symposium (EMS). Madrid, Spain **2015**, 148–153 (2015). https://doi.org/10.1109/EMS.2015.32
26. C. K. Lakde and P. S. Prasad, "Navigation system for visually impaired people," 2015 International Conference on Computation of Power, Energy, Information and Communication (ICCPEIC), Melmaruvathur, India, 2015, pp. 0093–0098.doi: https://doi.org/10.1109/ICCPEIC.2015.7259447
27. S. Alghamdi, R. van Schyndel, I. Khalil, "Safe trajectory estimation at a pedestrian crossing to assist visually impaired people,", , Annual International Conference of the IEEE Engineering in Medicine and Biology Society. San Diego, CA, USA **2012**, 5114–5117 (2012). https://doi.org/10.1109/EMBC.2012.6347144
28. Hangrong Pan, C. Yi and Y. Tian, "A primary travelling assistant system of bus detection and recognition for visually impaired people," 2013 IEEE International Conference on Multimedia and Expo Workshops(ICMEW), San Jose, CA, USA, 2013, pp. 1- 6.doi: https://doi.org/10.1109/ICMEW.2013.6618346
29. R. Tapu, B. Mocanu, A. Bursuc and T. Zaharia, "A Smartphone-Based Obstacle Detection and Classification System for Assisting Visually Impaired People," 2013IEEE International Conference on Computer VisionWorkshops, Sydney, NSW, Australia, 2013, pp. 444- 451.doi: https://doi.org/10.1109/ICCVW.2013.65
30. N. Mahmud, R. K. Saha, R. B. Zafar, M. B. H. Bhuianand S. S. Sarwar, "Vibration and voice operated navigation system for visually impaired person," 2014 International Conference on Informatics, Electronics & Vision (ICIEV), Dhaka, Bangladesh, 2014, pp. 1–5.doi: https://doi.org/10.1109/ICIEV.2014.6850740
31. Noorithaya, M. K. Kumar and A. Sreedevi, "Voice assisted navigation system for the blind," International Conference on Circuits, Communication, Control and Computing, Bangalore, India, 2014, pp. 177–181.doi: https://doi.org/10.1109/CIMCA.2014.7057785:

32. S. Abdullah, N.M. Noor, M.Z. Ghazali, "Mobility recognition system for the visually impaired, in IEEE 2nd International Symposium on Telecommunication Technologies (ISTT). Langkawi, Malaysia **2014**, 362–367 (2014)

Sentiment Analysis of Zomato and Swiggy Food Delivery Management System

Anand Upadhyay, Swapnil Rai, and Sneha Shukla

Abstract In today's world, millions of people are using micro-blogging sites for expressing their views. So, a large number of data are available on these micro-blogging sites. Understanding this kind of data helps in determining what a business needs to improve on. Our objective is to identify a better food delivery management system between Swiggy and Zomato, by performing sentiment analysis on Twitter data. Sentiment analysis is a part of AI which analyses the subjective information in a sentence, it can be a perception, appraisals, emotions, or attitude towards a topic, person, or entity. In this paper, we have extracted the Twitter data through Twitter API for performing sentiment analysis on tweets related to Zomato and Swiggy. We have calculated the subjectivity and polarity of a sentence to know whether the reviews are positive, negative, or neutral. This calculation helped us in determining which food delivery system is better among Zomato and Swiggy.

Keywords Twitter · Micro-blogging · Swiggy · Zomato · Sentiment analysis · Polarity · Subjectivity

1 Introduction

Sentiment Analysis is a process of analyzing sentiment with the help of Natural Language Processing, by analyzing text and performing statistics on the calculated values. Every business needs feedback for improvement in the product and services provided by them to customers. Sentiment Analysis helps businesses to better understand the sentiments of consumer, their opinion, their approach, and their feelings.

A. Upadhyay (✉)
Thakur College of Science and Commerce, Mumbai 400101, India

S. Rai
Thakur Global Business School, Mumbai 400101, India

S. Shukla
Guru Nanak Vidyak Society Institute of Management, Mumbai 400037, India

A. K. Luhach et al. (eds.), *Second International Conference on Sustainable Technologies for Computational Intelligence*, Advances in Intelligent Systems and Computing 1235,
https://doi.org/10.1007/978-981-16-4641-6_4

Over the past few years, we have witnessed the rise in the use of social media platforms. Nowadays people are more willingly sharing their thoughts and views on micro-blogging sites. These sites can be considered as a source of secondary data for analysis purposes. By automatically analyzing customer feedback such as survey answers and social media interactions, brands may understand why people find certain factors or products relevant or irrelevant. This enables them to modify goods and services to the requirements of the customer. We have gathered 3000 tweets of swiggy and zomato user's feedback from Twitter. In the second part, we have cleaned the data by performing the Data Pre-processing method. After the pre-processing part, we found polarity and subjectivity. Based on polarity and subjectivity, we got the positive, negative, and neutral feedbacks of respective food delivery services.

2 Literature Review

Mayank Nagpal et al. had used sentiment analysis for different types of food delivery apps are available for the delivery of the food. Here the authors collected the data from various digital platforms i.e. WhatsApp, Facebook, and Twitter, the collected data were used for the analysis of various issues using sentiments [1]. Shrawan Kumar had used Twitter sentiment analysis to create awareness regarding the use of social media to gather the data and reviews from the customers and helps to gain more customers if compared to its competitors. He has gathered all these tweets from Twitter and these data were further analyzed [2]. Devipriya et al. in their research aimed at various aspects that influence the customers to use online services like food delivery apps. How these online service providers have changed the way a consumer reflects towards their services and uses their services [3]. Vikas Gupta et al. have shared his case study about the various risks and the advantages of choosing the right OFDAS. The paper is all about the perspective of a consumer in choosing the OFDAS and what makes them re-order from the same service provider [4]. Arghya Ray et al. in this article had research and also had used a survey method to know about why people are more inclined towards the use of FDA's. What is making them using the FDA's platform [5]. Anamika Sinha et al. in her research paper has gathered information through various sources like reports through different sources, meeting the CEO's, data analyst, etc. The main purpose of the research was to find out or to predict the growth of FDAs in India in the future [6]. Anuj Pal Kapoor et al. in their article have shared how the OFA's have changed people's perspective. People Nowadays are more Inclined towards ordering food online and not going to restaurants. In his paper, he has shared the various factors that influence customers to order online [7]. Reena Jain et al. in their research paper is stating about the bullwhip effect which is a type of supply chain which is nowadays used by the FDA. How E-supply chain has changed the way a customer orders his food online and how do these apps use the advantages to the fullest [8]. Swarnalatha et al. in the research paper have done some research about the existing FDA's. What are the problems in the current FDA platform and which changes are to be done to enhances customer satisfaction and can increase the number of demands [9]. Rudy Prabowo and Mike Thelwall in their research have shared their learnings on sentiment analysis and its importance. We can gain the sentiments of consumers through different modes like reviews, comments, and later by different tools and methods we can extract the relevant contents and the business can improvise itself accordingly [10]. Clare Grover et al. in their paper have discussed extracting the datasets from the social media network Twitter. They have used different methods and models to find out the relevancy of the tweets [11]. Wilson and Theresa Ann discuss share their perspectives on detecting private state manifestations and their features, including the origins and amplitude of the private states. They've tried to detect the strength, polarity, and perspectives of private states [12].

3 Methodology

3.1 Sentiment Analysis

Sentiment analysis, generally known as opinion mining, is a methodology for identifying the positivity, negativity, or neutrality of data through natural language processing. Sentiment analysis is frequently used to textual data to help enterprises monitor brand and product sentiment in customer feedback and better explain customer's requirements. For example, using sentiment analysis to evaluate 3000+ reviews about your product could help you figure out whether consumers are satisfied with your pricing and customer service. It aids in the storage of large amounts of data, real-time analysis, and the application of consistent criteria. Also, humans fail to analyze emotions correctly, so emotion interpretation remains one of the most difficult challenges in natural language processing. Data science is improving its ability to create more reliable sentiment classifiers, but there is still much work to be done (Fig. 1).

3.2 Data Acquired

Data Acquiring is also known as collecting data. The collection of data is an essential process. With the help of Twitter API, the tweets are extracted in the CSV file for Preprocessing.

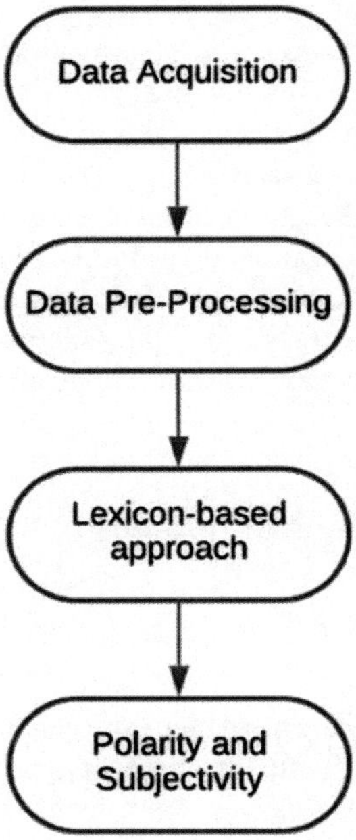

Fig. 1 Steps of sentiment analysis

3.3 Data Pre-processing

After acquiring the data Pre-processing method removes the unnecessary words or they can be called stop words, repeating words. In Pre-processing elimination of emoticons, elimination of URLs, etc. things take place.

3.4 Lexicon-Based Approach

In the Lexicon-based approach, each term's semantic orientation and magnitude are compared to a well before defined dictionary classification that divides words into positive, negative, and neutral subcategories. A text message is nothing more than a string of characters. After giving scores to every word, the sentiment of a complete sentence is calculated by taking an average of all sentiments or using some other pooling operation.

3.5 *Polarity and Subjectivity*

Polarity is knowing whether the sentence is positive, negative, or neutral by seeing the values. The Polarity of any sentence lies between [1, −1]. where 1 means positive, −1 means negative, and if the value is coming as 0 it means the sentence is neutral. Subjectivity refers to the opinion of a person their emotions or acumen. Subjectivity value lies between 0 to 1.

4 Result

The Polarity of a sentence is calculated by averaging the scores given to the words by comparing them with the pre-defined dictionary. Subjectivity gives the idea about the perception, tone, and the emotion of sentence. The sentences with more opinion, emotions, or acumen will give more Subjectivity value (Figs. 2 and 3).

Calculating the positive, negative and neutral Tweets of Swiggy and Zomato food delivery system:

We have collected 3000 data for each delivery system. To know which food delivery system is better than the other we have found the percentage of positive, negative, or neutral tweets of Swiggy and Zomato food delivery system separately. The positive tweet percentage is obtained by dividing the number of positive tweets by total tweets and then multiplying with 1000. In the same way, we have found the negative and neutral tweets percentage of both the food delivery system (Figs. 4 and 5).

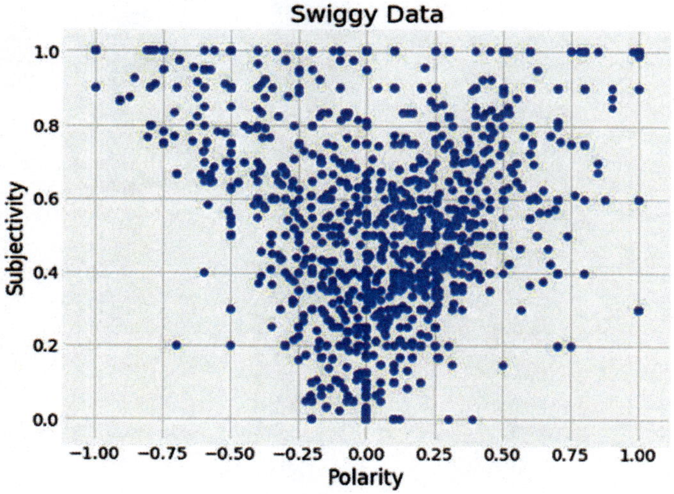

Fig. 2 Polarity and subjectivity graph of Swiggy's tweet

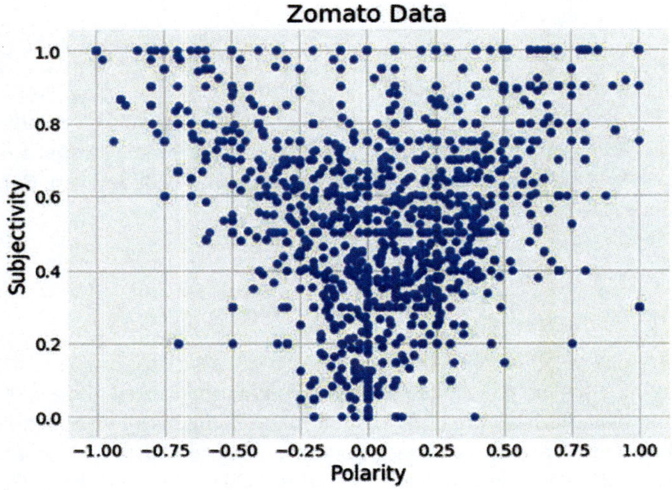

Fig. 3 Polarity and subjectivity graph of Zomato's tweet

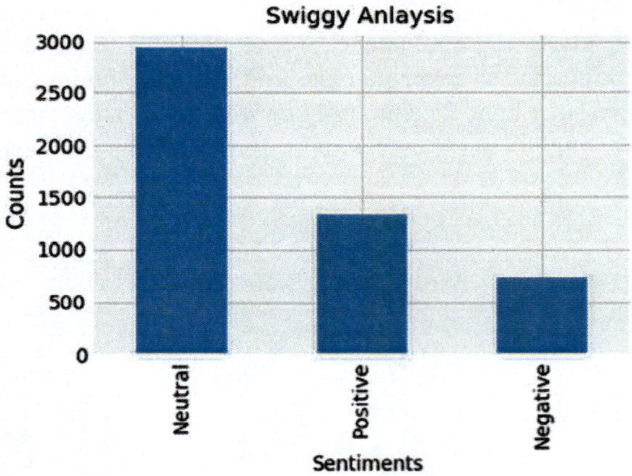

Fig. 4 Sentiment analysis graph of Swiggy

Let's assume,

Positive Tweet = PoTw
Negative Tweet = NeTw
Neutral Tweet = NuTw

Categorizing tweets of Swiggy data:

PoTw percentage = (PoTw/total tweets)*100 = 26.6%
NeTw percentage = (NeTw/total tweets)*100 = 16.6%
NuTw percentage = (NuTw/total tweets)*100 = 50.8%

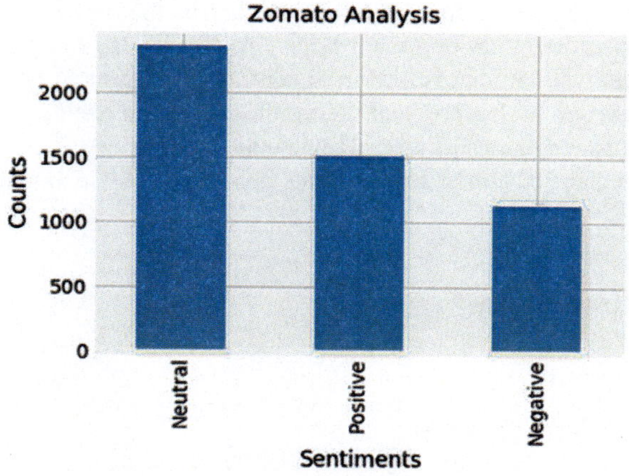

Fig. 5 Sentiment analysis graph of Zomato

Categorizing tweets of Zomato data:

PoTw percentage = (PoTw/total tweets)*100 = 30.2%
NeTw percentage = (NeTw/total tweets)*100 = 22.2%
NuTw percentage = (NuTw/total tweets)*100 = 47.6%

5 Conclusion

As we can see from the above calculation that Zomato has more positive tweets than swiggy. But by observing individual data we can see that the Swiggy has less negative feedback than the Zomato. This implies that the Swiggy customers are more satisfied with their services as compared to the Zomato food delivery system.

By manual reading of the tweets, we have seen that Swiggy's idea of delivering the new categories like medicines, groceries, etc. has satisfied the customers more. Also, Zomato provides more offers and cashback in comparison to Swiggy. Which attracts customers more for preferring Zomato over Swiggy. So we can conclude that Zomato provides better offers and Swiggy provides better services to gain their customers.

6 Future Work

In this research, we have used a fine-grained type of sentiment analysis which determines the polarity and subjectivity of customer reviews to know their positive and negative review. It doesn't give any idea about the reason behind the positive and negative feedback. In the future, we can use aspect-based sentiment analysis to know which are the aspect liked and disliked by the customers so that companies can improvise their services accordingly. ABSA is also known as abstract based sentiment analysis. ABSA is a form of sentiment analysis in which the text is analyzed, categorizes by feature, and the sentiment behind each one is determined. ABSA assists businesses in sorting and reviewing consumer data as well as automating procedures such as customer service tasks in order to obtain valuable insights.

References

1. M. Nagpal, K. Kansal, A. Chopra, N. Gautam, V.K. Jain, Effective approach for sentiment analysis of food delivery apps, in *Soft Computing: Theories and Applications* (Springer, Singapore, 2020), pp. 527–536
2. S.K. Trivedi, A. Singh, Twitter sentiment analysis of app based online food delivery companies. Glob. Knowl. Mem. Commun. (2021)
3. P.J. Devipriya, J. Mohan, A.B. Gowda, The influence of various factors on online food delivery services. IUP J. Supply Chain. Manag. **17**(2) (2020)
4. V. Gupta, S. Duggal, How the consumer's attitude and behavioural intentions are influenced: a case of online food delivery applications in India. Int. J. Cult. Tour. Hosp. Res. (2020)
5. A. Ray, A. Dhir, P.K. Bala, P. Kaur, Why do people use food delivery apps (FDA)? A uses and gratification theory perspective. J. Retail. Consum. Serv. **51**, 221–230 (2019)
6. N. Meenakshi, A. Sinha, Food delivery apps in India: wherein lies the success strategy? Strat. Dir. (2019)
7. A.P. Kapoor, M. Vij, Technology at the dinner table: ordering food online through mobile apps. J. Retail. Consum. Serv. **43**, 342–351 (2018)
8. R. Jain, M. Verma, C.K. Jaggi, Impact on bullwhip effect in food industry due to food delivery apps. Opsearch **58**(1), 148–159 (2021)
9. P. Swarnalatha, N. Pandey, G. Agrawal, N. Mathur, S. Pandey, An Insight on existing online food delivery applications in India and proposition of a new model. Int. J. Prog. Sci. Technol. **17**(2), 80–85 (2019)
10. R. Prabowo, M. Thelwall, Sentiment analysis: a combined approach. J. Inf. **3**(2), 143–157 (2009)
11. C. Llewellyn, C. Grover, B. Alex, J. Oberlander, R. Tobin, Extracting a topic specific dataset from a Twitter archive, in *International Conference on Theory and Practice of Digital Libraries* (Springer, Cham, 2015), pp. 364–367
12. T.A. Wilson, Fine-grained subjectivity and sentiment analysis: recognizing the intensity, polarity, and attitudes of private states. PhD dissertation, University of Pittsburgh, 2008

Text Similarity Identification Based on CNN and CNN-LSTM Model

Rohit Beniwal, Divyakshi Bhardwaj, Bhanu Pratap Raghav, and Dhananjay Negi

Abstract Natural Language Processing is an active area of research where new challenges and issues such as text similarity, semantic analysis, paraphrase identification, citation matching, and language translation, etc. foster the need to further exploit this area. Among the above issues, text similarity is one of the key operative challenges which is being explored by many researchers. In text similarity, extracting the meaningful word, phrase, and sentence patterns from texts requires an effective way that can be evaluated based on various standardized metrics. Therefore, we propose a model based on Convolutional Neural Networks (CNN) and Convolutional Neural Networks- Long Short Term Memory (CNN-LSTM) to find the text similarity as image recognition. Our results show that we achieved better classification results with the precision of 61.3% and recall of 57.9% with MatchGridTower as compared to the earlier MatchPyramid model having a precision of 54.0% and recall of 53.1%.

Keywords Convolution neural network · Long short term memory · Machine learning · Natural language processing · Text matching · Text similarity

1 Introduction

"Natural Language Processing (NLP) is an area of research and application that explores how computers can be used to understand and manipulate natural language text or speech to do useful things." [1]. In this area, there are numerous research problems such as text matching, similarity analysis, citation matching, and language translation, etc. Among them, text matching is one of the key problems where different researchers are actively working on it [2–4]. Therefore, for extracting the meaningful word, phrase, and sentence patterns from texts, we require an effective model that can evaluate the text similarity scores based on various standardized metrics. In this regard, various models have been explored based on the Deep Convolutional Neural Network, Deep Structured Semantic model, MatchPyramid model, etc. Similarly, we

R. Beniwal · D. Bhardwaj (✉) · B. P. Raghav · D. Negi
Department of Computer Science & Engineering, Delhi Technological University, Delhi 110042, India

A. K. Luhach et al. (eds.), *Second International Conference on Sustainable Technologies for Computational Intelligence*, Advances in Intelligent Systems and Computing 1235,
https://doi.org/10.1007/978-981-16-4641-6_5

also propose a model based on Convolutional Neural Networks (CNN) and Convolutional Neural Networks- Long Short Term Memory (CNN-LSTM) to find the text similarity as image recognition. However, our model provides better results when compared to the earlier available models.

As far as the text similarity process is concerned, it is widely seen and recognized that rich interaction structures are taken into account to make a good matching decision. These interactions between words, sentences, and phrases are identified as matching patterns. In image recognition, CNN along with LSTM networks provides an effective way to grasp various levels of matching patterns of an image.

In order to implement our model, we formed the matching matrix in which the similarity between the different words is represented by their elements. The matching matrix is further viewed as an image. Subsequently, with the help of CNN along with the LSTM layer, we captured matching patterns from the composed images and got the requisite results.

The rest of the paper is organized as follows: Sect. 2 discusses the related work; Sect. 3 introduces the research approach followed by Sect. 4 which explains the implementation of the research approach along with the discussion on results and evaluation. Finally, Sect. 5 concludes the research paper.

2 Related Work

As far as the text similarity problem is concerned, many researchers have worked in this area. Many of them worked on text similarity based on a single text. These researchers attempted to find a better representation of single text and used simple scoring functions for obtaining the matched results. Recently, new approaches came, which mainly focussed on modeling interaction between words and sentences. In this regard, Huang et al. [2] created Deep Structured Semantic Models (DSSM). These models are used for searching the web. They used word hashing techniques. These models were trained by maximizing the conditional similarity of the clicked documents. These models achieved the Mean Average Precision of 49.5% and Normalized Discounted Cumulative Gain of 53.6%.

On this subject, Hu et al. [3] created ARC-I and ARC-II models based on Deep Convolutional Neural Network. Regarding ARC-I, it was based on two texts interacting at the end of the process. Whereas ARC-II was based on the interaction of two texts at the beginning of the process and made abstraction on this basis. This basis made ARC-II to directly capture sentence-level interactions. As a result, their model achieved the Mean Average Precision of 56.9% and Normalized Discounted Cumulative Gain of 54.7%.

Similarly, in this area, Pang et al. [4] created the MatchPyramid (MP) model for text matching as image recognition. This model extracted meaning at three levels of abstraction for solving fundamental natural language tasks using CNN. As an output, their model achieved the Mean Average Precision of 53.2% and Normalized Discounted Cumulative Gain of 57.2%.

Now, when comparing the earlier works, we first concluded that Hu et al. [3] in ARC-II used a sum operation due to which interactions were not clear. Secondly, Pang et al. [4] in the MatchPyramid model did the representation of the text matching into an image matrix using CNN with the Mean Average Precision of 53.2% and Normalized Discounted Cumulative Gain of 57.2%. However, in this paper, we are proposing a text similarity identification model based on CNN and CNN-LSTM to further improve the earlier results.

3 Research Approach

The research approach for text matching as image recognition using CNN and CNN-LSTM is divided into four phases namely the data collection phase, data preprocessing phase, model design phase, and evaluation phase. Figure 1 depicts the research approach of text matching as image recognition using CNN and CNN-LSTM, which is as follows.

3.1 Data Collection Phase

In the elementary phase, first, we will import the required libraries. Then we will download the suitable dataset. The required properties for the suitable dataset are that it should have a set of questions and their answers. Moreover, for each question, there should be many answers, and among them only one is correct.

3.2 Data Preprocessing Phase

In this phase, we will preprocess the collected data. It includes five steps. The first step performs the tokenization of the dataset i.e., questions will be converted into words. "Tokens are the smaller units, they can be words, subwords, or characters, and the process of separating a piece of text into tokens is called tokenization" [5]. In the second step, all the words will be converted to lowercase because lowercase words are considered differently as compared to uppercase if recognized as images. The Third step will include the removal of all non-ASCII characters. In the fourth step, stop words will be removed. "Stop words are common English words such as the, am, their, which do not influence the semantic of the review. Removing them can reduce noise" [6]. Lastly, in the fifth step, n-grams will be computed. "N-grams of texts are extensively used in text mining and natural language processing tasks. They are basically a set of co-occurring words within a given window and when computing the n-grams we typically move one word forward" [7].

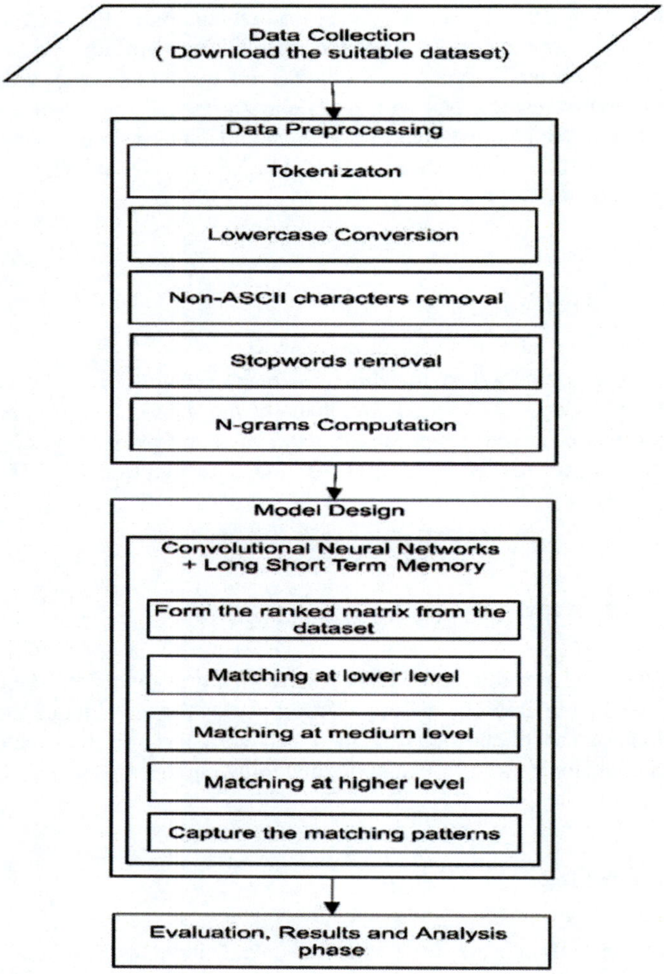

Fig. 1 Overview of research approach

3.3 Model Design Phase

In this third phase to build our model, we will use CNN class for image classification along with LSTM networks. A CNN [8] begins with an image and various filters are applied to produce a map. It applies the Rectified Linear Unit (ReLU) function and pooling layer to every feature map which is flatted to a one-dimensional vector. The vector is input into an artificial network and the features are processed in the neural network. The training process includes forwarding propagation and backpropagation for the defined number of epochs. In the end, we will get a CNN with trained weight

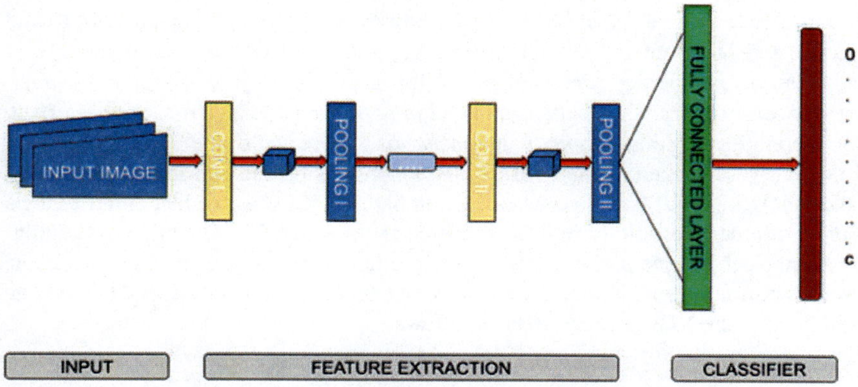

Fig. 2 Image recognition using CNN

and features produced by the feature extractor. The resultant class is a vote of layers, which we want to achieve. Figure 2 depicts image recognition using CNN.

LSTM [9] is a sophisticated Recurrent Neural Network that uses gate and memory cells to acquire dependencies between a sequence. We will add the LSTM layer into our model after adding the pooling layer and then adding the multilayer perceptron layer. Finally, the output result will be evaluated by integrating these relationships through a dynamic pooling layer along with a multilayer perceptron.

Therefore, our model will consist of five steps. Firstly, the ranked matrices will be formed from the dataset. In the second step, matrices interaction matching will be done at the lower level. Thirdly, matching will be done at a medium level. In the fourth step, matching will be done at a higher level. Lastly in the fifth step, various levels matching patterns will be captured and formulated to get the required results.

3.4 Evaluation, Results, and Analysis Phase

Lastly in the fourth phase, we evaluate the results using standard metrics such as precision, recall, and F1 score [10]. They are calculated as follows:

$$Precision = \frac{True\ Positive}{True\ Positive + False\ Positive} \tag{1}$$

$$Recall = \frac{True\ Positive}{True\ Positive + False\ Negative} \tag{2}$$

$$F1Score = 2 \times \frac{Precision \times Recall}{Precision + Recall} \tag{3}$$

The results will be calculated in the form of Mean Average Precision (MAP) and Normalized Discounted Cumulative Gain (NDCG). "MAP for a set of queries is the mean of the average precision scores for each query" [11]. The MAP evaluation process comprises the sorting and coupling of each question–answer pair. Each corresponding question–answer pair consists of a score and a label. The comparison for equality is checked on the threshold, which is also a parameter for our evaluation function. NDCG [12] function takes the number of results to consider and the labels along with the threshold of relevance degree as the input. The Discounted Cumulative Gain metric generates a value for the predicted scores of each document. The result is normalized with the metric value of each document over the true labels. The equation for the NDCG function is as follows.

$$MAP = \frac{\sum_{q=1}^{Q} AveP(q)}{Q} \tag{4}$$

$$nDCG_p = \frac{DCG_p}{IDCG_p} \tag{5}$$

where,

$$IDCG_p = \sum_{i=1}^{|REL_p|} \frac{2^{rel_i} - 1}{\log_2(i+1)} \tag{6}$$

4 Implementation

To implement the research approach, we pick the WikiQA Corpus dataset as it matches the requirements as prescribed in the research approach. We then use the Python programming language to further implement our research model. Now, the implementation for text matching as image recognition using CNN and CNN-LSTM model is as follows.

4.1 Data Collection Phase

In this preliminary phase, first, we downloaded the WikiQA Corpus dataset using Python programming language. In this dataset, we have several different questions and their answers. For each question, there are many answers, and among them only one is correct.

4.2 Data Preprocessing Phase

Next in the second phase, we performed data preprocessing using Python programming language. This phase included five steps. In the first step, the dataset was tokenized i.e. we converted the questions of the dataset into words. Secondly, we performed the lowercase conversion. In the third step, we removed the non-ASCII characters using the Unicodedata Python library. Fourthly, we removed stopwords using NLTK. At last, in the fifth step, we computed n-grams.

4.3 Model Design Phase

In this phase, we built our model with CNN layers for image classification along with LSTM layers. We formed ranked matrices from the dataset. From these matrices, interactions were matched at different levels to get the required pattern. At last, patterns were captured and formulated to get the required results.

4.3.1 Form the Ranked Matrix from the Dataset

The standard representations of text and images used in machine learning applications are different from each other. The texts are 2D words while the images are usually a two-dimensional grid in which each cell consists of a pixel value. Now to deal with text and images, we converted the text into a 2D matrix such that it represented the image in which each pixel value represents the similarity between the two texts.

This implementation comprised aggregation interactions at three hierarchical levels, which gave the overall score for the pixel in the produced image matrix. The model attempted to innovate visualization of text similarity as an image matching problem.

The MatchGrid consisted of 2D grids of Matching Matrix, 2-D convolution, and 2-D pooling. The layered architecture constituted the lower, middle, and higher matching levels that form the tower of MatchGridTower (MGT).

4.3.2 Matching at Lower Level

Matching at the lower level refers to one to one matching within the two texts. Both the identical words as well as words that have similar meanings are included and can contribute to all-over textual similarity.

With reference to the given example as shown in Fig. 3, We found identical matching in "long–long", "time–time", "ago–ago", "letters–letters", "and–and", "telegrams–telegrams", and "were–were", however, we also got the mapping in words that convey similar meaning such as "preferred–favored" and "mode–modes".

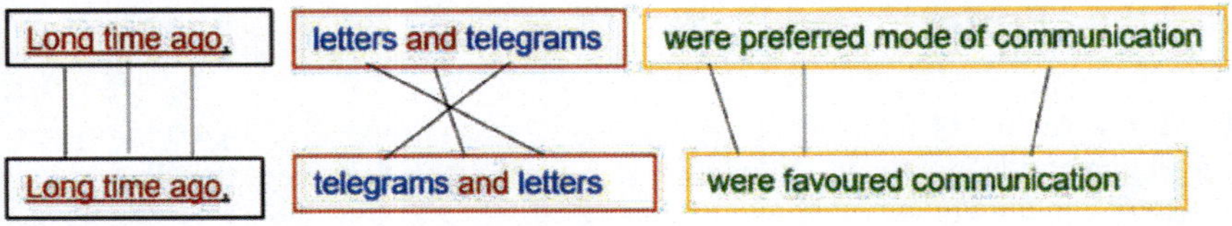

Fig. 3 Sentences to show how interaction works

4.3.3 Matching at Middle Level

Matching at the middle level refers to the mapping between groups of words. Some matching can be identified as an N-gram match, in which the successive order of words are also exactly similar.

With reference to Fig. 3 as shown above, we found the n-term matching in which the order of words could be different but they delivered a similar meaning. Hence, the group of words that were considered, they represented the same meaning. For e.g. "(letters and telegrams)–(telegrams and letters)", and "(were preferred mode of communication)–(were favored communication modes)".

4.3.4 Matching at Higher Level

Here, matching is done at the sentence level to get a better understanding of the text at a higher level. This matching comprises the earlier mentioned successive matchings at each level. Example sentences as shown in Fig. 3 would match at sentence level as their overall meaning conveys a similar meaning. The matching was a sentence-level matching, but we can consider paragraphs as a group of sentences only. Once we consider the paragraphs in the above fashion, we can analyze the overall text similarity.

4.3.5 Capture the Matching Patterns

The MatchGrid was a normal CNN that was designed to extract matching patterns at various levels. As described by the earlier sub-sections, matching patterns were extracted just as in the case of image recognition.

For the initial layer of CNN, the kernel $w^{(1,k)}$ traced the entire matrix to produce the feature map. The following equation represents the feature map generation.

$$z_{i,j}^{(i,k)} = \sigma\left(\sum\nolimits_{s=0}^{r_k-1}\sum\nolimits_{t=0}^{r_k-1} w_{s,t}^{(1,k)} .z_{i+s,j+t}^{(0)} + b^{(1,k)}\right) \tag{7}$$

where $z^{(1,k)}$ is the feature map and r_k gives the size of the kth kernel. Activation function sigma denotes Rectified Linear Unit (ReLU) [13]. The output of the ReLU function will be the same as input when the input is positive, otherwise, the output is 0. It is a piecewise activation function used in neural networks. The value inside the sigma function is the input to one neuron. This feature map value is calculated at every index i, j.

As text could be of different sizes, we incorporated the text size variation scenarios, which helped us in obtaining the fixed-sized feature maps. The following Eq. 8 represents the Dynamic pooling strategy [14] used to solve the problem of length variability.

$$z_{i,j}^{(2,k)} = \max_{0 \le s \le d_k} \max_{0 \le t \le d'_k} z_{i.d_k+s,j.d'_k+t}^{(1,k)} \tag{8}$$

The length of texts determines the d_k and d'_k. d_k and d'_k are the corresponding length and width of our square kernel and can be calculated using the size of feature maps and kernels.

4.4 Evaluation, Results, and Analysis Phase

We used the WikQA dataset [15] that is trained, validated, and tested using the above built model.

The following Table 1 shows the comparison of the MatchGridTower and MatchPyramid with their associated precision, recall, and the F1 score.

From Table 1, we can infer that we achieved better classification results with MGT having a precision of 61.30%, F1 score of 56.8, and recall of 57.9% as compared to the MatchPyramid model having a precision of 54.0%, F1 score of 45.7% and recall of 53.1%. These results are also shown in Fig. 4 via bar graph representation.

As far as this text matching as image recognition problem is concerned, different researchers have worked with various models to provide their solution. The same has been discussed in the related work section. Therefore, here we also compare our model results in the form of Mean Average Precision (MAP) and Normalized Discounted Cumulative Gain (NDCG) with the models of other authors such as [2–4]. The following Table 2 shows the related comparison.

The Fig. 5 as shown below shows the above tabular results via bar graph representation where the green bar shows the MAP and the orange bar depicts the NDCG.

From the above tables and graphs, it can be concluded that our model outperforms the other models by a considerable margin. Hence, the inclusion of the LSTM layer shows considerable improvement in the identification of matching question–answer pairs.

Table 1 Comparison of MatchGridTower and MatchPyramid

	Precision	Recall	F1 score
MatchGridTower	61.3	57.9	56.8
MatchPyramid	54.0	53.1	45.7

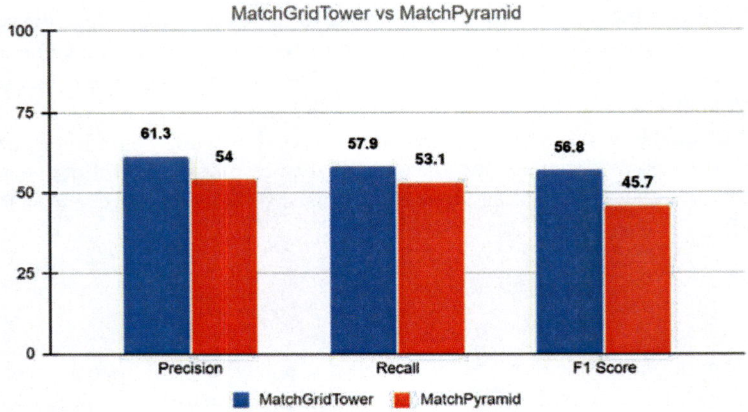

Fig. 4 Comparison graph of MatchGridTower and MatchPyramid

Table 2 Results comparison of MatchGridTower and previously published models

Model	Mean average precision	Normalized discounted cumulative gain
MGT	78.3	74.5
MGT + LSTM	80.8	88.3
MP	53.2	57.2
ARC	56.9	54.7
DSSM	49.5	53.6

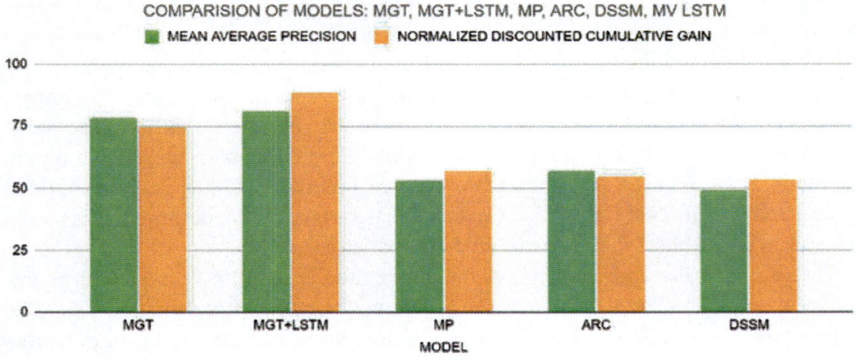

Fig. 5 Bar graph to compare the relative performance of machine learning models on WikiQA dataset

5 Conclusion

In this research paper, we proposed a text similarity identification model known as MatchGridTower based on CNN and CNN-LSTM. In our model, we considered text similarity as image recognition using CNN along with LSTM. To further implement the MatchGridTower model, we picked the WikiQA Corpus dataset. This MatchGrid-Tower model resembled the hierarchical matching of words, phrases, and sentences while identifying text similarity. The model captured these phrases at different levels of the tower and provided the corresponding output. As an output, our results have shown that we achieved better classification results with the precision of 61.3% and recall of 57.9% with MatchGridTower as compared to the earlier MatchPyramid model having a precision of 54.0% and recall of 53.1%.

References

1. G. Chowdhury, Natural language processing. Annu. Rev. Inf. Sci. Technol. **37**(1), 51–89 (2005). Available: https://doi.org/10.1002/aris.1440370103
2. P. Huang, X. He, J. Gao, L. Deng, A. Acero, L. Heck, Learning deep structured semantic models for web search using clickthrough data, in *Proceedings of the 22nd ACM International Conference on Conference on Information & Knowledge Management - CIKM'13*, no. 20131027 (2013), pp. 2333–2338. Available: https://doi.org/10.1145/2505515.2505665
3. B. Hu, Z. Lu, H. Li, Q. Chen, Convolutional neural network architectures for matching natural language sentences. arXiv.org (2015). Available: https://arxiv.org/abs/1503.03244v1
4. L. Pang, Y. Lany, Jointly considering Siamese network and MatchPyramid network for text semantic matching, in *IOP Conference Series: Materials Science and Engineering*, vol. 490 (2019), p. 042043. Available: https://doi.org/10.1088/1757-899x/490/4/042043
5. "Tokenization". Nlp.stanford.edu (2021). Available: https://nlp.stanford.edu/IR-book/html/htmledition/tokenization-1.html. Accessed 25 Mar 2021
6. "Stopwords", W. Maalej, H. Nabil, Bug report, feature request, or simply praise? On automatically classifying app reviews, in *2015 IEEE 23rd International Requirements Engineering Conference (RE)* (2015), pp. 116–125. http://dx.doi.org/https://doi.org/10.1109/RE.2015.7320414
7. "N-grams", https://kavita-ganesan.com/what-are-n-grams/#.YGXeFK8zZPY
8. P.Y. Simard, D. Steinkraus, J.C. Platt, Best practices for convolutional neural networks applied to visual document analysis, in *2013 12th International Conference on Document Analysis and Recognition*, vol. 2 (IEEE Computer Society, 2003) , pp. 958–958
9. S. Wan, Y. Lan, J. Guo, J. Xu, L. Pang, X. Cheng, A deep architecture for semantic matching with multiple positional sentence representations. arXiv.org (2021). Available: https://arxiv.org/abs/1511.08277
10. "Precision and recall". En.wikipedia.org (2018). https://en.wikipedia.org/wiki/Precision_and_recal
11. "Mean Average Precision". En.wikipedia.org, https://en.wikipedia.org/wiki/Evaluation_measures_(information_retrieval)#Mean_average_precision
12. "Normalised Discounted Cumulative Gain", https://towardsdatascience.com/normalized-discounted-cumulative-gain-37e6f75090e9
13. "A Practical Guide to ReLU". Medium (2021). Available: https://medium.com/@danqing/a-practical-guide-to-relu-b83ca804f1f7

14. R. Socher, E.H. Huang, J. Pennin, C.D. Manning, A.Y. Ng, Dynamic pooling and unfolding recursive autoencoders for paraphrase detection, in *Advances in Neural Information Processing Systems* (2011), pp. 801–809
15. Microsoft Research WikiQA Corpus, https://www.microsoft.com/en-us/download/details.aspx?id=52419

Survey Based on Configuration of CubeSats Used for Communication Technology

Gunjan Gupta and Robert Van Zyl

Abstract CubeSat misssions grown progressively competent with increased complexity, since their first dispatch. Moderately accurate appropriation rates and advancement in innovation permit mission designers to look over changed orbital heights, setups of the CubeSat, and commercial off-the-shelf (COTS) subsystems. In order to satisfy specific mission necessities, creators have additionally evolved custom subsystems. In this investigation, a review of the CubeSat mission for communication is done. It is seen that the mission type affects picking the control, configuration, and launching. The survey reflects the most common 3U configuration and less used 6U configuration. The research motivation is also provided with respect to modulation technique. Bar plots are used to show the trends in launching, types of configuration.

Keywords CubeSats · Satellite · Nanosatellite

1 Introduction

In 1957, the USSR launched the very 1st satellite named SPUTNIK into an elliptical low earth orbit [1]. It weighed less than 100 kg. The expenses connected with the physical materials, parts, development of the labor, the fuel of the launch vehicle, are directly proportional to the size of the satellite. The smaller the size of the satellite, the lesser the expenses and the shorter the development cycle. The main advantage is that the satellite can be reconfigured in accordance with the mission needs. As time progresses, lots of development has been made in this field. Jordi Puig-Suari and Bob Twiggs from California Polytechnic State University and Stanford University, respectively, proposed the concept of the first CubeSat in the year 1999. The weight of the CubeSat limits to 10 kg. It has been adopted by numerous academic, government, civil bodies, commercial users, and military within 20 years time. CubeSats are gaining

G. Gupta (✉) · R. Van Zyl
Department of Electrical, Electronics and Computer Engineering, French South African Institute of Technology, Cape Peninsula University of Technology, Cape Town, South Africa
e-mail: guptag@cput.ac.za

A. K. Luhach et al. (eds.), *Second International Conference on Sustainable Technologies for Computational Intelligence*, Advances in Intelligent Systems and Computing 1235,
https://doi.org/10.1007/978-981-16-4641-6_6

popularity because of their fast development time starting from design till their launch and operations. Due to the advancement and rising popularity in CubeSats, several COTS subsystems are available for use to undergraduate and postgraduate students, academic, and several other fields. The CubeSat's mission is categorized into six categories, namely, Communications, Educational, Earth imaging, Military, Technology Demonstration, and Science. The communication satellites build the communication between the two points, some of the examples are amateur radio service, AIS tracking. The educational or the E-class missions satellite have no science or technology value. They do the health check and send data to the ground station. Earth Imaging satellites send images of the earth which are used for commercial and research purposes. Military mission satellites have military relevance. The Science satellites collect data for scientific research, including Earth science, atmospheric science, space weather, etc. This data is for specific research. Technology demonstration satellites' mission involves the first flight of new technology and checks the readiness levels. In this paper, a survey of all the communication satellites is presented.

All CubeSats are not the same. They are divided into several categories depending upon the purpose for which they are used. Some CubeSats are considered as "Hobbyist" and are used in Universities and secondary schools, where there is an opportunity to learn. Lower cost resources and risky approaches are used. These CubeSats can have high failure rates. The "Crafters", also known as "SmallSat", have enough experience and capabilities to understand how to build and test space vehicles. Industrial missions uses standard practices to build high-performance spacecraft. They may have high costs and preferably long development time. The "Constellations" use dozens of spacecraft and use distributed data services to achieve a mission due to which their performance depends on the other missions. The most visible CubeSat constellations are Planet and Spire.

The survey provided contains updated data of all the communication satellites launched as of March 2019. The survey Table 1 given in [2] includes the launch date, CubeSat name and nation, organization, configuration, mass, orbit details, source of power, application, and sources [3–7]. Another section includes the statistical data of all the communication CubeSats launched in the following years and the study of the data according to their configurations.

2 Data Collection

Data is collected with the help of the online database created by Prof Michael Swartwout. He works at the Space Systems Research Laboratory (SSRL) of Saint Louis University. He developed an online database of each and every launched and failed CubeSat. Data comprises of the updated information of launch date, name, release date, contractor, class, subtype, ejector, mission type, mission status, orbit status, and launch vehicle. To date, a total of 1184 CubeSat's are launched, out of which 407 are decayed and 777 are still in orbit, that is, around 65.6% of the total launched

CubeSats. Around 870 of the launched CubeSats are developed by the US and the remaining 314 are developed by the rest of the world.

Data is also collected from Gunter's Space Page, which is created by Gunter Krebs (dipl.phys.). It is a very useful online source that provides information on spaceflights, like their launch vehicles, satellites, space shuttle, and Astronautics. Nanosats Database is also very helpful in collecting information about the application, configuration, sources, mass, and orbit study of the CubeSats. This website has data of around 2500 nanosats and Cubesats. The database does not include information about Femtosatellites, Chipsats, and suborbital launches.

Nanosats database, Gunter's Space page, and Michael Swartout's online database provide a complete list of all the launched and failed CubeSats missions. Some of the particular information was taken as a base for this study [3, 4]. The study includes the name, launch date, organization, configuration, mass, orbit, power, application, and sources. The main focus of this study is the equipment used in the designing of the CubeSats along with the operational orbit, configuration, and power of the CubeSat.

The data mentioned in this study is taken from the official space websites or related academic publications. Some CubeSat missions did not have uploaded the present information related to their equipment used and orbital configuration. So, this data is labeled as "Not Available, i.e., N/A". This study contains all the communication satellites launched from the year 2000 till 2019.

3 CubeSats Information

Falcon-ODE [8] is an educational mission with 1U configuration built by the US Air Force Academy to provide learning about real DoD space missions and to provide radar calibration and optical targets for ground-based space sensors. It is launched on an Electron KS rocket on the STP-27RD mission. The Astrocast 0.2 was built in Switzerland, in the year 2019, with 3U configuration. It provides global L-band M2M (machine-to-machine) services. The satellite demonstrates multi-satellite operations and functionalities of the satellites. Astrocast is basically a network of nanosatellites and estimates to build, launch, and operate a 64 CubeSats constellation for a low cost. It is launched on an Indian PSLV-QL [9]. Aistechsat-3, also known as Danu Pathfinder was launched in early 2019 on an Indian PSLV-QL. The satellite is built for aeronautical, maritime, and aircraft racking via an ADS-B receiver. Aistech Organisation plans for 25 such Danu satellites initially, but now 102 such satellites are envisioned by 2022. AISTECHSAT 1 is developed by Aistech [10] and is a 2U configuration. The main focus is on aeronautical tracking as a prototype for a larger constellation. Various versions of AISTECHSAT were launched. AISTECHSAT 3 was launched in early 2019 on an Indian PSLV-QL. It is also named as Danu Pathfinder. The NEXUS, Next Generation X Unique Satellite [11] is designed at Nihon University in collaboration with Japan AMSAT Association. It is of 1U Configuration and was launched on Epsilon (2) CLPS vehicle. The main purpose of the satellite is to demonstrate next-generation amateur satellite communication technology. The linear transponder was installed in this conventional ultra-small satellite. The RAAF M1 [12] was built by the University of New South Wales, Australia. This is a 3U-CubeSat whose objective is to test and validate the capabilities of Australian SSA by utilizing engineering design for the tracking of the LEO spacecraft. It also focusses on the demonstration and development of the building blocks and the CONOPS ADF space capabilities in the future. The satellite was launched on a FALCON-9 V1.2 (Block 5) rocket. Lemur-2 is built by SPIRE and is the initial constellation of LEO satellites. These satellites were carrying two payloads, one for meteorology and another for ship traffic tracking which is named as STRATOS GPS radio occultation payload and AIS receiver. The STRATOS payload tracks ships worldwide with the help of received AIS signals, and the GPS satellite signals are listened by using a SENSE payload. The LEMUR-2 satellites are launched in small batches as secondary payloads [13]. The PSLV-XL rocket was used to launch the first four Lemur-2 satellites. The 2nd batch was launched on Cygnus CRS-6 vehicle. The 3rd batch of four satellites was launched on a NanoRacks NRCSD-E deployer on the CygnusCRS-5 cargo craft and a 4th batch was launched on H-2B-304 [14]. There are still many batches of Lemur-2 which were launched in batches on different launch vehicles. Spire announced a second series of Lemur satellites to be launched in early 2018, which will feature ADS-B payload to track airplanes. D-Star one is designed and built by German Orbital systems and is a 3U configuration. It is used as a technology demonstration for CubeSat communication constellation. The satellite is equipped with two D-Star communication modules onboard. The satellite was launched on a Soyuz-2-1 b Fregat-M rocket on 28 November 2017. Two more D-Star One satellites, called SPARROW and iSAT were successfully launched in 2018 [15]. Two more D-Star One satellites named as EXOCONNECT and LightSat were also launched in 2019. CAT, also known as CubeSat Assessment and Test, has two 3U CubeSats, developed by the Johns Hopkins Applied Physics Laboratory for the Department of Defense (DoD) to demonstrate communication technologies. It is a low-cost CubeSat mission. CATA and CATB were launched to the ISS in late 2018 and were deployed on 31 January 2019. These satellites use two commercial off-the-shelf spacecraft to support government furnished equipment communications. Survey Satellite Technology Ltd (SSTL) has signed a contract with Honeywell Aerospace to build VESTA. This project is funded by the UK Space Agency. It was launched on a Falcon-9 v 1.2 (Block 5) Rocket. One of the Indian companies Exseed Space launched ExseedSat1 of 1U Configuration. ExseedSat1 is an amateur communication satellite with a multi-function UHF/VHF narrow band frequency modulation. It has also a digital feature for UHF uplink and VHF downlink with APRS. The satellite life depends upon the battery life, it is expected to be 2 years, and after that, it de-orbits naturally. The rocket, Falcon-9 v1.2 is used to launch the ExseedSat1. Hiber 2 is a 6U configuration, built by ISIS [16]. It is a constellation of around 24 CubeSats and was also launched on a Falcon-9 v1.2 rocket. These satellites are basically designed to provide interlinks to IoT devices. The Fox-1C CubeSat was developed by AMSAT. Fox-1C is a research CubeSat developed by AMSAT, which is a radio amateur and technology and is hosting a payload developed by several universities. Fox-1C is based upon 1U CubeSat and is designed to operate in LEO Orbit built using the flight spare of

Fox 1A. It has two whip antennas of 2 m and 70 cm. With the scarcity of launching sites which are affordable for amateur communications, AMSAT developed a small CubeSat which can carry payloads for amateur and scientific communications, which made these satellites qualify for launches in sponsored programs for, e.g., NASA's Educational Launch of Nanosatellites (ELaNa) [17]. Fox-1C serves as a relay for communications for the worldwide amateurs which makes it possible to run at the same time both experiment and communication missions. AMSAT and Spaceflight Inc. came together for the integration and launch, which uses Spaceflight's SHERPA system on a Falcon-9 v1.2 in the 3rd quarter of 2015 as a launch vehicle to a sun synchronous orbit. The same mission was repeated on the SSO-A cluster launch also in 2017–18. The Polar Scout or Operationally Responsive Space 7 (ORS 7) [18] mission consists of satellites that were able to efficiently detect transmissions from emergency positions which indicates radio beacons (EPIRBs), which carry onboard vessels which are used in distress to broadcast their positions. They are based on 6U CubeSat form factor. This was done considering that each Polar Scout CubeSat will travel after every 90–100 min over the North Pole. The orbit time of CubeSats around the earth is around 15 or 16 times in a day which allows a possible search of more than 3 h and for rescue operations every day in the Arctic. The satellites were launched on Spaceflight Industry's SSO which is a multi-satellite launch on a rocket of Falcon-9 v1.2 (Block 5). The four SpaceBEE which was formerly known as Basic Electronic Elements (BEEs) made picosatellites which were based on the 0.25U CubeSat form factor which demonstrated two-way satellite communications and data relay for Swarm Technologies Inc. The mission is intended to test the smallest satellites of the world that can provide two-way communications in order to provide cost-effective services to low data rate IoT networks. This connectivity is a solution for mobile and remote sensors. The experimental space deployment setup consists of four satellites, each of them with a 0.25U form factor which employs the radar signature enhancement technology, which allows passive tracking of the satellites and use of VHF band frequencies. These small or tiny satellites have a radar cross section which is very small, which leads to complications in tracking. Considering this complication, a GPS device is installed in them which broadcasts its position when required. The U.S. Navy's Space and Naval Warfare Systems Command developed a passive radar reflector which is used to cover the four smallest faces of the satellites, which were supposed to increase the satellites radar profile by a multiple of 10 as per the FCC application. The Swarm's application was dismissed by FCC and on an Indian PSLV-XL rocket, the satellites were launched in January 2018 without getting a license with the name SpaceBEE. The ownership of the SpaceBEEs was not concealed at that time which was later revealed by an IEEE Spectrum. Fleet Space Technologies' planned network uses Centauri 2 and 2 as pathfinder satellites for connectivity to the Internet of things (IoT). These Centauri satellites are based upon 3U CubeSat form factor. Two satellites were launched in 2018 for the purpose of secondary payloads. SpaceX Falcon-9 v1.2 (Block 5) rocket and an Indian PSLV rocket were used to launch Centauri 1 and Centauri 2, respectively. Integrated Communications Extension Capability (ICE-Cap) is an experimental communications based on 3U CubeSat developed for the US Navy. The US Navy's PEO Space

Systems, in coordination with Systems Center Pacific (SPAWAR) developed ICE-Cap satellite [19–21]. It was launched in 2016. The ICE-Cap was able to relay the communications between users, one near the North Pole and another somewhere in a different part of the world.

The main objectives of the mission were:

- Signify a cross-link from LEO to MUOS WCDMA in Geosynchronous Orbit
- Polar UHF SATCOM relay with CubeSats
- Developed and smaller radio antenna for the UHF SATCOM missions.

Compact Spectral Irradiance Monitor Flight Demonstration (CSIM-FD) [22] is a nanosatellite project of the Laboratory for Atmospheric and Space Physics (LASP) to make a 6U CubeSat which will take Sun observations. The main purpose of the mission is to use solar spectral irradiance to get insight into the effect of solar variability on the Earth's climate and to confirm the validation of the climate model sensitivity on solar forcing which is spectrally varying. LASP will facilitate and integrate to test the CSIM-FD payload using BCT's 6U spacecraft bus. The functional bus testing was conducted by BCT and the environmental testing was conducted by LASP before launch. The satellite was launched in November 2018 into a 500 km × 500 km, 52° orbit. Cost-effective High E-Frequency Satellite (CHEFsat) [3] is a 3U CubeSat to test and prepare a consumer communications technology for utilization in space. CHEFSat will test new, emerging millimeter wave components in space. Low cost, high performance, reliable IC devices that operate in E-band are now readily available. The main mission objective is to better understand the effects of weather and atmospheric conditions on E-band links. After ejection from the dispenser, CHEFSat will be in low power mode for 45 min. At the 45 min mark, CHEFSat will deploy the S-band telemetry flip-out patch antenna and the UHF folded dipole antennas. The bus will remain in low power tumble mode until the first contact over the ground station. CHEFSat will be commanded to de-tumble. Once the tumble rate drops below an acceptable level (<1°/min), CHEFSat will be commanded to deploy solar panels during the next available ground station contact. Once deployed, CHEFSat will enter into a coarse sun-pointing mode to increase power generation. Post deployment, spacecraft checkout activities will commence to verify the functionality of the remaining subsystems. Once the bus systems have been verified, the payload checkout activities will begin. A reflight, CHEFSat 2 [23] utilizes Cygnus CRS-10 cargo mission board for launch and was deployed using the external deployer of space crafts. Fleet Space Technologies used Proxima 1 and 2 as pathfinder satellites for the planned network of Centauri satellites connectivity to the Internet of things (IoT). These Proxima satellites were based on 1.5U CubeSat form factor. Both of these satellites were launched in 2018 as secondary payloads using an Electron KS rocket. Changshagaoxin or TY 4-01 [3] is another amateur radio satellite developed by Tianyi Research Institute independently. The satellite was equipped with an amateur radio payload. It is planned to conduct on-orbit ground communication test for amateur radio enthusiasts. At the same time, the satellite will verify the next-generation satellite platform independently deployed by Tianyi Research Institute. The weight of the satellite was 8 kg and was built on the 6U CubeSat form factor. It was launched in

October 2018 as a co-passenger to CFOSAT on a CZ-2C (3) rocket. Tongchuan 1, Zhaojin-1 or TY 4-02 [3] is the first verification test satellite of the Tiange Project. It is an astronomical exploration satellite jointly developed by Tianyi Research Institute and Tsinghua University Astrophysics Center and Tongchuan City of Shaanxi Province. This satellite, as the first verification satellite of the Tiange project, will lay the foundation for the development of the subsequent Tiange plan. At the same time, the star is equipped with a new type of high-sensitivity X-ray polarization detector to carry out a new astronomical observation test. In addition, the satellite will also verify the inter-satellite communication technology and conduct a natural-ground link test with the satellite ground station of Shaanxi Xingyi Space Technology Co., Ltd. The weight of this satellite is also 8 kg and is built on the 6U CubeSat form factor. It was launched in October 2018 as a co-passenger to CFOSAT on a CZ-2C (3) rocket. Bird was a small technological built on 1U CubeSat. BIRDS-2 [24], is launched in 2018. EnduroSat One is 1U CubeSat mission of Bulgaria's first. It's a 1U CubeSat from ISS with goals to encourage Amateur Radio's contribution to space research and for communication purposes in Bulgaria. The satellite has two beacons: a CW (A1A) beacon which is used to transmit its own call sign with TLM data by coding in Morse code and another beacon is a telemetry which is AX.25 operated in datagram mode. EnduroSat One was launched on 21 May 2018, which is to begin the cargo resupply mission Cygnus CRS-9. TRICOM 1 is developed at the University of Tokyo as an experimental 3U Cubesat. The weight of the satellite was 3 kg, it has features to store and forward information and an imaging camera for the earth. Link Space Aerospace Technology developed FengMaNiu 1 which is a 3U CubeSat. It has deployable solar panels. The critical mission of the satellite is to test the performance of added components, for example, two cameras in space. It also served as a repeater for amateurs worldwide with the help of the onboard transponder system. In 2018, it was launched along with Zhangheng 1 and other several small satellites using a CZ-2D (2) booster. Fox-1D has almost the same configuration and properties which Fox-1C with additional repeaters with a VHF downlink with analog voice using FM and data using FSK and L-band for uplink. SHARC (Satellite for High Accuracy Radar Calibration) is a picosatellite built by the Air Force Research Laboratory (AFRL) in order to provide calibration using radar for radars situated on the ground. It is built to the non-standard 5U (1 × 5U) CubeSat form factor. SHARC is to express the capability to perform important calibration of around 120 TriService C-Band radars. The calibration is required to do the tracking of the orbital objects. It is to show the data delivery at low latency. It is also used to investigate the performance of Hypervisor on-orbit for DARPA. It features a GPS system and a C-band return transponder. SHARC features the EyeStar communications system, tested on GEARRS, which uses the Globalstar satellite communications network for command and control. SHARC also carries equipment, especially a highly precise GPS system, to be tested for the Biarri-Squad constellation and is designated Biarri-Point or Biarri Risk Mitigation Flight in this capacity. NRO refers to this mission as Square Dance. SHARC was launched in April 2017 on board of a cargo craft to the International Space Station from where the satellite was deployed on 17 May 2017. SHARC reentered on 3 or 4 May 2019. Singaporean experimental nanosatellite mission is known as VELOX-II. It is a 12 kg 6U CubeSat that carries satellite-based communication hardware developed by Addvalue Innovation Pte Ltd. The unique payload of which transmits data that is collected by a geostationary satellite named as Inmarsat-4 L-band. The SMDC Nanosatellite Program (SNaP) are e 3U CubeSats which were built for the mission for the development of software defined radio (SDR) for providing coverage to users at remote locations. They have five times the data rate of the SMDC-ONE satellites. On the zenith side, the satellites have four deployable solar arrays, while the nadir side features four pop-up UHF antennas. Fox 1A is a 1U CubeSat that operates in LEO. Same as Fox 1C and 1D. Fox 1A was accepted into the NASA ELaNa program in February 2012. It was launched on the ELaNa-12 mission. Three more units will be constructed with a different science payload. The first of these is satellite is named Fox 1B (RadFxSat). Fox 1C and 1D have been assigned a radiation effects experiment and a high energy radiation instrument, respectively. AAUSAT 3 was built using no Cubsat-kit. It got an Automated Identification System (AIS) receiver tested in orbit. AAUSAT 4 tested AIS receiver built by a student. AAUSAT 5's primary mission is to receive AIS beacons from ships using a new design which is an extension of AAUSAT 3. DCBB or Chinese Amateur Radio satellite 3G (CAS 3G) is a 2U Cubesat. It is launched in a cluster of 20 satellites on the maiden flight of the CZ-6 rocket. The Perseus-M satellites are small maritime surveillance satellites developed by Canopus Systems US, an US American independent affiliate of the Russian Dauria Aerospace. The Perseus-M satellites are built as pathfinder satellites to the CubeSat (6U) standard and feature an Automatic Identification System (AIS) receiver provided by LuxSpace similar to Dauria's DX 1 prototype satellite. Originally an imaging payload was planned. Dauria and Canopus Systems also planned to launch eight optical Perseus-O earth observation satellites in late 2015, but this was canceled. HumSat-D spacecraft is based on the XaTcobeo and it is built on Pumpkin's 1U CubeSat. The problem associated with it is that it has no attitude or propulsion control. The satellite was launched from a Dnepr booster and is deployed from the Italian unisat 5 satellite. Triton 1 and 2 are both 3U CubeSat nanosats which were developed by Innovative Solutions In Space BV (ISIS-BV) in order to validate new technologies in the space and also to test AIS receiver. The South African ZACUBE 1 or TshepisoSat is a 1U CubeSat which is developed by students of Cape Peninsula University of Technology (CPUT), Cape Town, South Africa. In order to characterize Earth's ionosphere and for calibration of South African National Space Agency (SANSA's) auroral radar installation at the SANAE-IV base in Antarctica, the main payload used is a high frequency (HF) beacon transmitter. The Dnepr booster is used for the launch of this satellite in November 2013 and the launch was provided by ISI Space. Hoʻoponopono 2 is a 3U CubeSat mission is developed for radar calibration and is developed at the University of Hawaii by students. The second CubeSat Hoʻoponopono 3 used results of the Hoʻoponopono 2 mission to place the satellite in the longer orbit for a longer time. TurkSat-3USat is a communication 3U-based CubeSat built in the Istanbul Technical University (ITÜ), Istanbul. The payload of this is used for voice communication using a VHF/UHF transponder. It is also equipped with a small camera. A deorbiting system is also included to deorbit the satellite following its end of life. The nanosatellite has sufficient volume for all the systems in case of failures. In September 2012, SMDC-ONE [21] spacecraft were launched aboard an ATLAS V rocket as an auxiliary payload which allowed more detailed testing. The two SMDC-ONE satellites which were launched on OUTSat were displayed as SMDC-ONE 2.1 (Able) and SMDC-ONE 2.2 (Baker). The most important mission of the SMDC-ONEs was to demonstrate voice and data communications through a LEO satellite using military standard radios. Both of them Able and Baker have provided important test data since their deployment in the orbit. Aeneas is a nanosatellite project of the University of Southern California/Space Engineering Research Center (USC/SERC). A 3U (30 cm × 10 cm × 10 cm) is modified at National Reconnaissance Office (NRO)-specified Colony I CubeSat bus to address a required research for the Department of Homeland Security (DHS). An improved version of which was planned as PTecSat. RAFT1 is one of two small CubeSats which were launched from a PLA-2 Launcher from the Space Shuttle as a constellation.

RAFT1 Hardware Requirements are:

- For hearing NSS radar ping a 217 MHz downconverter/detector is required.
- The ping detected in above step is transmitted to USNA through an FM audio downlink.
- NSSS radar fence require a 217 MHz 4 mW Oscillator.
- For the Amateur Satellite Service, the communication requires a digital transponder.

STS-116 on 10.12.2006 as part of the RAFT project launched two CubeSats in a constellation of two satellites one of which is the MARSCOM satellite [25].

The secondary mission of the RAFT and MARSCOM is to provide relays for the packet digital communications used for mobile units in order to transmit their GPS coordinates and messages through the satellites. CUTE-1.7 (Cubical TITech Engineering Satellite) [26] is mainly developed by students of the Laboratory for Space Systems (LSS). The main requirements from this are:

- Amateur radio service
- Avalanche Photo Diode (APD) sensor demonstration
- Deployment of Tether.

It was launched as a sub-payload on the 3rd stage of Japanese solid rocket M-5 in February 2006.

4 Data Analysis

CubeSat is gaining demand day by day due to many benefits. As the data can be seen from Table 1 [2], the number of launched communication satellites in the year 2018 is 65 which is a pretty high amount as compared to the previous years. The data is recorded till early 2019 where only 4 satellites were launched till date. Numbers are expected to be more than 2018. In 2017, the number of communication satellites was

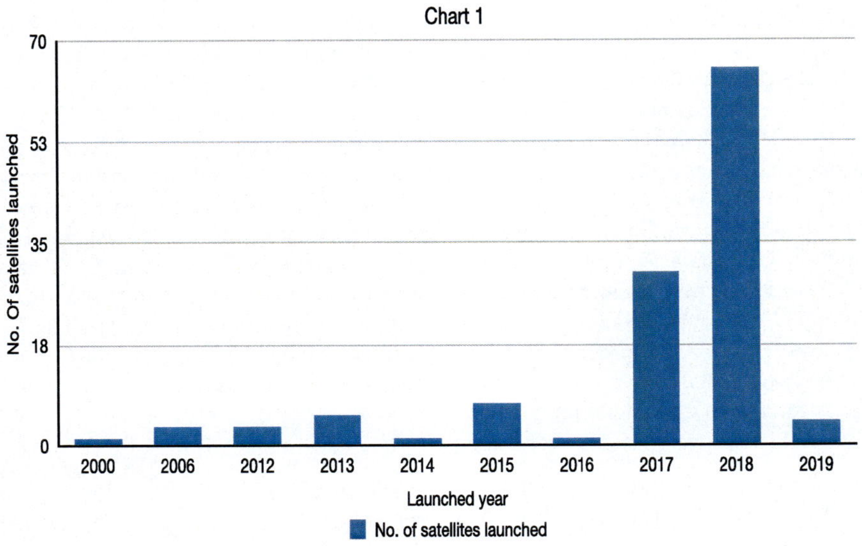

Fig. 1 Number of launched satellites (communication) missions versus launch year

30, which has contacted the base station. In Fig. 1, a bar plot is used to show how many satellites were launched every year from 2000 to 2019. Some more satellites were launched but were a failure. Among all six different CubeSat configuration sizes—0.5U, 1U, 1.5U, 2U, 3U, and 6U—the 3U configuration, with 76 missions, is the most often selected configuration as shown in Fig. 2. The 3U provides a larger volume, therefore, allowing packing more components (subsystems or payload). 1U CubeSats are limited in size and volume as compared to 3U which are free from these limitations. Many CubeSat start-up companies or Universities are developing and commercializing the 3U CubeSat platform. It is also interesting to see the distribution of CubeSat missions according to the configuration sizes in a particular year shown in Fig. 3. In the year 2018, the 3U configuration outnumbered all the other configurations due to its less limitations. The number of 3U satellites in the year 2018 was 38 as compared to 24 in the year 2017.

5 Research Motivation on CubeSats

In order to ensure communication coverage in remote areas like mountains, sea, etc., satellites are used, these satellites are usually in LEO orbits. Long range (LoRa) is a low power wide area network modulation technique which is gaining attention to be implemented for CubeSats. Researchers across the globe are including

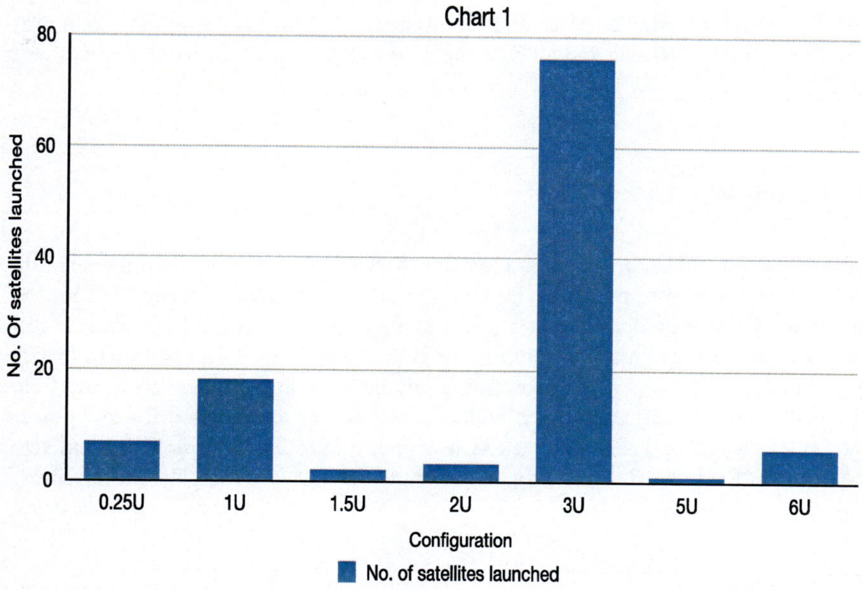

Fig. 2 CubeSat configuration size distribution

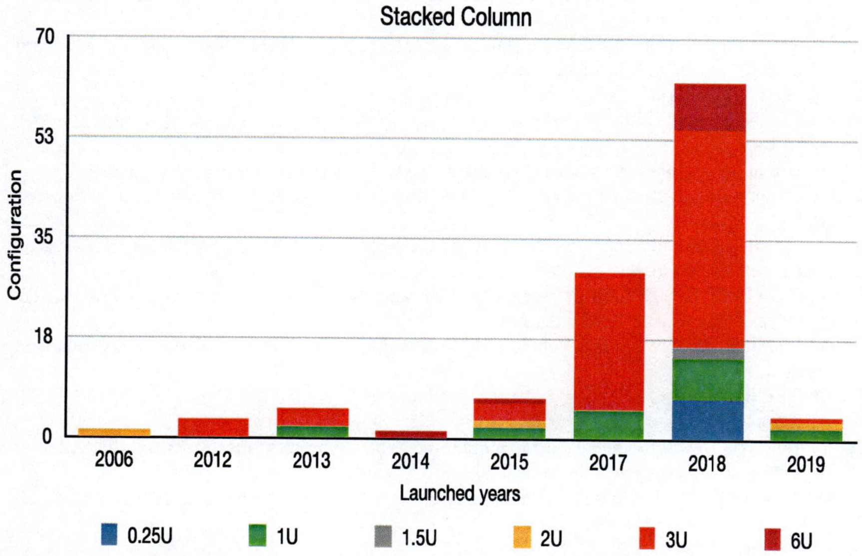

Fig. 3 CubeSat configuration sizes with launch years

LoRa in different aspects of CubeSats mission [27–30]. Other similar technologies like Sigfox, NB-IoT, and LTE-M are also suggested, LoRa outperforms these technologies [28].

6 Conclusion

Expanding interest in the CubeSat idea shows itself in the dispatch numbers and in the wide assortment of missions. Despite the fact that numerous recently proposed CubeSats utilize the 6U arrangement, the quantity of dispatched 6U CubeSats is still tiny, with the 3U design overwhelming the CubeSat populace. The pattern to bigger sizes has likewise helped the reception of dynamic control. Research in the field of CubeSat is ongoing and new modulation techniques are adopted for low power and better reception. LoRa emerges as a winner due to lesser complexity and size requirement for Long Range communication, which is a key for CubeSat missions.

References

1. J. Bouwmeester, J. Guo, Survey of worldwide pico- and nanosatellite missions, distributions and subsystem technology. Acta Astronaut. (2010)
2. G. Gupta, CUBE SATS used for communication technology (researchgate.net, 2021). https://doi.org/10.13140/RG.2.2.33627.90402
3. Gunter's Space Page. Available from: https://space.skyrocket.de/
4. CubeSat Database. Available from: https://sites.google.com/a/slu.edu/swartwout/home/cubesat-database
5. Nanosatellite & CubeSat Database. Available from: https://www.nanosats.eu/database
6. ESA. Earth Observation Portal (2016). Available from: https://directory.eoportal.org/web/eoportal/
7. C. Lafleur, Spacecraft Encyclopedia (2016). Available from: http://claudelafleur.qc.ca/Spacecrafts-index.html#Table-1
8. (1) F-O (AFOTEC). Falcon Orbital Debris Experiment (2019). Available from: https://space.skyrocket.de/doc_sdat/falcon-ode.htm
9. PSLV. Polar Satellite Launch Vehicle (2019). Available from: https://www.isro.gov.in/launchers/pslv
10. AISTECHSAT. AISTECHSAT 1 (2018). Available from: https://space.skyrocket.de/doc_sdat/aistechsat-1.htm
11. T. Tomoaki, Development of Next Generation X-band digital transponder. Available from: https://www.isas.jaxa.jp/e/forefront/2004/toda/
12. M1 R. RAAF M1 3U-CubeSat. Available from: https://space.skyrocket.de/doc_sdat/raaf-m1.htm
13. 2 L. Lemur-2 Nanosatellite Constellation of Spire Global. Available from: https://directory.eoportal.org/web/eoportal/satellite-missions/l/lemur
14. NanoRacks. Portal to Space. Available from: https://nanoracks.com/
15. ISAT D and. D-STAR ONE Sparrow and ISAT. Available from: http://www.orbitalsystems.de/d-star-one-sparrow-and-isat-ready-for-launch/
16. ISIS. ISIPOD CubeSat Deployer. Available from: http://www.isispace.nl/brochures/ISIS_ISIPOD_Brochure_v.7.11.pdf

17. ELaNa. Small Satellite Missions. Available from: https://www.nasa.gov/mission_pages/smallsats/elana/index.html
18. ORS. ORS 7 (Polar Scout 1, 2 / Yukon, Kodiak). Available from: https://space.skyrocket.de/doc_sdat/ors-7.htm#:~:text=The Polar Scout or ORS, their position if in distress.
19. ICAP. ICE cap. Available from: https://space.skyrocket.de/doc_sdat/ice-cap.htm
20. N.S. Alhassoun, M.Y. Sarwar Uddin, N. Venkatasubramanian, Context-aware energy optimization for perpetual IoT-based safe communities. Sustain. Comput. Inform. Syst. **22**, 96–106 (2019). https://doi.org/10.1016/j.suscom.2019.01.016
21. SMDC-ONE. Space & Missile Defense Command-Operational Nanosatellite Effect (2009). Available from: https://earth.esa.int/web/eoportal/satellite-missions/s/smdc-one
22. E. Richard, D. Harber, G. Drake, J. Rutkowsi, Z. Castleman, M. Smith, J. Sprunck, W. Zheng, P. Smith, M. Fisher, A. Sims, B. Cervelli, M. Fowle, M. Miller, M. Chambliss, T. Woods, P. Pilewskie, J.L. Christ, Compact spectral irradiance monitor flight demonstration mission, in *SPIE Optical Engineering + Applications*, vol. 11131 (2019), p. 1113105
23. SAT2. CHEFSat 2 (2018). Available from: https://www.nanosats.eu/sat/chefsat
24. BIRDS-2. Unveiling QSL Card of BIRDS-2 CubeSats (2018). Available from: https://birds2.birds-project.com/
25. RAFT1 Mars. RAFT1 (NO 60, Navy-OSCAR 60) (2006). Available from: https://space.skyrocket.de/doc_sdat/raft1_marscom.htm
26. 1.7 C. Cubical Tokyo Tech Engineering Satellite (2003). Available from: https://directory.eoportal.org/web/eoportal/satellite-missions/c-missions/cute-17-apd
27. G. Gupta, Z.R. Van, Energy harvested end nodes and performance improvement of LoRa networks. **14**(1), 1–15 (2021)
28. P. Anantachaisilp, M. Muangkham, N. Punpigul, M. Thammawichai, SSC20-WKVIII-07 store and forward CubeSat using LoRa technology and private LoRaWAN-server (66) (2020)
29. P. Techavijit, P. Sukchalerm, N. Wongphuangfuthaworn, A. Plodpai, S. Manuthasna, S. Chivapreecha, Internet network in space for small satellites: concept and experiments. Sens. Mater. **30**(10), 2295–2311 (2018)
30. L. Fernandez, J.A. Ruiz-De-Azua, A. Calveras, A. Camps, Assessing LoRa for satellite-to-earth communications considering the impact of ionospheric scintillation. IEEE Access **8**, 165570–165582 (2020)

Ontology Driven Software Development for Better Understanding and Maintenance of Software System

Rohit Beniwal, Kumar Abhijeet, Kushal Kumar, and Mrigank Sagar

Abstract The last few decades have seen tremendous growth in the IT industry where offshoring software has become a general phenomenon. Due to this offshoring, development and maintenance teams are generally located in different time zones along with at distinct locations. This separation of teams at different locations leads to the problem of sharing information in a consistent manner. In this respect, the Semantic Web plays an essential role as it helps in sharing and reusing already available information. Due to the emergence of the Semantic Web, it has been applied in various Software Engineering tasks and issues. Among those various issues, one is to provide a model or system, which can produce automatic documentation for every version of a software system for their better understanding and maintenance. Therefore, this research paper proposes an *ontology driven software development approach for better understanding and maintenance of software system.* The research approach consists of a multi-phased framework in a cyclic manner that generates the various versions of software and their associated documentation. To implement and demonstrate the research approach, we picked a case study. The results of the implementation demonstrate that the paper successfully generated all versions of the software along with its automatic documentation. Moreover, the documentation is accessible in machine-understandable together with human-understandable form, which leads to the generation of consistent information without any ambiguity. Thus, the research work helps in better understanding and maintenance of software systems.

Keywords Documentation · Maintenance · Ontology · Semantic Web · Software development · Software Engineering

R. Beniwal · K. Abhijeet · K. Kumar · M. Sagar (✉)
Department of Computer Science & Engineering, Delhi Technological University, Delhi 110042, India

A. K. Luhach et al. (eds.), *Second International Conference on Sustainable Technologies for Computational Intelligence*, Advances in Intelligent Systems and Computing 1235,
https://doi.org/10.1007/978-981-16-4641-6_7

1 Introduction

The modern era of technology has seen the prevalent presence of outsourcing techniques. This system often results in inconsistent and conflicting information, and it leads to the development of imprecise, incorrect, and unsought software in the absence of a proper knowledge-sharing mechanism. Besides, reusing already available information and knowledge helps in improving efficiency and limits the efforts during the development and maintenance phase. Hence, it becomes a crucial operative challenge to increase the reusability and shareability of software development knowledge [1].

Considering the above-stated problem, the 'Semantic Web' has played an indispensable part as "it provides a common framework that allows data to be shared and reused across application, enterprise, and community boundaries" [2]. Aside from it, the Software Engineering is "the application of a systematic, disciplined, quantifiable approach to the development, operation, and maintenance of software, and the study of these approaches; that is, the application of engineering to software" [3]. Thus, integrating the knowledge domain of Semantic Web [4] and Software Engineering can resolve the issue of shareability and reusability of software development knowledge in case the development and maintenance teams are dispersedly located. Consequently, the above-stated integration provides different problems and challenges, retaining this integrated field exuberant for the forthcoming years.

Among the existing challenges, there is an issue of updating and maintaining the software system. As the size of software increases, so does the task of maintaining along with analyzing how a component has emerged and evolved and what change has occurred, and how. Therefore, maintenance becomes reasonably cumbersome after several versions of software. Hence, a model, which can produce documentation for every version in the machine and human-understandable form will solve this problem. Therefore, this research article provides an ontology driven software development approach for better understanding and maintenance of software systems. Here, "Ontology is a formal and explicit specification of a shared conceptualization" [5]. Our approach consists of a multi-phased framework in a cyclic manner that generates the various versions of software and their associated automatic documentation.

The organization for the rest of the research paper is as follows: Sect. 2 mentions and discusses the work related to our research; Sect. 3 explains the proposed research approach followed by Sect. 4, which expounds its implementation along with the discussion on result & analysis. Finally, the research paper is concluded in Sect. 5.

2 Related Work

To enhance the description and quality of software documentation, a wiki page with annotations was created via lightweight Software Engineering ontology [6] by Tang et al. [7] and de Graaf [8]. The limitation in their work is that the lightweight ontology had a limited number of classes and associated properties. Moreover, their work did not generate any documentation of the software. Later, to develop the software architecture documentation, Lopez et al. [9] created an ontology-based tool known as Toeska Rationale Extraction. However, this tool generated documentation for the software architecture part only.

In this regard, Bhatia et al. [10] provided an ontology driven approach for automating the documentation of a software system. However, it was also limited to generating documentation of a single version software system. Hence, the research gap is still present for documenting the software evolution for maintenance of software systems. Therefore, this research article brings forth an ontology driven software development approach for better understanding and maintenance of software systems. In our approach, we provide a multi-phased framework in a cyclic manner that generates the various versions of software and their associated automatic documentation. Moreover, we generate the documentation in machine-understandable together with human-understandable forms, thus, leading to documentation without any ambiguity. The next section discusses the proposed research approach in detail.

3 Research Approach

The research approach for ontology driven software development for better understanding and maintenance of software systems consists of a multi-phased framework that generates the various versions of software and their associated documentation. Each cycle of software development goes through six phases, namely domain ontology phase, Software Requirement Specification ontology phase, design phase, source code phase, source code listing and cross reference listing ontology phase, and test case ontologies phase. The following Fig. 1 illustrate the research approach.

As per the above research approach, each version of the software is developed using the above-mentioned six phases. The rest of this section describes what happens in each of these phases.

In the elementary phase also known as the domain ontology phase, the major concepts and features related to the domain under consideration will be recorded. Usually, this process is done manually starting from scratch. However, this process may be automated using Artificial Intelligence/Machine Learning techniques. Nonetheless, 100% automation is still not possible as it requires help from experts to understand various terminologies related to the concerned domain. After this ontology is created, it will be converted to its HTML form. Thus, the generated ontology is accessible in both human-understandable together with machine-understandable form.

The second phase known as Software Requirements Specification (SRS) phase, requires the improvisation in domain ontology so that it can be converted to SRS Ontology. This ontology will be organized as per IEEE standard guidelines for organizing the SRS [3]. After this ontology is created, it will also be transformed to its HTML form.

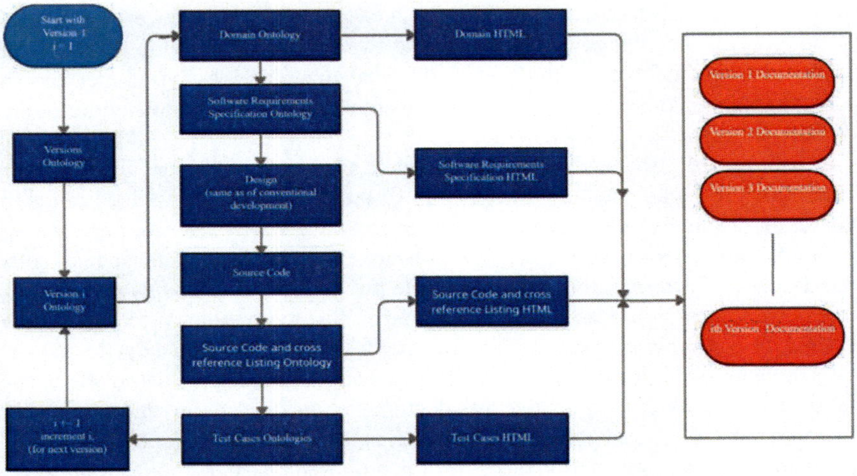

Fig. 1 The research approach for ontology driven software development with 'n' versions of documentation

In the third phase, the design phase, the design for the concerned software system will be developed. Here, we will follow conventional UML design to create the end product known as Software Design Document (SDD).

In the fourth phase, also known as the source code phase, the software or actual program will be written through conventional/object-oriented programming practices. Hence, this phase requires the actual implementation of the concerned software system.

In the fifth phase, source code and cross reference listing ontology is created. This phase aims to explain the purpose and working of variables, functions, classes, etc. used in the source code. This phase provides a great understanding of for what purpose various components of source code are written, thus, giving details about them and their relationships with each other. After this ontology is built, it will also be converted to its HTML form.

Lastly in the sixth phase known as the 'test cases ontologies phase', requires the generation of test cases ontologies for the concerned software system. The testing using the test cases ensures the proper functioning of the software system. It verifies and validates the requirements as per SRS and customer's expectations. After development, this ontology will also be converted to its HTML form. In all of the above six phases, once ontologies are built, we will check their consistencies using reasoners in order to avoid any ambiguity.

After all these phases are over, we will have the first version of the software system along with its associated documentation. Let us call the documentation of this version software as 'version 1' documentation. Once this version of the software system is delivered to the client or customer, the software goes under the operation and maintenance system. Moreover, after the system goes into the operation and maintenance system, it requires a periodic update as per the changing environment. Therefore, for succeeding updates, the software will go through the same cycle of development, leading to succeeding versions and their documentation. Hence, this entire development process takes place in a cyclical manner where each successive iteration generates the next version of the software and its associated documentation. Therefore, at last, our approach requires the creation of version ontology that handles and manages the documentation belonging to various versions of software systems. Thus, this approach of ontology driven software development will help in better understanding and maintenance of software systems.

4 Implementation, Result, and Analysis

To demonstrate the research approach's effectiveness, we selected a case study as used by Bhatia et al. [10]. This case study is based on a software named 'Result Management System' developed for the Delhi Technological University (DTU). The Result Management System (RMS) software includes tasks such as student registration, semester course registration, generating mark-sheets for each semester as well as consolidated view, updating students' marks, etc. The first version of RMS was initially developed by Bhatia et al. [10] via the 'ontology driven Software development' approach as described in the previous section. Once the software was deployed, it came into the operation and maintenance phase. During the maintenance phase, many changes occurred in the examination and marking scheme of DTU. Hence, the second version of RMS is required to be developed. The snapshot of the updated problem statement for Version 2 of the RMS is shown below (Fig. 2).

Therefore, in order to develop the RMS version 2, we implemented our research approach. For this purpose, we used version 5.5.0 of software known as Protege to develop and update the existing ontologies. "Protégé is a free, open-source platform that provides a growing user community with a suite of tools to construct domain models and knowledge-based applications with ontologies" [11]. Protégé has many plugins such as HermiT reasoner [12, 13], FaCT++ reasoner [14], Owl difference plugin, Change Tracker, OWLDoc plugin, etc. where each of them provides the unique functionality. Here, HermiT reasoner and FaCT++ reasoner plugins are used to check whether or not the ontologies are consistent. OWLDoc [15] plugin is used to automatically convert ontologies to their associated HTML pages.

First of all, we picked the RMS ontology and then created a version class under it. Under this version class, we created two sub classes known as version 1 and version 2. In the version 1 sub-class, we put all the existing ontologies as created by Bhatia et al. [10]. After that, we followed our research approach, which has six phases to develop the version 2 ontologies so that they can reflect the updated examination and marking scheme of DTU.

In the first phase, we updated the domain ontology where all major concepts and features related to RMS are recorded as per the updated examination and marking scheme of DTU. After this, we converted this domain ontology to its HTML form.

Delhi Technological University (DTU) conducts an 8-semester B.Tech program in various engineering disciplines. The students are offered different subjects which may or may not have practical evaluation during each semester. The subjects offered in these semesters are either categorized as 'Core' or 'Elective, Core subjects do not provide any alternative, whereas elective subjects do provide choice among various offered subjects. Thus, a student has a choice to select any subject available from the list of an elective subject.

The subjects are divided into two categories: (1) those having a practical component and (2) those that don't have any practical component. For the first category, i.e. those having a practical component, out of total 100 marks, the midsems are of 30 marks, practical labs are of 15 marks, internals are of 15 marks, and end sems are of 40 marks. For the second category, i.e. those that don't have any practical component out of total 100 marks, the midsems are of 25 marks, internals are of 25 marks, and the end sems are of 50 marks.

Every subject has some credit points assigned to it. If the total marks of the student are >=35 in a subject, he is considered 'Pass' in that subject. Otherwise, the student is considered Fail in that subject. If the student passes in a subject, he earns all the credits points assigned to that subject, but if the student fails in a subject, he does not earn any credit points in that subject. At any time, the latest information about the subject being offered in the various semesters and their credit points can be obtained from the university's website.

It is required to develop a system that will store and manage information about subjects being offered in various semesters, students enrolled in various semesters, electives opted by various students in different semesters, whether or not those subjects contain practical components, marks and credit points obtained by students in different semesters, etc. The system should also have the ability to generate printable mark sheets for each student. Semester wise detailed mark lists and student performance reports also need to be generated.

Fig. 2 Problem statement for updated result management system

In the second phase known as the SRS phase, we improvise the domain ontology so that it is converted to SRS Ontology. This ontology is organized as per the IEEE standard requirements and guidelines for organizing the SRS [3]. Again, the SRS ontology is transformed to generate its HTML form. The following Fig. 3 shows the HTML page of the index of SRS ontology from version 2.

In the third phase, the design phase, we amended the design of the RMS to reflect the updated SRS configuration, thus, formed the updated SDD. In the fourth phase, also known as the source code phase, the RMS software is updated using PHP and Visual Basic programming language. In the fifth phase, source code and cross reference listing ontology is also updated and converted to its HTML form. Lastly in the sixth phase known as the 'test cases ontologies phase', we updated the test cases ontologies for RMS in order to make sure the proper functioning of RMS. At last, these test case ontologies are also converted to their corresponding HTML form.

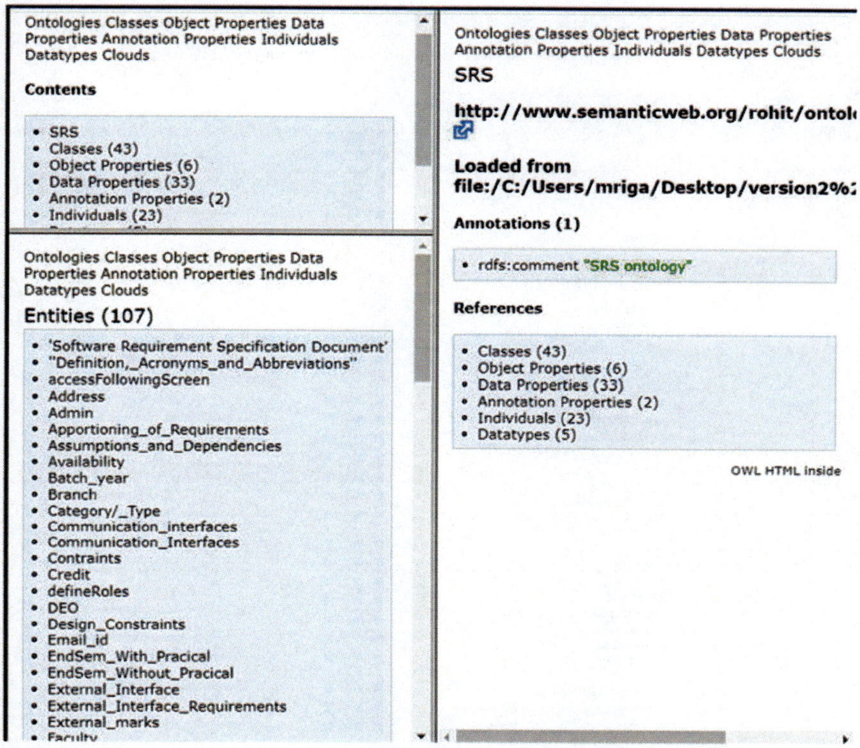

Fig. 3 HTML of the index page of version 2 SRS ontology

Once the above six phases of development are over, we have RMS version 2 software available with us along with its version 2 documentation. The Fig. 4 shown below depicts the protege snapshot consisting of ontologies of version 1 as well as of version 2.

After completion of one cycle of development, it leads to the next version of the software. Moreover, the RMS software is again deployed and it again goes into the operation and maintenance phase. As maintenance is a continuous process, our research approach will be followed again and again in a cyclical manner to generate the succeeding versions. Therefore, as a result, we will always have updated software with updated documentation. Thus, our approach will always help in better understanding and maintenance of software systems where all versions of the documentation are available in one umbrella which facilitates quick understating of the concerned software system.

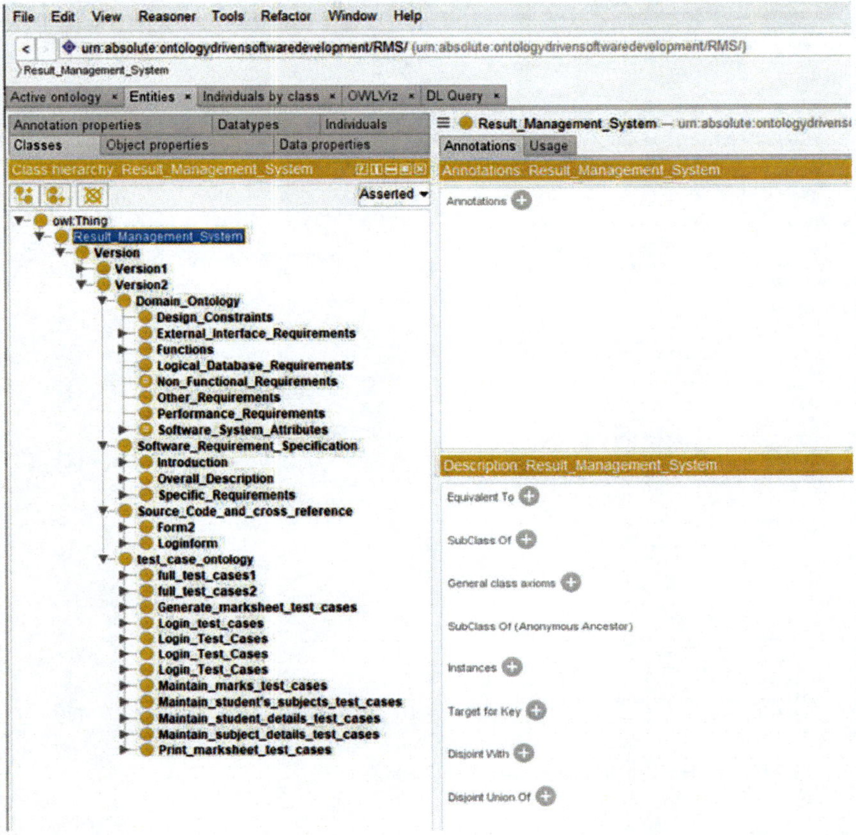

Fig. 4 Protégé snapshot for result management system

5 Conclusion

This paper suggested a research approach for *ontology driven software development approach for better understanding and maintenance of software systems*. For implementation, we picked the RMS case study of DTU. The results of the implementation demonstrate that the research approach successfully generated the succeeding RMS version along with its corresponding documentation. Moreover, the documentation is accessible in the machine-understandable form together with human-understandable form, which leads to the generation of consistent information without any ambiguity. Therefore, this approach stands very useful in comparing different versions of the software along with quick understanding for the maintenance purpose. Furthermore, the approach helps in tracking changes, monitoring new and obsolete features, and understanding how gradually the system developed from its primitive version to its current version. Overall, the research approach will always help in better understanding and maintenance of software systems as it shows all versions of documentation under one umbrella for quick understating of the concerned software system.

References

1. M. Bhatia, A. Kumar, R. Beniwal, SWOT analysis of ontology driven software engineering. Indian J. Sci. Technol. **9** (2016)
2. W3C Semantic Web Activity Homepage. W3.Org (2021), https://www.w3.org/2001/sw/. Accessed 1 Jan 2021
3. *610.12-1990 IEEE Standard Glossary of Software Engineering Terminology* (IEEE, Los Alamitos, 1990)
4. M. Bhatia, A. Kumar, R. Beniwal, Ontologies for software engineering: past, present and future. Indian J. Sci. Technol. **9** (2016). https://doi.org/10.17485/ijst/2016/v9i9/71384
5. R. Studer, V.R. Benjamins, D. Fensel, Knowledge engineering: principles and methods. Data Knowl. Eng. 25(1–2), 161–197 (1998)
6. A. De Nicola, M. Missikoff, A lightweight methodology for rapid ontology engineering. Commun. ACM **59**, 79–86 (2016)
7. A. Tang, P. Liang, H. Vliet, Software architecture documentation: the road ahead, in *2011 9th Working IEEE/IFIP Conference on Software Architecture*, Boulder, CO, USA (2011)
8. K. de Graaf, Annotating software documentation in semantic wikis, in *International Conference on Information and Knowledge Management* (Association for Computing Machinery, New York, 2011), pp. 5–6
9. C. López, V. Codocedo, H. Astudillo, L. Cysneiros, Bridging the gap between software architecture rationale formalisms and actual architecture documents: an ontology-driven approach. Sci. Comput. Program. **77**, 66–80 (2012)
10. M. Bhatia, A. Kumar, R. Beniwal, Ontology driven software development for automated documentation. Webology **15** (2018)
11. Research, Stanford. "Protégé", https://protege.stanford.edu/. Accessed 10 Jan 2021
12. "Hermit Reasoner: Home", http://www.hermit-reasoner.com/. Accessed 10 Jan 2021
13. "Hermit - Protege Wiki", https://protegewiki.stanford.edu/wiki/HermiT. Accessed 10 Jan 2021
14. D. Tsarkov, I. Horrocks, FaCT++ description logic reasoner: system description, in *Automated Reasoning. IJCAR 2006*, ed. by U. Furbach, N. Shankar. Lecture Notes in Computer Science, vol. 4130 (Springer, Berlin, 2006). https://doi.org/10.1007/11814771_26
15. "OWLDoc - Protege Wiki", https://protegewiki.stanford.edu/wiki/OWLDoc. Accessed 10 Jan 2021

Application of Genetic Algorithm (GA) in Medical Science: A Review

Rahul Karmakar

Abstract Medical diagnosis is the key prerequisite for any medical treatment. To get that optimized result of any diagnosis, several tests have been proposed in a cost and time-effective manner. Metaheuristic algorithms are used in many fields; especially in medical science, it has a huge impact. With the help of these algorithms, many models have been developed to get accurate results during diagnosis. In this paper, we are elaborating on the Genetic Algorithm (GA). It is a well-known metaheuristic algorithm. We categorically represent the GA applications in medical science. The genetic algorithm finds its way in different fields of medical science like Cancer Treatment, Image Segmentation, Gynecology and Obstetrics, Cardiology, Personalized Health Care, Plastic Surgery, Disease Diagnosis, Radiology, Radiotherapy, and Diabetes Prediction. We discuss how the genetic algorithm principle is successfully applied in these applications. We also try to make a comparative discussion among the selected applications on different parameters like diagnosis time, cost, and many more in a lucid manner and find the research gaps.

Keywords Genetic algorithm · Health care · Metaheuristic algorithm

1 Introduction

Nowadays, computers have a huge impact on our day-to-day life. It has covered all the fields from education to business, entertainment, shopping, banking, and even medical science. In medical science, computers are the backbones of different medical research, monitoring, imaging, treatments, diagnosis, maintaining patient records, etc. To get all these advantages of technologies, many devices are designed, and their hardware and software are designed in such a manner that they can manage or handle those tasks. Software is based on some sets of programs. Programs follow a proper algorithm to perform their tasks. The algorithm is a set of rules that are

R. Karmakar (✉)
Department of Computer Science, The University of Burdwan, Burdwan, India
e-mail: rkarmakar@cs.buruniv.ac.in

A. K. Luhach et al. (eds.), *Second International Conference on Sustainable Technologies for Computational Intelligence*, Advances in Intelligent Systems and Computing 1235,
https://doi.org/10.1007/978-981-16-4641-6_8

Fig. 1 Block diagram of GA

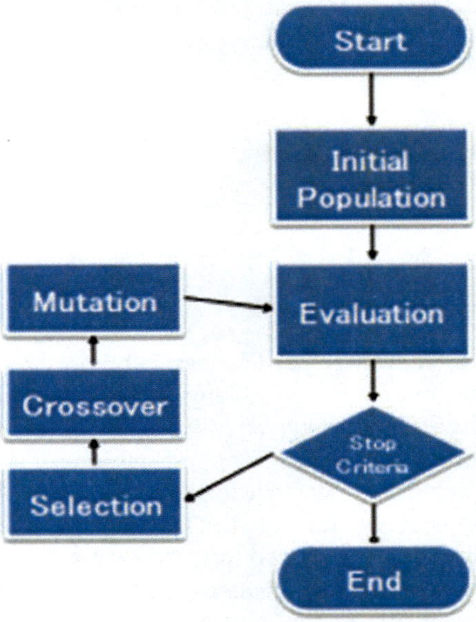

followed for solving problems and calculations. There are two types of algorithms; heuristic and metaheuristic. Heuristic methods are problem-dependent techniques. On the other hand, metaheuristic methods are best for solving complex problems faster and finding the optimal solution. The genetic algorithm is one of them. This paper is all about the genetic algorithm and its application in medical science.

The Genetic Algorithm (GA) is a metaheuristic approach, inspired by the laws of genetics. It was first discovered by John Holland and his student David Goldberg in 1981 [5]. In this method, genes of other creatures are combined and a better creature is formed that is optimization. This optimization is obtained in three different steps [7]. Selection: Two fittest parents are selected among the population with the help of the fitness function. Crossover: The new offspring is generated by the crossover operation performed between the selected parents. Mutation: The genes or the chromosomes of individual parents are randomly changed to create a child. Through these three steps, the genetic algorithm proceeds [16] (Fig. 1).

The selection operator helps to pick fitter individuals from the population. It also probabilistically selects the better solution and removes the bad. The recombinations of genes are performed in the crossover. For the next generation, a new individual is produced by two parents.

The rest of the paper is organized as follows: After the introduction in Sect. 1, different GA-based medical applications have been discussed in Sects. 2.1–2.10. The discussion is presented in Sect. 3 followed by the conclusion in Sect. 4.

2 Applications

2.1 *Cancer Diagnosis*

The human body is composed of numerous cells, tissues, and bones; skin; and different organs. In the natural process, these cells grow, reproduce, and finally die. When this natural process goes wrong, there is an abnormal growth of cell division that occurs which then leads to cancer. It may occur in different places of the human body such as lung cancer, breast cancer, and esophageal cancer. For the destruction of cancer cells, doctors suggest chemotherapy, surgery, etc. Two local control of esophageal cancer is the Artificial Neural Network (ANN) and Genetic Algorithm. ANN is used to predict the treatment result. The human body has 20,000–25,000 genes approximately, each of them has a particular sequence. The process by which the instructions in our DNA are converted into a functional product such as protein by two steps of transcription and translation is called Gene Expression [6]. It first uses Information Gain (IG) for feature selection. Then Genetic Algorithm (GA) is employed for feature reduction. Finally, it uses Genetic Programming (GP) for the classification of cancer disease. The feature selection technique is used to select the cells with different expressions. The GA helps to reduce the number of features effectively. It modifies the chromosome population for better optimization of the problem. Fitness values have been assigned to each of the chromosomes. Lung cancer patient status is predicted by the Support Vector Machine (SVM) and Artificial NeuralNetwork (ANN) classifiers. These models are powerful tools been predicting lung cancer using the features (gene) selection method [2].

2.2 *Plastic Surgery*

Plastic surgery helps in the enhancement of beauty by correcting feature anomalies and treating the aging process [18]. Feature extraction is started with 50 chromosomes. All bits are labeled as 1. The UCLBP is used as a feature extractor for all others 40 face granules. The SURF is used as a feature extractor. A chromosome with identification accuracy is used to assign the proper weights to the face granule. It is used as an initial chromosome for this purpose. Then crossover operation is performed on both chromosome types 1 and 2. A crossover operation is performed on 10 best performing chromosomes to populate a new generation of 100 chromosomes. With the probability of 0.02, the mutation operation has been performed. For the type 1 chromosome, a random mutation is performed. The bits are randomly inverted with the chromosome with a chance of 2%. For the type 2 chromosome, the mutation operation is completed with the changes in weights. It is changed to one or more weights by a factor of the standard deviation of the predecessor generation [15, 20].

2.3 Disease Diagnosis

In the field of medical imaging, Computer-Aided Diagnosis (CAD) is flowering very fast. The feature selection technique is used to improve the rate of disease detection accuracy. The main disadvantage of feature selection is not retrieving the removed features. The GA can be used as a solution to this. GA is very useful for evolving new features from the selected features. The initial population is called the candidate solution, which is the start point of GA. A fitness function is assigned to each candidate solution. Chromosomes having the highest fitness value are preceded by the next generation. After that crossover operation is performed, a child solution is generated. The mutation process gives genetic diversity from one generation to another. Several models are proposed based on GA with SVM and ANN to predict different diseases [22]. Heart disease: The GA and Support Vector Machine (SVM) are the suggested machine learning algorithms in this field. Hongmei Yan et al. have developed a real coded genetic algorithm. It is used to ignore irrelevant and redundant features and gather significant relevant features. This improved algorithm results that acquire an accuracy rate of 94.2% [13, 22]. Lung disease: Jamshid Dehmeshki et al. have presented a method of modules by shape-based Genetic Algorithm Template Matching (GATM) based on geometric shape features. The lungs' phantom nodule images have been used as a reference. With the help of this method, it is possible to detect non-spherical nodules. It is also flexible by nature. This method gives an accuracy rate of 90% [24]. Diabetes: Sapna et al. have proposed an approach to diagnosing diabetes using a fuzzy neural genetic algorithm. It touches the accuracy rate of 92.24% [22]. Breast cancer: Dheeba et al. have proposed a model to detect breast cancer based on the CAD system. It selects the features of cancerous tissues and normal tissues and helps to give an accurate result of 97.8% [19].

2.4 Cardiology

The Artificial Neural Network (ANN)is a machine learning approach to solve complex problems. The hidden neurons of the Neural Network (NN) represents the error on the nodes. The outputs are connected with the nodes. To overcome this problem, an approach is developed that minimized the number of nodes. Here, GA is combined with NN from Genetic Neural Approach which is used to predict heart disease. A backpropagation algorithm is used to learn and train the NN for this hybrid system. The nodes in the hidden layer are calculated to optimize the NN and handle the overfitting problem. The initialization of the Neural Network weight is a blind process. It makes it critical to find the initial weight, and the network output would run toward optimum value. Using the GA, the weights of the NN are initialized. It also calculates the NN layers along with the total number of weights. The is bias used to initialize the network [3]. For each chromosome, the fitness function is also calculated based on the mean square error. After applying GA, chromosomes having lower adaption are replaced with better ones, and the better fitter chromosome that is related to interconnection of weights and thresholds of the neural network is generated. Thehe difference between the desired and actual output error is calculated. The weights are modified randomly to reduce errors. This process is repeated. The Genetic Algorithm-based Neural Network Approach for Cardiovascular disease prediction is used to predict whether the patient has heart disease or not [26, 27].

2.5 Diabetes Prediction

There are two types of diabetes. The type1 is Insulin Dependent Diabetes Mellitus (IDDM) or Juvenile Onset Diabetes Mellitus. The pancreas stops creating insulin because the pancreas undergoes an autoimmune attack by the body itself. This type is very common in children and young adults. Type 2 is called Non-Insulin Dependent Diabetes Mellitus (NIDDM) or Adult On Set Diabetes Mellitus. In this type, the patient produces adequate insulin but the body is unable to use it. The lack of sensitivity to insulin by the cells of the body is the main reason behind this [24]. The proposed system uses the GP part of GA to predict diabetes. GP initially selects the individual by the tournament selection method. The individual is carrying out over the entire population space. The fittest result is obtained through the fitness function. Then the new generation is formed and forwarded to the new generation. In crossover, two programs are selected and copied to the mating pool. And the resultant program is copied to the next generation. In the mutation operator, a single program is selected and copied to the mating pool. And a new program is obtained as a result. The function set is typically directed by the nature of the problem domain. The primitive sets include the specialized function and terminals which are designed to solve the problem. Many research have been performed on diabetes like Rahman and Nandi designed a classifier that diagnoses diabetes patients as diabetic or non-diabetic. The proposed system classifies the patients in any of the three classes pre-diabetic, non-diabetic, and diabetic [10, 20].

2.6 Image Segmentation

Image processing refers to the detection, feature extraction, recognition, and classification of an object. It truly depends on the quality of the segmentation techniques. Image segmentation is a process of partitioning a digital image into a set of pixels [8]. The process of partitioning can be done by edge detection and region extraction. A different object of an image has the same color shape contours after they are partitioned. The entire image is formed by the union of sets of regions. The noisy image is to be segmented of any size. The fitness function is assigned to each chromosome. The individual in the mating pool produces offspring in the crossover technique. Then each bit (pixel value) in the candidate is trapped. Or we may define a mutation rate and randomly flip bits. A maximum no. of iterations are performed. These techniques are applied to every sub-images. The resultant segmented image is more accurate even though it may have a misconnection between two objects. Segmentation methods used artificial intelligence that further includes the Artificial Neural Network (ANN) technique. It is combined with GA to get an optimized result. GA is used for image segmentation. It modifies both the parameters of the existing segmentation algorithms and the pixel-level segmentation [23]. GA determines an optimal number of regions in the segmented image. A quality image of image segmentation of each region must be homogeneous and each and every pixel in an image must belong to a non-overlapping region. It can be done by several other methods such as Edge-based, Region-based, and Clustering-based. The edge detection technique helps to detect the boundaries of different regions of an image. In the case of the region splitting, technique pixels are classified into regions. In the Clustering-based technique, clustering separates the image into various classes without any prior information. GA is used in Image segmentation that helps to divide two major classes. They are parameter selection and Pixel-level segmentation. Parameter selection is the process in which GA is used to modify the parameters of an existing image. In the case of pixel-level segmentation, region labeling is done by a genetic algorithm. Between these two methods, the first method is used because most of the image segmentation methods have many parameters and they need to be optimized [12, 23].

2.7 Gynecology

A Fatal Electrocardiogram Signal (FECG) is an important tool in monitoring and diagnosis of the fetus. It checks the high-risk condition of the fetus and Arrhythmias. It is difficult to record fatal electrocardiogram signals and others from a mother's body during pregnancy. The FECG signals are recorded at the mother's abdominal surface. It is dominated by MECG. The FECG signals range between 5 and it is 1000 times smaller in intensity than an adult. A new mathematical model of the FECG signal has been proposed. The FECG template is obtained with adaptive elimination of MECG interference. Eight skin electrodes were used on different parts of the abdomen for collecting the FECG signals. The recording was done for 5 s. Two signals namely Abd1 and Thr3 are placed as input for Least Means Square (LMS) noise canceller. These MECG Abd1 and Thr3 are eliminated and a relatively well filtered FECG signal is generated. To overcome such problems, a mathematical model is proposed for FECG using GA to optimize its coefficient. Here GA is used to derive the 10 unknown factors and the signal strength. Each FECG signal is encoded in the form of chromosomes [28]. A fitness function is assigned to each chromosome. The basic three operators of GAs are used to manipulate the genetic composition that is selected is done on the chromosome (signals) and fittest are selected. The crossover operation is performed on the selected chromosomes. The mutation is done occasionally with a lower probability. A fitness function is defined for each chromosome. The proposed mathematical model is used to calculate the fitness function. With the help of the

Adaptive Noise Canceller mixture of MECG and FECG, signals are separated. The proposed model minimized the FECG template and the artificially generated signals with the help of the Sum Square Error (SSE) method [14].

2.8 *Radiology*

The Fuzzy Clustering Method (FCM) has been used effectively in the segmentation of Fuzzy boundaries. Though it has the disadvantage of convergence to the local minimum point while finding the value of the centroid clustering. The combination of GA and FCM can be used to get the value of the global centroid [14]. Brain imaging is also very helpful in detecting brain tumors, stroke, and paralysis. All these problems are detected by the clustering technique. It is a task of grouping a set of objects such that objects of the same group are more similar to each other. Several techniques are applied to segment the MRI cerebral tissues such as White Matter (WM), Gray Matter (GM), and Cerebrospinal Fluid (CSF). They are K-Means, Fuzzy, and C-Means (FCM) and Genetic Algorithm-based FCM. It is a local search optimization algorithm that is combined with GA to develop an optimized fuzzy segmentation technique. Clustering can be defined as a group or a set of similar objects. Brain imaging is also useful in detecting brain tumors, stroke, and paralysis. It can be detected by clustering techniques. Different clustering techniques have been applied to cerebral tissues like the White Matter (WM), Gray Matter (GM), and Cerebrospinal Fluid (CSF). The clustering techniques can be the K-Means, Fuzzy and C-Means and GA-based FCM. This local search-based algorithm can be amalgamated with GA for optimized and improved fuzzy segmentation. The authors proposed an improved algorithm for MRI brain tissue segmentation. Initially, all the parameters are set. Cluster centers can be represented with the chromosomes. The population size is 100 and each chromosome is of size 3. Using this algorithm, we can get three clusters for the chromosomes. The fuzzy membership function is used to assign pixels to each tissue. As a result, the cost function is reduced and the fuzzy-ness is controlled. The membership function is used to define the probability of a pixel for the cluster. Then, for all chromosomes, the fitness is computed. The fittest chromosomes are selected for the next generation. The parents are selected using the roulette wheel technique. A new chromosome is produced by the single point crossover succeeded by mutation. The fitness function is saved in the last generation. Depending on the termination criteria, the population is selected or moved to the next iteration. Finally, the results for each cluster are determined. The algorithm accurately segregates the cluster tissues of MRI [17].

2.9 *Personalized Health Care*

The authors in [25] presented a personalized health care system. It is a GA-based system that monitors cardiac patients. They have selected different functions like fitness, severity, overall abnormality, etc. Then the Genetic Algorithm operations are performed. Half of the chromosome pairs are selected from the population for the next. Two offsprings are generated by the 4-point crossover to get the abnormal and lesser abnormal genes. To prevent the local optima, the author used genetic operators to alternate one or more genes. The main goal is to detect the cardiac abnormality of a patient. Using population replacement and the incremental replacement technique, management of the population is done. The offspring of the new population is considered in population replacement, and replacements of one parent with one offspring are done in incremental replacement. The chromosomes that have a fitness value of more than 60% are carried forward to the next generation using the Rank-based selection technique. There is an increase of 10% threshold fitness value for each generation. Thus, the chromosomes that have more than 90% fitness are promoted to the fifth generation. The extremely abnormal chromosomes are detected.

2.10 *Radiotherapy*

The treatment of radiotherapy is done either manually or controlled by a computer. In computerized methodology, the shaping of the field and the intensity is measured. The device is called a multi-leaf collimator [9]. The GA is included in the radiotherapy treatment to generate the primary set. Without breaking any of the constraints, the maximum strength of the radio beams for each side is calculated. Then the crossover is performed to the two randomly selected beams. The multiplication is the new crossover is saved if it is accepted. Now the whole set is stored. The primary set is determined by the best solutions. Two elements are randomly selected from the primary set and the mutated. The resulting set is produced by adding both of the elements having half of their energies. The randomly selected beams are now altered by the amount of 10 to its strength. Then the multiplication numbers are checked. The strengths of all the new beans are halved. If the multiplication factor is not valid, then the products are discarded. Otherwise, it is included in the training set and the process will continue till the training set is completed [1]. Then the resultant set is saved and the topmost element is accepted as the primary set. In this way, radiotherapy treatment doses are calculated for each patient based on their severity.

3 Discussion

GA can be implemented in different sectors, and only a few of its applications in medical science are discussed above. Among a large number of features, specific features are selected with the help of the feature selection method to make the prediction error-free. In this paper, a comparative study is given between different applications of GA in medical science. In Table 1, a comparative study on different processing techniques with different parameters like the author, year of invention, required time, cost, and accuracy of that technique is given. GA can be implemented in different sectors and only a few of its applications in medical science are discussed above. GA is combined with ANN and SVM to select the fittest features from the population, over that crossover and mutation are performed to give optimal results in cancer diagnosis, image segmentation, gynecology lungs disease diagnosis, etc., and it gives an accurate result on or above 90%. Eventually, it is also combined with the Fuzzy Clustering system to predict diabetes, and also in the field of medical imaging it is used. Using the feature selection method, applying GA facial granules are also selected. BSN is combined with GA to give proper results in a personalized health care system. In the case of radiotherapy, the strength of radiation dose of X-Ray is measured based on their disease severity. However, there is also an effective reduction in the time and cost of those treatments. In this paper, a comparative study is given between different applications of GA in medical science. In the case of breast cancer diagnosis using a genetic algorithm costs $\leq$40 but without GA it costs near about 100. So it's a huge difference in cost.

There are other works where [4, 21] lung cancer and Parkinson's disease have been diagnosed using radiomic approach and chaotic firefly decision-making model. Further, for the correctness of the models, formal methods like Event-B [11] can be used.

4 Conclusion

In this paper, we have discussed the Genetic Algorithm and some of its applications in medical science. The GA is a metaheuristic algorithm used to solve a complex problem and give an optimal result within a reasonable time. Medical science is getting advanced rapidly and is also able to offer a satisfactory service to the patients. Several research works are performed using GA in medical science to obtain the optimized output. The lack of efficient interaction between computer scientists and physicians and the unfamiliarity of physicians with the complex mathematical formulas are responsible for the inherent complexity of medicines. To overcome these problems, an efficient interaction is needed between physicians, computer scientists, and engineers.

Table 1 Comparison on different parameters

Application	Author	Year	Technique	Treatment time	Cost	Accuracy
Cancer diagnosis [2]	Hossein Ghayoum I Zadeh	2012	Model based on GA, SVM, and ANN	228 days	< =40	86
Plastic surgery [15, 20]	Singh et al.	2010	Feature selection based on GA and feature extractor	2–3 week	2000–3500	47.8
Disease diagnosis [10, 20]	Jamshid Dehmeshki	2007	Feature selection based on GA Neural Network (NN) and Support Vector Machine (SVM)	0.09 s		93.25
Cardiology [26, 27]	Sumit Bhatia	2008	Computer-aided diagnosis system based on GA, ANN, and data mining	22.6322 s	= 50	94.17
Diabetes [10, 20]	Karaganda	2011	Model based on GA and fuzzy system		= 20	87
Image segmentation [8, 12, 23]	F. Meyer	1992	Model based on ANN and GA	0.046 s		81.41
Gynecology [14]	Liang Xu	2013	Feature selection using GA and SVM		$290	98
Personalized health care [25]	Vincent et al.	2013	Body sensor network based on GA	440 ms		77
Radiology [1, 17]	R.M. Patton	2007	Fuzzy clustering method, fuzzy C-means and L-means clustering	20 s	20	95
Radiotherapy [1, 9]	K. Raja Kumar	2013	Feature selection based on GA	3–7 weeks	<13.70	74.2

References

1. N.H. Barth, An inverse problem in radiation therapy. Int. J. Radiat. Oncol. Biol. Phys. **18**(2), 425–431 (1990). https://doi.org/10.1016/0360-3016(90)90111-v
2. H.S. Bhatt, S. Bharadwaj, R. Singh, M. Vatsa, Recognizing surgically altered face images using multiobjective evolutionary algorithm. IEEE Trans. Inf. Forensics Secur. **8**(1), 89–100 (2013). https://doi.org/10.1109/TIFS.2012.2223684
3. K.V. Dalakleidi, K. Zarkogianni, V.G. Karamanos, A.C. Thanopoulou, K.S. Nikita, A hybrid genetic algorithm for the selection of the critical features for risk prediction of cardiovascular complications in type 2 diabetes patients, in *13th IEEE International Conference on BioInformatics and BioEngineering*, November (2013), pp. 1–4. https://doi.org/10.1109/BIBE.2013.6701620
4. S. Dash, A. Abraham, A.K. Luhach, J. Mizera-Pietraszko, J.J. Rodrigues, Hybrid chaotic firefly decision making model for Parkinson's disease diagnosis. Int. J. Distrib. Sens. Netw. **16**(1), 155014771989521 0 (2020). https://doi.org/10.1177/1550147719895210
5. K. De Jong, Learning with genetic algorithms: an overview. Mach. Learn. **3**, 121–138 (1988). https://doi.org/10.1007/BF00113894
6. J.M. Diaz, R.C. Pinon, G. Solano, Lung cancer classification using genetic algorithm to optimize prediction models, in *IISA 2014, The 5th International Conference on Information, Intelligence, Systems and Applications*, July (2014), pp. 1–6. https://doi.org/10.1109/IISA.2014.6878770
7. D. Goldberg, *Genetic Algorithms in Search Optimization and Machine Learning* (1988). https://doi.org/10.5860/choice.27-0936
8. S. Jansi, P. Subashini, Modified FCM using genetic algorithm for segmentation of MRI brain images, in *2014 IEEE International Conference on Computational Intelligence and Computing Research*, December (2014), pp. 1–5. https://doi.org/10.1109/ICCIC.2014.7238461
9. P. Kallman, B. Lind, A. Eklof, A. Brahme, Shaping of arbitrary dose distributions by dynamic multileaf collimation. Phys. Med. Biol. (IOP Publishing) **33**(11), 1291–1300 (1988). https://doi.org/10.1088/0031-9155/33/11/007
10. A. Karegowda, A. Manjunath, M.A. Jayaram, Application of genetic algorithm optimized neural network connection weights for medical diagnosis of PIMA Indians diabetes. Int. J. Soft Comput. (IJSC) **2** (2011). https://doi.org/10.5121/ijsc.2011.2202
11. R. Karmakar, B. Biman Sarkar, N. Chaki, System modeling using event-B: an insight. SSRN Electron. J. (2019). https://doi.org/10.2139/ssrn.3511455, https://www.ssrn.com/abstract=3511455
12. A. Kos, A. Skalski, T.P. Zielinski, D. Gomes, V. Sá, P. Kedzierawski, T. Kuszewski, Feature selection for automatic CT-based prostate segmentation, in *2016 IEEE International Conference on Imaging Systems and Techniques (IST)*, October (2016), pp. 243–248. https://doi.org/10.1109/IST.2016.7738231
13. H. Kumar, R. Kumar, J. Yadav, A. Rani, V. Singh, Genetic algorithm based PID controller for blood pressure and cardiac output regulation, in *2016 IEEE 1st International Conference on Power Electronics, Intelligent Control and Energy Systems (ICPEICES)*, July (2016), pp. 1–6. https://doi.org/10.1109/ICPEICES.2016.7853680
14. K.K.H. Kunasekaran, R. Sugumaran, Exploratory analysis of feature selection techniques in medical image processing, p. 5
15. M. Mun, A. Deorankar, Implementation of plastic surgery face recognition using multimodal biometric features **5**, 5 (2014)
16. A. Olusesan, Review of feature selection methods in medical image processing. IOSR J. Eng. **4**, 01–05 (2014). https://doi.org/10.9790/3021-04140105
17. R.M. Patton, B.G. Beckerman, T.E. Potok, Learning cue phrase patterns from radiology reports using a genetic algorithm, in *2009 1st Annual ORNL Biomedical Science Engineering Conference* (2009), pp. 1–4. https://doi.org/10.1109/BSEC.2009.5090446
18. E. Sumathi, M.P.R. Rajeswari, Genetic algorithm based recognizing surgically altered face images for real time security application. IJSRP (2013), http://www.ijsrp.org/research-paper-1213.php?rp=P242086

19. H. Salem, G. Attiya, N. El-Fishawy, Gene expression profiles based human cancer diseases classification, in *2015 11th International Computer Engineering Conference (ICENCO)*, December (2015), pp. 181–187. https://doi.org/10.1109/ICENCO.2015.7416345
20. S. Sapna, D. Tamilarasi, M. Kumar, Implementation of genetic algorithm in predicting diabetes. Int. J. Comput. Sci. Issues **9** (2012)
21. S. Sharma, P. Nanglia, S. Kumar, A. Luhach, Detection and analysis of lung cancer using radiomic approach, pp. 13–24 (2019). https://doi.org/10.1007/978-981-13-6295-8-2
22. S. Sindhiya, S. Gunasundari, A survey on genetic algorithm based feature selection for disease diagnosis system, in *Proceedings of IEEE International Conference on Computer Communication and Systems (ICCCS14)*, February (2014), pp. 164–169. https://doi.org/10.1109/ICCCS.2014.7068187
23. Singh, V., Misra, A.K., Varsha, Cardiac image segmentation using simulated genetic algorithm, in *2015 International Conference on Advances in Computer Engineering and Applications*, March (2015), pp. 1024–1027. https://doi.org/10.1109/ICACEA.2015.7164857
24. R. Sonawane, S. Patil, Diabetes detection using genetic programming. Int. J. Comput. Appl. (Foundation of Computer Science (FCS), NY) **127**(10), 12–16 (2015), https://www.ijcaonline.org/archives/volume127/number10/22764-2015906503
25. V. Vaidehi, K. Ganapathy, V. Raghuraman, A genetic approach for personalized healthcare, in *2015 IEEE 28th Canadian Conference on Electrical and Computer Engineering (CCECE)*, May (2015), pp. 196–201. https://doi.org/10.1109/CCECE.2015.7129185. ISSN: 0840-7789
26. N.P. Waghulde, N. Patil, Genetic neural approach for heart disease prediction (2014), https://www.semanticscholar.org/paper/Genetic-Neural-Approach-for-Heart-Disease-Waghulde-Patil/48edb7e31e049cc0a22c2af7717d9be647a7e2d9
27. S.R. Warhade, U.W. Hore, Intelligent prediction of heart disease diagnosis using ANFIS classification model **6**(5), 5 (2015)
28. L. Xu, A. Georgieva, C.W.G. Redman, S.J. Payne, Feature selection for computerized fetal heart rate analysis using genetic algorithms, in *Annual International Conference of the IEEE Engineering in Medicine and Biology Society*, vol. 2013 (2013), pp. 445–448. https://doi.org/10.1109/EMBC.2013.6609532

Designing a Machine Learning Model to Predict Parkinson's Disease from Voice Recordings

Jaya Singh, Ranjana Rajnish, and Deepak Kumar Singh

Abstract Parkinson's disease is one of the most prevalent neurodegenerative disease among the older population more than 60 years, it directly affects the brain cells, and it keeps deteriorating the health of the patient without being cured. It has been found in our research work that there is no cure to Parkinson's disease, so it must be detected early to prevent the fatal rate. Machine learning based approach has been used by many researchers to get a breakthrough for the early prediction of Parkinson's disease. PD cannot be diagnosed through traditional diagnostic methods, so there must be some other method for diagnosing it through cheaper and faster diagnostic tool. This paper proposed a predictive model through which PD can be diagnosed attaining highest accuracy and precision employing a dataset which is composed of voice recordings of the healthy subjects and PD patients. The results obtained from the experiment showed that the Two-Class SVM is the best model for the available dataset with an accuracy of 99%.

Keywords Machine learning · Disease prediction · Parkinson's disease

1 Introduction

Parkinson's disease is a long-term deteriorating disease, and it worsens the health of the patient over time if not treated or if there is delay in time. It affects the central nervous system which causes movements of the body, including tremors. Parkinson's disease is the second most spreading neurodegenerative disease around the globe after Alzheimer's disease. While giving treatment to Parkinson's disease patients, it is observed that only symptoms are being treated. This disease directly attacks the

J. Singh (✉) · R. Rajnish
Amity University, Lucknow, Uttar Pradesh, India

R. Rajnish
e-mail: rrajnish@lko.amity.edu

D. K. Singh
DK Infosolutions Pvt. Ltd, Lucknow, India

A. K. Luhach et al. (eds.), *Second International Conference on Sustainable Technologies for Computational Intelligence*, Advances in Intelligent Systems and Computing 1235,
https://doi.org/10.1007/978-981-16-4641-6_9

nerve cells in the brain, and it to a lesser extent affects the outside part of the brain. There have been many medications and devices that only improve the symptoms, which means that these medications helps in restoring the normal chemical balance in the brain through which various symptoms like rigidness, tremors, mobility, etc., but the process which is actually causing the damage has not been preserved or treated in any way.

Today diagnostic tests for Parkinson's disease patients are very limited. In other disease diagnosis there are blood tests and x-rays but in PD no such tests helped in the favor of the patient. Imaging technology like DaTscan can help to give insight about the patient's condition, but they are too expensive to afford and not efficient enough for the diagnosis. So scientists and researchers are looking for other alternative for the diagnosis of the PD. One of the most popular and efficient method which is creating valuable remarks is the Machine Learning models even without conducting clinical tests.

There are many symptoms of PD like tremors, walking difficulty, stiffness, and many more. One of the symptoms of PD which is not yet been researched on a very large-scale is its effects on vocal patterns. It has been researched that PD is predicted early even voice recordings recorded on smartphones. As we all know smartphones are available everywhere, so there exists a need to develop a model that is accurate, precise, and compact enough for the detection of PD through voice patterns from any kind of data available. In this paper, we attempted to develop a machine learning model which can diagnose PD with 99% accuracy even with coarse vocal patterns which is available on smartphones from the patients suffering from Parkinson's and regular healthy subjects. To get the best model, various algorithms were used for testing and then compared successfully, such as one-class SVM, two-class SVM, and Neural networks, and it was found that two-class SVM is the best suited model for diagnosing PD from voice patterns. Here, we used Machine learning algorithms for the diagnosis of PD patients, but this model can be used further in other fields where voice analysis is recently used like biometrics, since voice patterns of every individuals differ in some way or other, and it cannot be the same (Fig. 1).

2 Background

2.1 Machine Learning

Machine learning is a data science process which determines the results based on the past experiences that is data. Machine learning can be used to identify the diseases based on the previous data which is available. This would make the process of disease prediction even reliable, precise, and more productive through Machine learning models. In this paper we utilized supervised learning through the existing data available online.

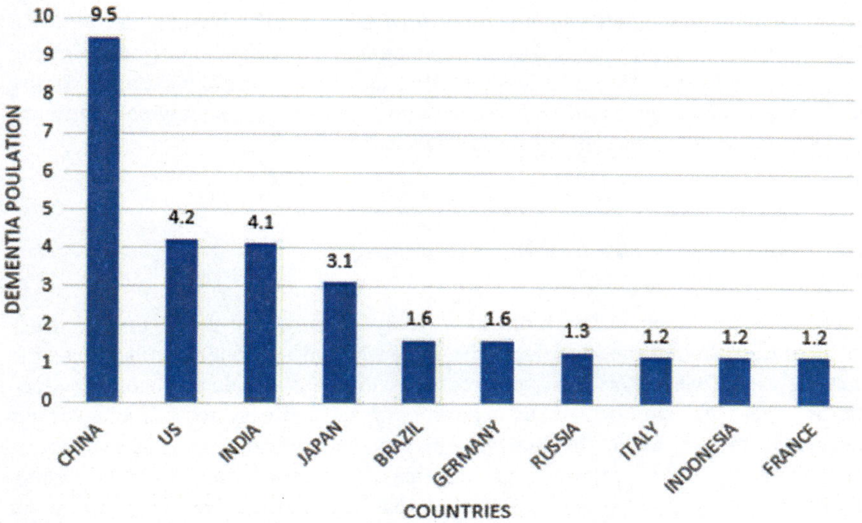

Fig. 1 Growth rate of neurodegenerative disorders in the world

2.2 *Parkinson's Disease Dataset*

The dataset used in our experiments was collected from UCI Machine learning repository where it is freely available for research purposes. As we could not have the access to real time data where we can observe and collect, so we chose this dataset which is already been used by many researchers for their project and research work. Parkinson's disease dataset is consisting of voice patterns from PD patients and healthy subjects out of which 23 are diagnosed with PD and remaining 8 are regular subjects. The data consists of each row with 196 voice recordings from every individual. To perform any experiment on this data firstly we have to differentiate the regular subjects from those who are diagnosed with Parkinson's disease, there is a status column which shows 0 for regular subjects and 1 for the one who is diagnosed with Parkinson's. The data is in ASCII CSV format. İn this dataset each person has six voice recordings each, which is distinguished by their name in the first column.

Here the dataset is composed of an individual voice recording which has various attributes and features which will be utilized in this model. Different attributes of the data include MDVP:Fo(Hz), MDVP:Fhi(Hz), and MDVP:Flo(Hz) which represents average vocal fundamental frequency, maximum vocal fundamental frequency, minimum vocal fundamental frequency, respectively, MDVP:Jitter(), MDVP:Jitter(Abs), MDVP:RAP, MDVP:PPQ, Jitter: DDP is several measures of variation in fundamental frequency, MDVP:Shimmer, MDVP:Shimmer(dB), Shimmer:APQ3, Shimmer:APQ5, MDVP:APQ, Shimmer:DDA is several measures of variation in amplitude, NHR, HNR are the two measures of ratio of noise to tonal components in the voice. RPDE, D2 is two nonlinear dynamical complexity measures, DFA is a signal fractal scaling exponent spread1, spread2, and PPE is a three nonlinear measures of fundamental frequency variation. Status health of the subject is labeled one for Parkinson's patient and zero for healthy subjects. Statistical analysis and filter-based feature selection is applied to select which attributes are more relevant for diagnosing Parkinson's disease.

2.3 Microsoft Azure Machine Learning

Microsoft azure Machine Learning studio analyses and speed-up the task of structuring and deploying the models more efficiently and accurately. In our machine learning model we utilize this studio for classification and data processing algorithms. The azure studio gives us a more compact and easy way to analyze our machine learning models. This tool contains many classification models like regression, neural networks, support vector machines, and many more. Machine learning studio can be adjusted to create a robust model for our predictive model. Common data processing modules include selecting features, deleting null columns, and splitting data before training. Figure 2 shows visual representation of this model used in this paper as followed in Microsoft Machine Learning Studio.

3 Methods

So as to achieve our machine learning model for the early prediction of PD we must take few steps for carrying out this experiment successfully. In the first step we have to do the pre-processing of the data collected from the online website in a proper format so that our machine learning experiments could be carried out on that data. In the dataset there are two kinds of data firstly Parkinson's patients and the other one is healthy subjects, in which the heath 'status' is used to categorize whether the patient is suffering from Parkinson's or not. If the status is 1 which means the patient is diagnosed with PD and status 0 symbolizes that the person is healthy.

3.1 Cleaning the Missing Data

In our foremost step of our experiment, we applied the pre-processing of the dataset, although it was just for the precautionary measures as our data is collected from reliable source, and it has been cleaned already while collection. There were no such missing rows, null values, or any other outlier that could change our data. While cleaning the data the outdated and useless information is removed which not only improves the data quality but also increases its overall productivity. Cleaning the

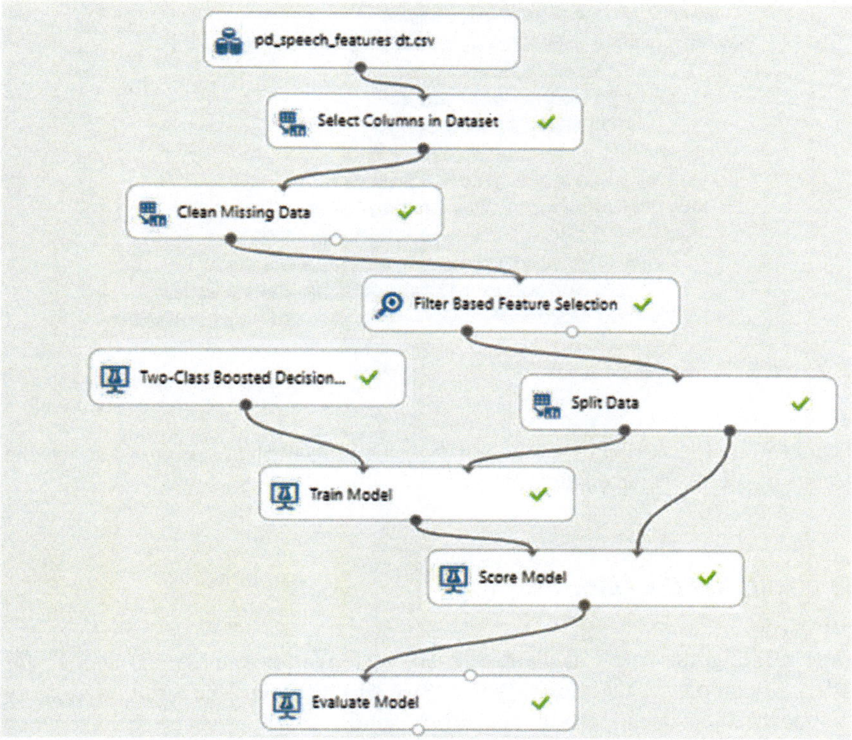

Fig. 2 Screenshot of visual algorithm

data is one of the most important step before starting any experiment or creating your model.

3.2 Filter-Based Feature Selection

In the next step, we needed to select columns for our experiment that could pertained to the problem. Here, we removed the column labeled as 'name' as it was of no use in our experiment. After removing the 'name' column as it was not that much productive for our model creation we chose the method for sorting the attributes which have higher weights for disease prediction. Then the remaining data was further placed for the next process in which it has to go through one of the Filter-base selection algorithm for deciding which columns can be more relevant to our machine learning model. We used SPSS software in which we applied Pearson's correlation algorithm to find out the weights of the attributes which were closely correlated. Spread1 was found to be the most weighted feature among all the given below attributes, so we

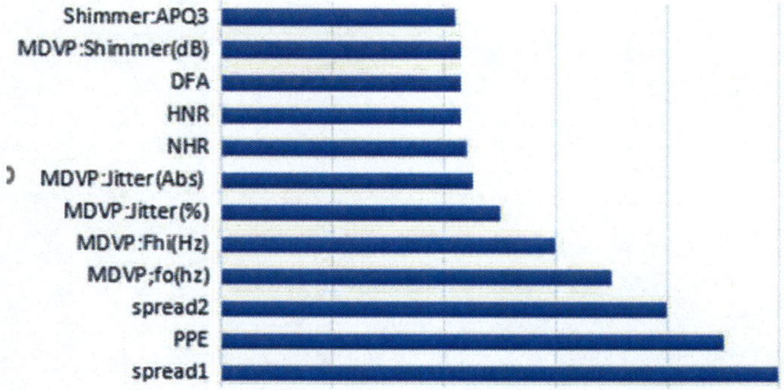

Fig. 3 Bar graph of feature weights

selected around 12 attributes, excluding the remaining attributes. Figure 3 shows the weights of the attributed calculated.

3.3 *Splitting the Data*

While splitting, the data was divided in two sets, one for training and another for testing. My ratio was 50% training and 50% testing as I found the results were more accurate through this proportion. The ratio of splitting the data can be changed when needed as it is up to the researcher how it chooses the model.

3.4 *Train Model*

To train a model in machine learning simply means settling the values for all the weights which attempts to find a model which minimizes loss or which has the minimum loss. The model whose "prediction" is almost flawless and absolute would be considered to be with zero loss otherwise greater loss. The goal of training the model is to determine the sets with minimum losses and which are independent of biasness that has minimum loss on average.

3.5 *Scoring the Model*

Scoring is referred as "prediction" in machine learning, and it generates values based on the trained machine learning model. The prediction made by the score can be used for future values, but it might represent likely outcome and category. The whole prediction is dependent on the type of the data you provide and the model you choose.

3.6 Evaluating the Model

Lastly, the most valuable step of our experiment was changing the model one by one and comparing their results. Many models were tested, compared, and represented in which some are one-class SVM, two-class SVM, BPM, and Random forests.

The various attributes under which it is being evaluated are given below.

Accuracy

Accuracy is the most accurate performance measure and is simply a well-predicted measure of any performance score. One would think that, the more accurate we are, the better our model will be. Yes, accuracy is a good measure but only if you have metrics data where the values of negative and positive elements are almost the same. Therefore, you should look at other parameters to test the performance of your model. For our model, we got 0.997 which means our model is approx. 99% accurate.

$$\text{Accuracy} = \frac{\text{correct predictions}}{\text{total predictions}}$$

Precision

Precision simply means the correct prediction of positive observations to the maximum observations predicted in any experiment. The precision of any experiment depends on the high positive rate. If there are no false positives, then the model had 100% precision. We have got 0.902 precision which is pretty good.

$$\text{Precision} = \frac{\text{true positive}}{\text{false positive} + \text{true positive}}$$

AUC

AUC stands for Area under the ROC Curve which basically means that it measures the entire two-dimensional area underneath the entire ROC Curve from (0,0) to (1,1). To understand better, AUC is the probability that a model will rank a random positive example more highly than a negative one. It ranges from 0 to 1 which means if a model's predictions are 100% wrong, then it will have an AUC value of 0.0 and if the predictions are 100% right then it will have a value of 1.0. It is desirable to use because it is scale-invariant, which means measures of predictions are ranked rather than their absolute values. It is also classification-threshold-invariant, as it measures the quality of prediction irrespective of the classification threshold used.

F1 Score

It is the harmonic mean of Precision and Recall which gives a better measure of incorrectly classified cases. It is also called the F Measure or the balanced F-Score. The formula for calculating the score is

$$2 * ((\text{precision} * \text{recall})/(\text{precision} + \text{recall}))$$

It takes both false positives and negatives into account and works better than Accuracy if you have an uneven class distribution. The best value of F1 Score is 1, and the worst is 0.

4 Results

The results obtained in this experiment concludes that machine learning model can be deployed on the given dataset even with less number of attributes and limited data. We tested and compared different models, namely, Decision Trees, Support Vector Machines, Regression, Neural Networks, and Decision Forests. Among all the experiments, the algorithm that outperforms with maximum accuracy is the two-class Support Vector Machines (SVM). SVM differentiates two available classes by dividing it into separate classes by creating a hyperlane. In this algorithm, the hyperplane keeps adjusting and changing until the optimal solution of the problem has been found. Recently SVM has been used in many biomedical experiments for disease diagnosis. It can also be used for other complex disease. The dataset used in this model is composed of numerous rows and columns which can be very lengthy for the machine learning models to outperform, and we minimized and reduced the data according to achieve the best results which best fits for our disease diagnosis. This method was reliable due to lesser use of parameters resulting it less complex. Without any extra modifications results obtained in our attempts are excellent in which the accuracy achieved was above 90% while precision, AUC (area under the ROC curve), and F-score were extremely better which shows that these results were much higher than expectations.

Table 1 shows the results of the experiment.

5 Future Improvements

In our experiment we were able to achieve the accuracy of 99% but still some algorithms do not perform as good as the SVM, so we can use cross validation method to achieve better accuracy. More statistical methods can be applied to remove the erroneous data and correlate it more strongly. One more issue was identified here was limited amount of data so we can improve our model if we have a larger and

Table 1 The results of the experiment carried out

Model	Accuracy	Precision	Feature	AUC
BPM	0.993	0.993	0.997	0.173
Boosted decision tree	0.967	0.997	0.993	0.492
Decision forest	0.997	0.902	0.999	0.941
One-class SVM	0.902	0.037	0.935	0.294
Two-class SVM	0.997	0.997	0.067	0.764
Decision jungle	0.089	0.002	0.032	0.31
Logistic regression	0.082	0.078	0.0789	0.088

more relevant data in future. The method proposed by this paper is not only limited to treating Parkinson's disease patients but also can be used by other model for the clearer view of voice production. There are many researchers who are working on treating voice disorders and help people learn through vocal skills, and this paper can be useful for such research.

6 Conclusion

Some important conclusions which can be included in our paper are as follows. Firstly, it is not impossible to diagnose Parkinson's from voice patterns as we earlier thought that it is not possible to diagnose Parkinson's. The data set was collected from reliable source, and it was already cleaned which made our model successful. The best model of machine learning through Microsoft Machine learning Studio was two-class Support Vector Machines with maximum accuracy of 99%. And we also concluded that the attributes spread1, spread2, and PPE are the most weighted attributes for the given data and can provide better results while diagnosing Parkinson's disease.

Prediction Techniques for Maintenance of Marine Bilge Pumps Onboard Ships

Abhimanyu Kumar and Arun Mishra

Abstract This paper endeavors to bring out the prediction techniques for maintenance of bilge pumps onboard ships and marine platforms. This work is oriented toward conversion of Planned Preventive Maintenance techniques to Condition based Preventive Maintenance (CBPM) for Bilge Pumps, by studying its present maintenance schedule, past defect history, etc. and creating a decision support system by applying the various Artificial Intelligence tools and techniques. The main motive is to reduce the present turnaround time required for the time-based maintenance of the pumps by incorporating CBPM technique and undertaking maintenance only when needed as the Decision support system predicts. Further, this will also ensure a high level of availability of the Marine platforms as well as improve the overall equipment life.

Keywords Microsoft Azure · Classification · Regression · PPM · CBPM · MAE · Confusion matrix

1 Introduction

A bilge pump is a centrifugal [1] rotating mechanism with six primary components that work together to keep the pump running smoothly. Pump casing, bearings, impeller, bearing frame, shaft, and mechanical seal are among the pump's components. It operates on the premise of mechanical energy being converted into pressure. During operation, the pump's shaft revolves, whirling the impeller, which accelerates the liquid, and the velocity of the liquid is transferred to pressure as the area of the pump casing expands. As a result, the pressured liquid is discharged from the pump discharge. This pump is used to remove water from the ship's bilge compartment. This pump frequently fails to sustain operational availability because the bilge contains a

A. Kumar (✉) · A. Mishra
Department of CSE, Defence Institute of Advanced Technology, Pune, Maharashtra, India

A. Mishra
e-mail: arunmishra@diat.ac.in

A. K. Luhach et al. (eds.), *Second International Conference on Sustainable Technologies for Computational Intelligence*, Advances in Intelligent Systems and Computing 1235,
https://doi.org/10.1007/978-981-16-4641-6_10

lot of undesirable particles as well as water. Its goal is to create a model/strategy that can forecast whether it will fail or live for a long time depending on the observed parameters. Various Machine Learning tools can be used to do this. The scope of work comprises gathering vast amounts of data samples, pre-processing the data, and labeling the samples for supervised learning. Test the model using known samples to see how well it works. Iterate the model from the data exploration step onwards. Determine which model, based on the data, produces the best results. Statistical data analysis can be used to compare the outcomes.

2 Present Maintenance Philosophy

Maintenance can be defined as the operations required to keep or restore a machine, piece of equipment, or system to the specified operable condition in order to extend its useful life. The primary goal of preservation is to ensure that the machine continues to perform as intended for the rest of its life. However, if equipment is not properly maintained, it can have a variety of consequences, including excessive equipment downtime, production losses, high product unit cost, poor quality control, and so on. Maintenance philosophy can be divided into three categories.

(a) **Reactive Maintenance** is what it is called when something goes wrong. It is also known as breakdown maintenance, which refers to repairs made after a machine has broken down in order to return the machine to its original functioning condition [2].

(b) **Preventive Maintenance** [2] is a term that refers to the maintenance that is done. It is done on a piece of equipment on a regular basis to lessen the chances of it failing. It is done while the equipment is still operational to ensure that it does not break down unexpectedly.

(c) **Maintenance that is planned ahead of time** The goal of predictive maintenance is to first forecast when equipment will fail and then to prevent the failure from happening by performing maintenance. Monitoring for potential failure allows maintenance to be planned ahead of time. Predictive maintenance, in theory, provides for the lowest possible maintenance frequency to avoid unforeseen reactive repair while avoiding the expense associated with performing too much preventative maintenance [2].

3 Methodology

A. Relevance of Present Methodology

Predictive Maintenance is a relatively new and well-known maintenance strategy. It was first launched in 1975 to improve the efficiency of planned preventive maintenance decision-making. The machine's operational condition is checked using several monitoring parameters (vibration, bearing temperature, lubricating oil condition, pollutants, and noise levels), and maintenance is performed only when necessary. Equipment is designed in such a way that certain vital parameters are monitored while it is in operation, and these values must fall within a predetermined range. Any deviation from this range indicates inappropriate conduct that will result in failure. As a result of this aspect, real-time parameters may be utilized to anticipate the equipment's health and detect any abnormalities, allowing for prompt maintenance and thereby preventing failure [3]. CBPM is motivated by the fact that nearly 99% of equipment failures are preceded by certain signs or circumstances, that a failure is imminent (Bloch & Geitner, 1983). As a result, Predictive Maintenance is required for better machinery health management, lower lifespan costs, and the prevention of catastrophic failures, among other things [4].

B. Proposed Methodology

*B1. **Data Collection Methodology***. The purpose of data collection is to provide a basis for analysis (Supervised learning). The initial information regarding the present maintenance schedule of pumps was gathered from repair organizations. The past defect history as well as the parameters at the time of occurrence of these defects were collected by undertaking visits to various ships at various ports. An interaction with crew was also carried out to take feedback on the existing maintenance policy. As the data is not available in public domain, data collection is a daunting task in this research work because of the nature of confidentiality. So, a lot of thought has gone in to decide the nature of data that can be used but still keeping the confidentiality intact. Based on this a comprehensive data set has been created containing 1027 instances of data. The main attributes of the dataset are Timestamp, Suction Pressure, Discharge Pressure, Vibration at 4 points, Bearing Temperature, and Pump Condition as (1-Normal Operation, 2-Alarming, and 3-Abnormal).

*B2. **Model Framing***. We need data on operating parameters to undertake predictive maintenance. The data needed for predictive maintenance should be in the form of a time series. A timestamp, as well as a collection of measurements taken at the same time as timestamps, must be included in the data. The goal of predictive maintenance is to predict whether or not equipment will break down in the near future at a time "t" using data collected up to that point. There are two techniques to formulate predictive maintenance:

(a) **Classification approach** [5]—this will classify the present condition of the pump whether it is normal, alarming, or abnormal.

   What is the present condition of the pump?

(b) **Regression approach**—this will predict the time that is remaining before the next failure or breakdown might occur. It can also be called as the Remaining Useful Life of the Equipment [5].

   How much time is left before the pump breaks down?

C. Microsoft Azure Machine Learning Studio [6]

Microsoft Azure Machine Learning Studio is a drag-and-drop collaborative platform for building, testing, and deploying predictive analytics solutions on data. Machine Learning Studio provides models as web services, which can then be accessed directly by custom applications like Excel. Azure Machine Learning Studio provides an interactive, visual workspace for quickly developing, testing, and iterating predictive analytical models. We can link datasets and analytic modules together to construct an experiment, which is then conducted in Machine Learning Studio, by dragging and dropping them into an interactive work of art. We can also turn our training experiment into a predicting experiment and then publish it as a web service so that others may use our model. We can simply drag and drop datasets and modules to create our predictive analysis model.

C1. Classification Approach [6]. Following steps are undertaken for the classification approach in Azure Machine learning studio:

1. **Uploading the dataset.**
2. **Creating a New Project and New Experiment.**
3. ***Preprocessing the data***
4. ***Feature Selection.***
5. ***Choosing and initializing the model.*** After adding the "Train Module" function we need to select our target column (output column) in the column selector which will send it to function as a parameter.
6. ***Evaluating the model.*** After training the module we would need to test the module with the 20% test data we set aside at the split data module. Using that test data and the output of the trained model we give input to the score model function to score the model on the basis of the test dataset shown in Fig. 1. After that we evaluate the model through evaluating the model function.

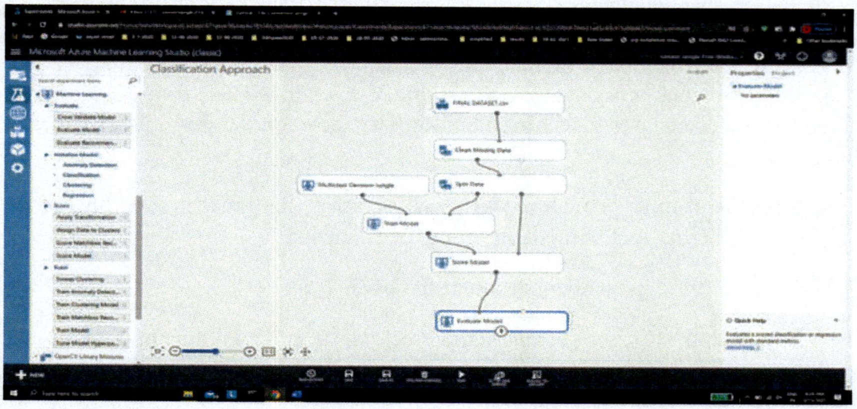

Fig. 1 Evaluation of model

We can create the same data flows for multiple models and train them to evaluate their results. Following model initializers were available on Azure machine learning studio:

Following are Binary Classification models [6]:

- Two-Class Averaged Perceptron
- Two-Class Bayes Point Machine
- Two-Class Boosted Decision Tree
- Two-Class Decision Forest
- Two-Class Decision Jungle
- Two-Class Locally Deep Support Vector Machine
- Two-Class Logistic Regression
- Two-Class Neural Network
- Two-Class Support Vector Machine

Following are the Multiclass models [6]:

- Multi-class Decision Jungle
- Multi-class Logistic Regression
- Multi-class Decision Forest
- Multi-class Neural Network

We will be using Multiclass classification models depicted in Fig. 2 because our output column has three distinguished classes referring to three types of Pump

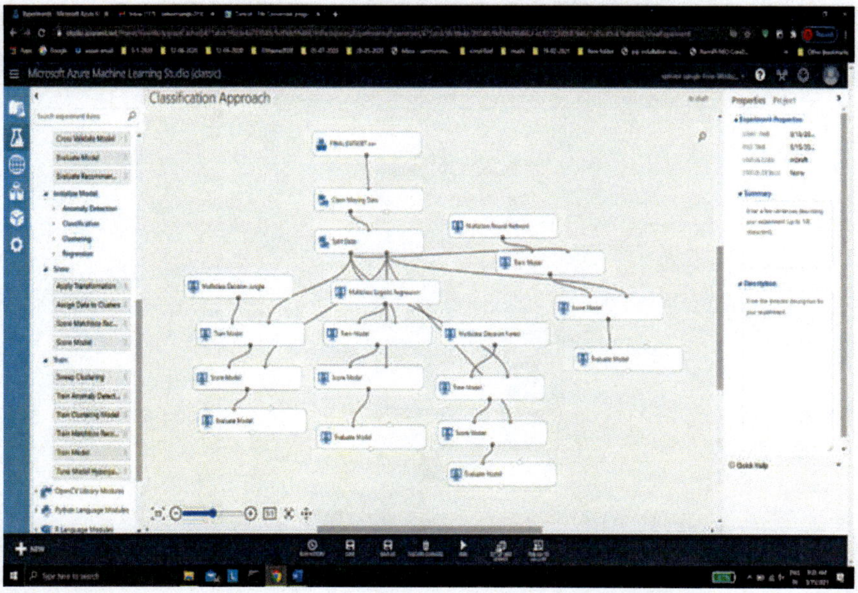

Fig. 2 Mulit-class model

condition failures. Whereas binary classification models are used to classify between two classes.

6.1. Evaluation Results:

After completing the above process and adding all the necessary functions in the process we run the script and evaluate the results for each module with the 20% test dataset we split in the Preprocessing data part.

Following are the evaluation results for the models trained on the input dataset.

- **Multi-class Decision Jungle**

For Multi-class Decision Jungle the Overall Accuracy was 0.99512 as shown in Fig. 3.

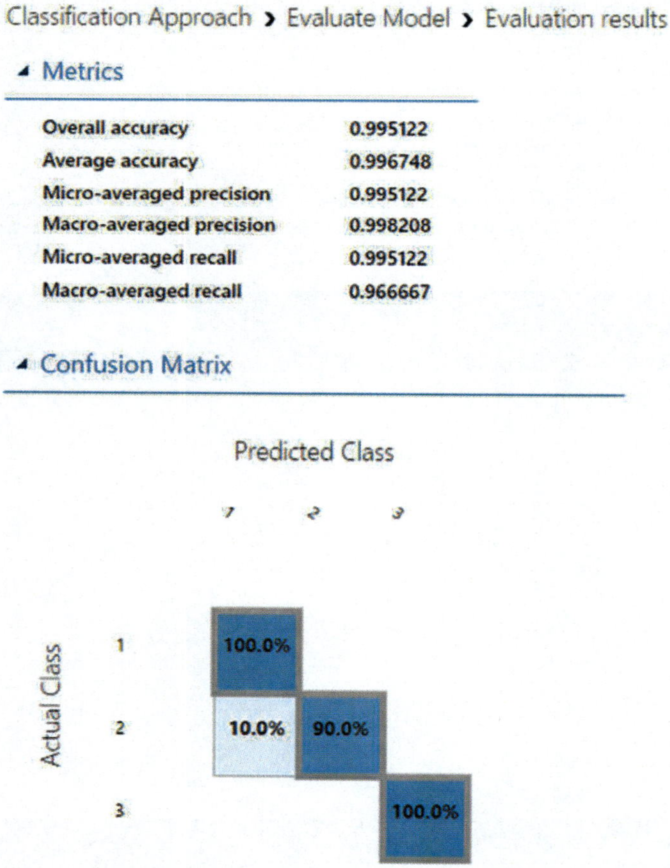

Fig. 3 Result

- **Multi-class Logistic Regression**

Similarly, for Multi-class Logistic Regression the Overall Accuracy is 0.907317 after evaluating the model on a test dataset.

- **Multi-class Decision Forest**

For Multi-class Decision Forest the Overall Accuracy is 0.980488 after evaluating the model on the test dataset.

7. Classification of Instance

To use the trained model, we would need to set it up as a web service. We would need to run it before setting it up as web service and then after finishing the run process. A web service flow will be created by automation to test it. You will need to run it once and then deploy as a web service. Once web service set up is complete we are free to test the trained Classification model with sample inputs which are not included in the training and test dataset. We can test the model in multiple ways by using the excel format and testing it on batches.

8. Result of Classification

For testing the trained model we used excel sheet and the azure web service portal. Figure 4 is the input dataset we used to feed the trained model; this dataset is separate from the Dataset used to train the model. This dataset is totally unknown to the model and is used to test its accuracy on unknown data.

As shown below in Fig. 5 the Multiclass Decision Jungle model got 6 out of 7 prediction correct and failing to predict the 4th one.

In Fig. 6 the Multiclass Logistic Regression model got 5 out of 7 predictions correct and failing to predict the 3rd and 5th one.

In Fig. 7 the Multiclass Decision Forest model also did well and got 6 out of 7 predictions correct and failing to predict the 4th one.

Time	Suction_pressure	Discharge_pressure	Bearing_Temp	Vibration_P1	Vibration_P2	Vibration_P3	Vibration_P4	Pump_condition
1	1.7	5.4	66	5.4	6.3	5.8	5.7	1
2	2.2	5.3	56	4.8	5.2	6.6	5.5	1
3	0.4	4.1	52	4.2	5.2	7.1	7.6	3
4	1.3	6.3	68	5.3	5.2	4.9	7.4	1
5	2.2	6.3	69	4.8	6.5	5.1	5.5	2
6	1.1	3.3	55	4.3	5.2	5.8	8.8	1
7	1.8	3.2	58	4.3	5.5	7.1	5.4	1

Fig. 4 Data set

Multiclass Decision Jungle													
INPUT								OUTPUT					Is Prediction Correct?
Time	Suction_pressure	Discharge_pressure	Bearing_Temp	Vibration_P1	Vibration_P2	Vibration_P3	Vibration_P4	Pump condition	Scored Probabilities for Class "1"	Scored Probabilities for Class "2"	Scored Probabilities for Class "3"	Scored Labels	
1	1.7	5.4	66	5.4	6.3	5.8	5.7	1	0.875	0.125	0.00E+00	1	YES
2	2.2	5.3	56	4.8	5.2	6.6	5.5	1	1	0	0.00E+00	1	YES
3	0.4	4.1	52	4.2	5.2	7.1	7.6	3	0	0	1	3	YES
4	1.3	6.3	68	5.3	5.2	4.9	7.4	1	0.125	0.875	0	2	NO
5	2.2	6.3	69	4.8	6.5	5.1	5.5	2	0	1	0.00E+00	2	YES
6	1.1	3.3	55	4.3	5.2	5.8	8.8	1	0.875	0	0.125	1	YES
7	1.8	3.2	58	4.3	5.5	7.1	5.4	1	1	0	0	1	YES

Fig. 5 Result

Multiclass Logistic Regression													
INPUT								OUTPUT					Is Prediction Correct?
Time	Suction_pressure	Discharge_pressure	Bearing_Temp	Vibration_P1	Vibration_P2	Vibration_P3	Vibration_P4	Pump condition	Scored Probabilities for Class "1"	Scored Probabilities for Class "2"	Scored Probabilities for Class "3"	Scored Labels	
1	1.7	5.4	66	5.4	6.3	5.8	5.7	1	0.89573	0.092035	1.22E-02	1	YES
2	2.2	5.3	56	4.8	5.2	6.6	5.5	1	0.94994	0.047138	2.92E-03	1	YES
3	0.4	4.1	52	4.2	5.2	7.1	7.6	3	0.63317	0.063036	0.303798	1	NO
4	1.3	6.3	68	5.3	5.2	4.9	7.4	1	0.80636	0.166586	0.027055	1	YES
5	2.2	6.3	69	4.8	6.5	5.1	5.5	2	0.88728	0.110112	2.61E-03	1	NO
6	1.1	3.3	55	4.3	5.2	5.8	8.8	1	0.82358	0.042762	0.13366	1	YES
7	1.8	3.2	58	4.3	5.5	7.1	5.4	1	0.95669	0.023349	0.01996	1	YES

Fig. 6 Result

Multiclass Decision Forest													
INPUT								OUTPUT					Is Prediction Correct?
Time	Suction_pressure	Discharge_pressure	Bearing_Temp	Vibration_P1	Vibration_P2	Vibration_P3	Vibration_P4	Pump condition	Scored Probabilities for Class "1"	Scored Probabilities for Class "2"	Scored Probabilities for Class "3"	Scored Labels	
1	1.7	5.4	66	5.4	6.3	5.8	5.7	1	0.875	0.125	0.00E+00	1	YES
2	2.2	5.3	56	4.8	5.2	6.6	5.5	1	0.875	0.125	0.00E+00	1	YES
3	0.4	4.1	52	4.2	5.2	7.1	7.6	3	0.25	0.125	0.625	3	YES
4	1.3	6.3	68	5.3	5.2	4.9	7.4	1	0.25	0.75	0	2	NO
5	2.2	6.3	69	4.8	6.5	5.1	5.5	2	0	1	0.00E+00	2	YES
6	1.1	3.3	55	4.3	5.2	5.8	8.8	1	0.875	0	0.125	1	YES
7	1.8	3.2	58	4.3	5.5	7.1	5.4	1	1	0	0	1	YES

Fig. 7 Result

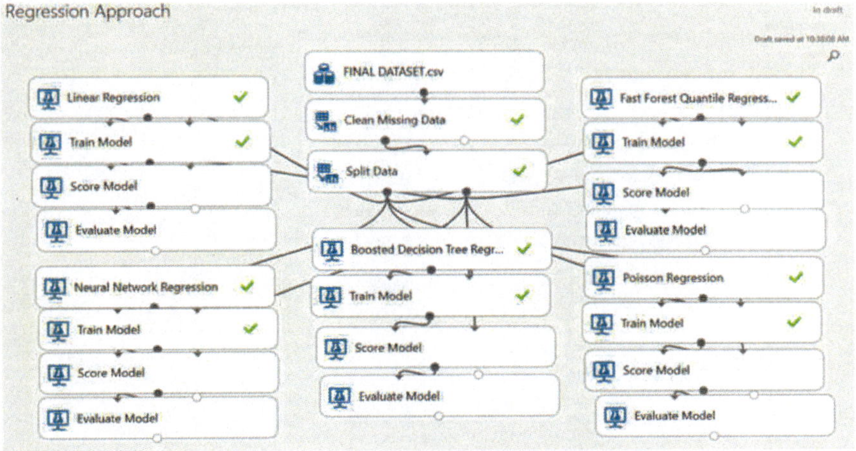

Fig. 8 Model

C.2. REGRESSION APPROACH [6]

In the Regression approach we follow similar steps as that of classifications. For regression approach, we will be training the model on following initializers:

Following are the models [6]:

- Boosted- Decision Tree
- Fast Forest Quantile-Regression
- Linear- Regression
- Neural -Network -Regression
- Poisson -Regression

After applying similar steps as in classification, we get the following model experiment as shown in Fig. 8:

After adding all the initializers, we add score model function and evaluate model function for evaluating our training results for each model separately as mentioned in the classification process. Then we proceed to run the experiment and evaluate the model.

1. **Evaluation results for the following model**: After evaluating the models with the 20% test dataset we kept aside for evaluations. Then we will visualize each of the results for the evaluate function.

- **Linear Regression**

Linear Regression model after training has an Mean absolute Error (MAE) score of 0.233725 and Root mean squared Error as 0.361113 as shown in Fig. 9.

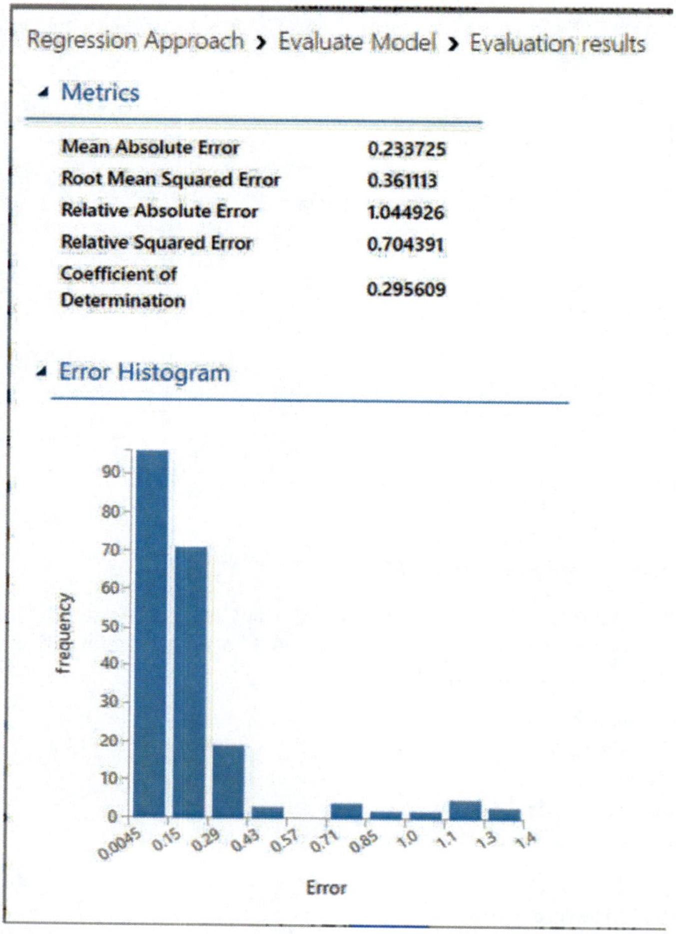

Fig. 9 Result

- **Neural- Network Regression**

Similarly, Neural Network Regression scored a very good MAE, i.e., 0.05103 means almost 99% accuracy.

- **Boosted Decision Tree Regression**

The Boosted Decision Tree Regression has an MAE of 0.399767 which is better than the Linear Regression model.

- **Fast Forest Quantile Regression**

Fast Forest Quantile Regression has shown good accuracy of 0.015 from the first Quantile loss and remains less volatile in the next 3 Quantile losses.

A web service flow will be created by automation to test it. You will need to run it once and then deploy it as a web service similarly as we did it in the classification approach so that we can test it with the unknown dataset as a web service. A web service flow will be created by automation to test it you will need to run it once and then deploy as a web service. Again for Testing the trained Regression model with sample inputs which are not included in the training and test dataset being totally unknown to the model will be used in evaluating the final results through web service portal.

2. Results of Regression. Figure 10 is the Input Dataset Table We Used to Evaluate the Model. This Data Set is Separate Dataset from the Original Data:

The Linear Regression model predicted 6 out of 7 of the labels correctly, and we can say it failed to predict or come closer to predict the 5th instance as shown in Fig. 11.

The Boosted Decision Tree Regression model predicted almost all of them correctly except the 5th instance as shown in Fig. 12.

The Fast Forest Quantile Regression was able to predict 5 out of 7 predictions correctly, but analyzing the scores for other quantiles, it has an accuracy of almost 80% as shown in Fig. 13.

Time	Suction_pressure	Discharge_pressure	Bearing_Temp	Vibration_P1	Vibration_P2	Vibration_P3	Vibration_P4	Pump_condition
1	1.7	5.4	66	5.4	6.3	5.8	5.7	1
2	2.2	5.3	56	4.8	5.2	6.6	5.5	1
3	0.4	4.1	52	4.2	5.2	7.1	7.6	3
4	1.3	6.3	68	5.3	5.2	4.9	7.4	1
5	2.2	6.3	69	4.8	6.5	5.1	5.5	2
6	1.1	3.3	55	4.3	5.2	5.8	8.8	1
7	1.8	3.2	58	4.3	5.5	7.1	5.4	1

Fig. 10 Data set

Linear Regression										
INPUT									OUTPUT	
Time	Suction_pressure	Discharge_pressure	Bearing_Temp	Vibration_P1	Vibration_P2	Vibration_P3	Vibration_P4	Pump condition	Scored Labels	Is Prediction Correct?
1	1.7	5.4	66	5.4	6.3	5.8	5.7	1	1.118878462	YES
2	2.2	5.3	56	4.8	5.2	6.6	5.5	1	0.823528675	YES
3	0.4	4.1	52	4.2	5.2	7.1	7.6	3	2.888020388	YES
4	1.3	6.3	68	5.3	5.2	4.9	7.4	1	1.281767693	YES
5	2.2	6.3	69	4.8	6.5	5.1	5.5	2	1.28189508	NO
6	1.1	3.3	55	4.3	5.2	5.8	8.8	1	1.039781136	YES
7	1.8	3.2	58	4.3	5.5	7.1	5.4	1	1.168452171	YES

Fig. 11 Result

Boosted Decision Tree Regression										
INPUT									OUTPUT	
Time	Suction_pressure	Discharge_pressure	Bearing_Temp	Vibration_P1	Vibration_P2	Vibration_P3	Vibration_P4	Pump condition	Scored Labels	Is Prediction Correct?
1	1.7	5.4	66	5.4	6.3	5.8	5.7	1	1.144126534	YES
2	2.2	5.3	56	4.8	5.2	6.6	5.5	1	0.98694402	YES
3	0.4	4.1	52	4.2	5.2	7.1	7.6	3	2.99784255	YES
4	1.3	6.3	68	5.3	5.2	4.9	7.4	1	1.436916828	YES
5	2.2	6.3	69	4.8	6.5	5.1	5.5	2	1.158388615	NO
6	1.1	3.3	55	4.3	5.2	5.8	8.8	1	1.014469981	YES
7	1.8	3.2	58	4.3	5.5	7.1	5.4	1	1.000419617	YES

Fig. 12 Result

Fast Forest Quantile Regression												
INPUT									OUTPUT			
Time	Suction_pressure	Discharge_pressure	Bearing_Temp	Vibration_P1	Vibration_P2	Vibration_P3	Vibration_P4	Pump_condition	Scores for quantile :0.250	Scores for quantile :0.500	Scores for quantile :0.750	Is Prediction Correct?
1	1.7	5.4	66	5.4	6.3	5.8	5.7	1	1	1	2	NO
2	2.2	5.3	56	4.8	5.2	6.6	5.5	1	1	1	1	YES
3	0.4	4.1	52	4.2	5.2	7.1	7.6	3	2	3	3	YES
4	1.3	6.3	68	5.3	5.2	4.9	7.4	1	1	1	2	NO
5	2.2	6.3	69	4.8	6.5	5.1	5.5	2	1	1	2	YES
6	1.1	3.3	55	4.3	5.2	5.8	8.8	1	1	1	1	YES
7	1.8	3.2	58	4.3	5.5	7.1	5.4	1	1	1	1	YES

Fig. 13 Result

4 Conclusion

Predictive Maintenance has a great relevance in the maintenance philosophy and having a system which can predict the condition of the equipment and also predict the remaining life which is useful and will help in deciding on when to undertake maintenance of the equipment. The artificial intelligence-based decision support system will help the operator in identifying the condition of the equipment at real time and then decide on the maintenance plan based on the prediction. This will ensure less downtime of equipment and also ensure highest operational availability thereby enhancing the overall equipment life feedback polynomials.

5 Future Scope

The same models can be again used to train and predict for different equipment and Predictive Maintenance can be undertaken on all the equipment onboard a ship. The main task for this will be collection of training data. With induction of ships with Integrated Platform Management System (IPMS), the collection of data will become much simpler as all the required data are being logged automatically in these systems.

References

1. D. Kernan, Pumps 101: Operation, Maintenance and Monitoring Basics[white paper], https://www.gouldspumps.com/ittgp/medialibrary/goulds/website/Literature/Whit%20Papers/ITT_white_paper_Pumps_101_Operation_Maintenance_and_Monitoring_Basics. Accessed 2 Sep 2018
2. Emerging trends in engineering, science and technology for society, energy and environment, in *Proceedings of the International Conference in Emerging Trends in Engineering, Science and Technology (ICETEST 2018)*, (2018)
3. A. Jardine, V. Makis, D. Banjevic, D. Braticevic, M. Ennis, A decision optimization model for condition-based maintenance. J. Qual. Maint. Eng. (1998)
4. A.K.S. Jardine, et al., A review on machinery diagnostics and prognostics implementing condition-based maintenance. Mech. Syst. Signal Process. **20**(7), 1483–1510 (2006)
5. https://en.wikipedia.org/wiki/Machine_learning
6. "Microsoft Azure", Azure ML Studio, https://docs.microsoft.com/en-us/azure/machine-learning/studio/what-is-ml-studio
7. D. Singh, A. Suhane, M.K. Thakur, The study of failure analysis of centrifugal pump on the basis of survey
8. R. Ahmad, S. Kamaruddin, An overview of time-based and condition-based maintenance in industrial application. Comput. Ind. Eng. Elsevier (2012)
9. E. Amaya, A.J. Álvares, Expert systems for power generation fault diagnosis using hierarchical meta- rules, in *Proceedings of IEEE 17th International Conference on Emerging Technologies & Factory Automation* (2012)
10. https://towardsdatascience.com/how-to-build-a-data-set-for-your-machine-learning-project-5b3b871881ac
11. A.K. Verma, A. Srividya, A. Goyal, P.G. Ramesh, Application of fuzzy logic for selection of machinery health monitoring strategies, in *Annual Meeting of the North American Fuzzy Information Processing Society (NAFIPS)* (2012)
12. C.E. Ebeling, *An Introduction to Reliability and Maintainability Engineering* (Waveland Press, Inc., 1997, reissued 2005)

A Systematic Literature Review on Software Development Estimation Techniques

Prateek Srivastava, Nidhi Srivastava, Rashi Agarwal, and Pawan Singh

Abstract Project Failure is the key target undergoing in today's scenario which is observed through the software project administrators. An ambiguity of the estimation is the major cause of this dilemma. According to tech project managers, project failure is the most common concern today. This issue arises as a result of measurement inaccuracy. When software's size and significance increase, its complexity makes it even more problematic to reliably estimate the cost of growth. This was the predicament of previous years. Software cost estimation (SCE) is the utmost significant subject in software design in modern times. Real calculation necessitates expense and commitment considerations when developing applications utilizing algorithmic or Artificial Intelligence (AI) techniques. In this paper, numerous predefined approaches for SCE are illuminated and their characteristics will be considered. This paper briefs numerous sessions of SCE models and methods. This study also claims that no particular procedure is superlative for all circumstances, software measures and metrics are also needed to enhance efficiency. This research paper focuses on Artificial Intelligence-based software cost estimation approaches, their implementation domain, and strategies for calculating software cost estimation. There are two kinds of SCE approaches: parametric/algorithmic models and non-parametric/non-algorithmic models. One of the most common algorithmic SCE models is the Constructive Cost Model (COCOMO). Basic COCOMO is ideal for fast, quick, and uneven estimates of the effort expected to manufacture software, but

P. Srivastava (✉) · N. Srivastava
Amity Institute of Information Technology, Amity University Uttar Pradesh, Lucknow Campus, Lucknow, India

N. Srivastava
e-mail: nsrivastava2@lko.amity.edu

R. Agarwal
Department of Information Technology, University Institute of Engineering and Technology, Chhatrapati Shahu Ji Maharaj University, Kanpur, India

P. Singh
Department of Computer Science & Engineering, Amity School of Engineering and Technology, Amity University Uttar Pradesh, Lucknow Campus, Lucknow, India
e-mail: psingh10@lko.amity.edu

A. K. Luhach et al. (eds.), *Second International Conference on Sustainable Technologies for Computational Intelligence*, Advances in Intelligent Systems and Computing 1235,
https://doi.org/10.1007/978-981-16-4641-6_11

its accuracy is restricted because of the lack of variables to account for cost-driven differences. Non-algorithmic approaches for overcoming the shortcomings of algorithmic models include expert Judgment, machine learning (ML), and price to gain. In checked studies, ML models were implemented to address the shortcomings of parametric estimation models.

Keywords Software cost estimation (SCE) · Estimation techniques · Artificial intelligence · Data mining · COCOMO · Software metrics

1 Introduction

The SCE is a method utilized by the company to approximate the cost of developing a software project. Estimation of the incomes and time required are particularly critical for all programs, especially for information systems, where the budget and timeline are normally overrun. Both measurement approaches must adhere to the metric scale of the software [1]. The evaluation of the software project presents difficulties relative to other sectors [2]. Current strategies strongly rely on the knowledge required for the project [3]. As the project moves on, the estimates are more precise because more reliable details are available. The method of estimation should be a continuous process, incorporating new material. This study demonstrates a method for calculating software development costs based on data pertaining to Artificial Intelligence (AI) techniques, defining the most influential causes and variables completed the cost of the software. The work is carried out based on a historical data collection of projects, considering the data assembled by the International Software Benchmarking Standards Group (ISBSG), which gathers knowledge from more than 2000 projects. This dataset includes both numerical values and categorical details. Due to the absence of certain standards in this dataset, some data mining techniques are used for the pre-processing of knowledge.

Many modern companies depend on software creation for their progress and survival in today's evolving IT market. Estimation of cost is a challenge for many companies. The expense on the program development can be calculated in such a manner that it does not sacrifice the accuracy, schedules, and performance. Algorithmic, non-algorithmic, and machine learning are the three main methods used in Software Cost Estimation. The previous approach relied on the source line of code, function points, and cost drivers for predictive simulation, which was based on previous project work. The main emphasis of the algorithmic standard is on architecture, research, execution, and optimization. The other emphasis is laid on the tentative assessment and the functional use of the program. The best often applied versions are COCOMO, the function point, and the Putnam model [4].

The COCOMO (Constructive Cost Model) [5] is a common method for estimating the SCE's effort and time as shown in Fig. 1. In present developments, a key position in software programs is performed by the COCOMO-II paradigm. Former data is used in non-algorithmic methodology to estimate the costs of ongoing

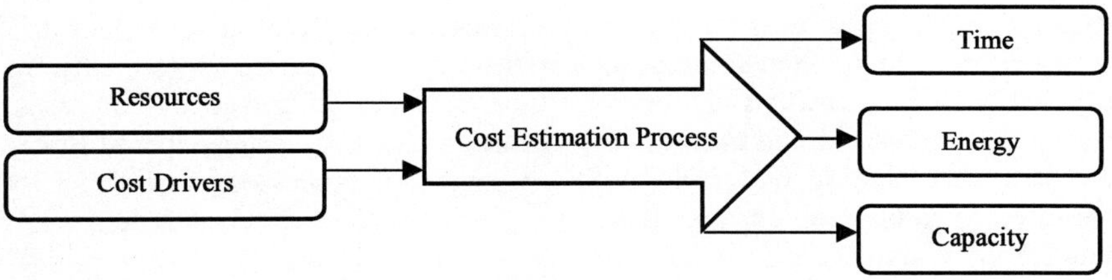

Fig. 1 Cost estimation process

programs with planned software projects. It is simple to read, and it should have the full detail on previous ventures. Current calculation approaches are dependent on analogue methodology, Judgment, win price, etc. Machine learning strategies expect accurate software attempts due to their essence of learning. It is merged with data processing techniques. The primary emphasis is on data processing and is regarded as unsupervised learning.

The goal of this review is to analyze software cost estimation methods and explore how they can be used as conceptual information in a software cost estimation expert database system. Expert database systems are addressed in terms of their different implementations. In almost all software development companies throughout the globe, a common thing found is the poor estimation of development effort and cost which results in vast kinds of glitches in all unsuccessful software development projects. The study aims to provide a better insight through an extensive review of the literature. The study about the efficient software cost estimation model is likely to forecast the total cost of projects. The majority of the research has concentrated on developing structured software effort estimation models. Earlier models were mostly depending on regression evaluation or arithmetically resulting from hypotheses through other fields. Hence, it is the need of time to study all the popular software estimation models.

2 Significance of Software Estimation

Software Estimation (SE) [6] is one of the main practices and challenges of a software enterprise. One of the company's objectives is to ensure that software goods are shipped on schedule and below the cost estimates. Underestimation of a project can contribute to inadequate resource control, poor efficiency, and reduced quality compliance, which can result in adverse business outcomes, competition, profitability, and market reliability. Overestimating the project also results in a bad division of resources, which raises costs, decreases profits, or often leaves the company in a competitive position; this often contributes to the lack of other business prospects, such as the starting up of a new project.

In the launching of every new software initiative, the proposal states that it is important to schedule and forecast all parameters to all customers and software firms to achieve win-win situations and thus mutually advantageous outcomes. Project management would naturally be involved with initiative and expense strategies, resource planning, budgeting, project organization processes, project success measurement, scheduling changes, product version control, reallocation of personnel, interactions within the organization and the customer. Hence, a software project manager will be interested to determine the project cost and the effort, which will be a basis to allocate the budget, decide the timeline, mitigation plan for the risks, resourcing, and others. Where there is no accurate forecast, project preparation and monitoring become a time-consuming operation. To quantify costs and commitment, the metrics of previous projects need to be accumulated and matched with comparable features of current projects.

3 Related Works

Qin et al. [7] pointed out that management is an important consideration. Resource, preparation, research, construction, coding, and checking are the essential requirements for estimating project costs. Several approaches are implemented into the SCE. These approaches are assessed at perfect cost and the adjustment is simple since each move is registered. Deshmukh et al. [8] argue that an unforeseen rise in the expenditure and a lack of execution of the project will reduce market opportunities in information technology. To solve this challenge, the taxonomy of data mining methods is applied to equate it with the COCOMO-II software cost estimation model. This model is a multiplier element that determines the cost generators of chronological content.

Arulmozhi et al. [9] used different approaches to measuring the expense of the software. It forecasts scale, cost, commitment, and design objects. A new data mining method deals with the Least Square-SVM method to enhance the efficiency of Software Cost Estimation. Regression function tries to keep the squared error to a minimum. The correct conclusion is the implementation of the AI techniques. Empirical approaches or expert-driven are useful for cost estimation. There are several approaches to increase total expense accuracy. Khalifelu et al. [10] have suggested an analytical position focused on data mining techniques. ANN, LR, KNN, and SVR are related to the COCOMO model and are applied through Artificial Intelligence methods to measure the correct expense of the program project. Training and testing datasets are applied to equate energy estimates and real energy. Using the NASA dataset and WEKA data mining tools, the effect is estimated using MRE, MAE, RMSE, and Correlation Coefficient.

According to Shekhar et al. [11], the cost estimation of software is algorithmic and non-algorithmic, according to two key software engineering methodologies. The Putnam model and the COCOMO are two examples of algorithmic models. Estimation is done using scientific calculations. Non-algorithmic approaches are expert judgment, analogy approach, and the price to win based estimation. It must include the earlier specification descriptions and should be based on historical sources. When the design is huge and the design experience is least, it is safer to select algorithmic methods. Rijwani et al. [12] have created a multi-layer feed forward neural network based on ML approaches applied to test the skills of the project. MMER and MSE are assessment techniques and are applied to minimize network counts, research, training databases, COCOMO datasets, and statistical instruments to increase economic development precision in machine learning algorithms and data mining.

Densumite et al. [13] have created an activity point approach for estimating the scale of the design using the activity diagram, which provides a graph showing the sequence of activities and the scheme's flow of control. TCF, ECF, and PRF are the modification variables applied to estimate the scale of the software. Sangwan et al. [14] proposed investigating machine learning methods used to approximate effort, like ANN, Analogy calculation, Fuzzy, (SVM), Genetic algorithm, and Bayesian Network. Accuracy metrics are being used to explain the discrepancy between predicted and realistic effort.

Whigham et al. [15] recommended a baseline model that could be applied to all activities including programmed effort estimation study. When evaluating a different or prevailing technique to analyzing the projection activity, it is critical to evaluate the outcomes to the baseline. Sarro et al. [16] proposed a new effort estimation algorithm that combined Confidence Interval Analysis and Mean Absolute Error estimation (MAE). The established algorithm was examined and assessed on three distinct variables, with positive results.

Masoudi-Sobhanzadeh et al. [17] recommended a novel software role selection model. It has the potential to be utilized in a variety of fields, including drug design, neuroscience, and image processing. The model is applied by choosing a subset of characteristics/aspects established on optimization algorithms, which are then allocated to classifiers or learners. Pospieszny et al. [18] developed an ML-based method for bridging the gap among recent research and real-world applications. The model suggested produced very accurate results and its creator has agreed that the software effort and duration calculations for realistic undertakings yield more rational results.

Al Asheeri et al. [19] conducted a review of the majority of the approaches and approaches used in computing cost estimation. He attempted to list the key benefits and disadvantages of both. The Author conveys the following qualitative considerations that should be used in assessment metrics: staff expertise, construction context, and culture; quantitative factors involve project scale and usable capital. Başkeleş et al. [20] conducted numerous software effort prediction experiments on three different databases utilizing a range of machine learning methods. As a consequence, they discovered that parametric models are not adequate for estimating scheduled initiative. Tunio et al. [21] proposed classical planning of AI techniques to promote quality and reduce the burden on crowdsourcing software development (CSD) platform and CSD developers by searching for open tasks and optimizing the matching process that CSD developers need to develop and provide a solution for that task.

4 Software Cost Estimation Methodologies

Effective software estimation is a crucial practice and a difficult challenge in the software industry [22]. The Cost estimate is a comprehensive method that impresses the project's commitment and timeline of the cost model parameters. Management of design involves seven actions:

- Establishing complete design of database
- Estimation of size
- Estimation of efficiency
- Distribution of Phase
- Sensitivity analysis and Risk
- Validation
- Calibration

Other methods implementing SCE models are in the field of sensitivity, risk analysis, and calibration. The characteristics in the program are dependent on the calculation of effectiveness as well as reliability needed for the implementation of the software. Cost drivers are the key action for estimating costs. In software industries, effective cost assessment is critical because it is necessary to define the resources matched to the specifications, enhance the software business plan with improved resources, determine the available resources, manufacture, validate, and verify resource-based development in software [23].

Practitioners have shared their dissatisfaction with their failure to correctly predict software development costs. The various methods for software cost estimation are shown in Fig. 2.

4.1 Algorithmic Methods

This methodology [24] employs sequential expense information to display project cost metrics (code lines). The cost is calculated using statistical equations. The main advantages are the ability to create repeatable estimates and the ease of which adjustments can be made using feedback data, as well as the ability to process and update formulas. Unrivaled situations are not possible to work with, certain knowledge and variables could not be measured, in the case of Algorithmic methods. The algorithmic approaches include the Constructive Cost Model, the Putnam method, and the Function Point method.

COCOMO

COCOMO is a major applicant for the figure of methods to be included in SCE depending on accessible information. Barry Boehm built the first model in 1981. COCOMO has three forms, namely, basic, intermediate, and detailed. Organic, semi-detached, and embedded are the three categories of systems established for

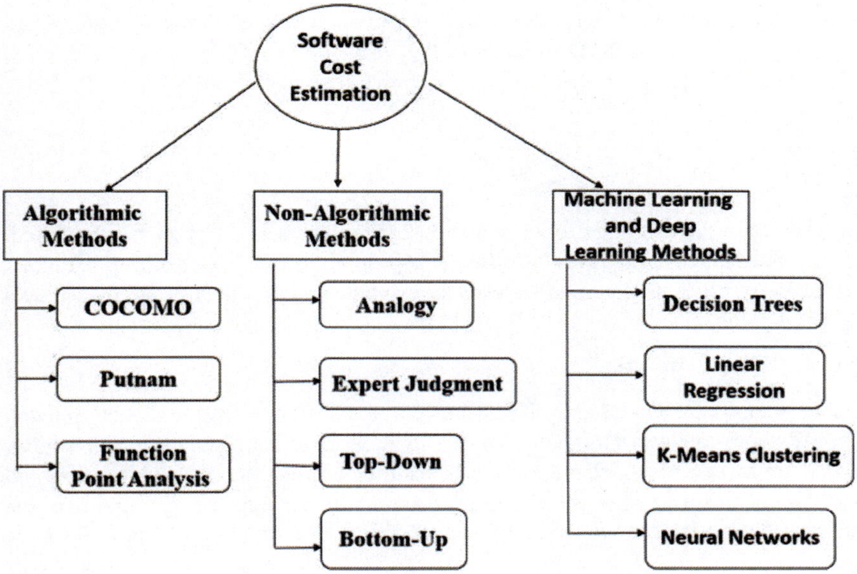

Fig. 2 Methods for software cost estimation

COCOMO. There are a lot of expense generators for COCOMO. The cost driver is a mechanism that improved the cost of few operations. The drivers are database capacity, program sophistication, memory limitations, resources, facilities, cost, project constraints, etc. Follow-up, assortment, and assessment for an estimation are all examples of project reviews. Estimation based on regression methods involve several traditional parametric estimation frameworks, such as COCOMO constructed on 63 chosen projects. The attempt is a metric to reflect the total operating time for the production of applications. The workforce is also a critical consideration in estimating computing costs.

Putnam Model

The macro approximation model is also known as the Putnam Model that calculates the commitment and time required by Rayleigh/Norden for large-scale ventures in the lifecycle management of applications. Tech study programs and human capital are the foundations for this model.

Function Point Analysis

This method calculates the model's operation and size. Complexity is minimized and Albrecht applies SLOC as a metric in 1983. The Cost of a single unit is measured in the initial stages [25]. The sophistication and weight of the program are determined through inputs–outputs from external and conceptual directories. By assessing the client's responsibilities and evaluating the magnitude of the problem, function count is used to quantify the program's complexity. FPA method of estimation is adopted by the models such as ESTIMACS and SPQR/20.

4.2 Non-Algorithmic Methods

To estimate initial design experience and design requirements, this approach is used. Expert assessment, price to win estimate, top-down estimate, bottom-up estimate, and analogy estimate can also be used to evaluate the previous program design's coherent paths.

Analogy Cost Estimation

It is easy to implement because it is handled at the system as well as function level. The efficiency rate is around 60%. Metrics include scale, sophistication, and weight. Needs are historical information records, estimated project costs, architecture, and functions. The key experts of this approach are the improvement of the quality of the software due to the knowledge of ancient evidence. The Analogy Process Measures are:

- Analogy Assortment
- The connections and discrepancies between plans
- Examining the analogy value
- Verifying evaluation.

Expert Judgment Method

Human professionals utilize their expertise in the creation of applications and device domains to forecast device costs and schedules. They utilize the experience of comparable efforts to explain the programs. They have a skill such as regarding the strength to create the app. The most popular Judgment process techniques are Work Breakdown Structure, Wideband Delphi, and PERT.

Top-Down Method

This approach divides the design's total cost into various device variables based on the overall properties. Detailed information is not accessible at the previous point, as the system-level operation costs are not missing, and minimum project specifics are required.

Bottom-Up Method

Designed for every element, the cost is calculated and merged to generate the project's overall principles. This approach uses tiny projects to approximate costs. The predicted errors are balanced in the different elements of the project.

4.3 *Machine Learning and Deep Learning Methods*

Machine Learning has become a cost estimating component that utilizes statistical analysis. The project is assessed using the training samples, which is usually split into two categories: supervised and unsupervised learning. For both the data of anticipatory model input and output, the supervised learning is the fastest and most optimal computational technique. It is subdivided into Regression and Classification. Various algorithms under this category are SVM, Linear Regression, Nearest Neighbor, Decision Trees, Neural Networks, and Naïve Bayes. Unsupervised Learning works with unlabeled data sets. Cluster analysis is most widely used to discover hidden correlations in massive unlabeled datasets.

Decision Trees

The internal nodes of a decision tree signify the features of a dataset, the branches symbolize the decision laws, and every leaf node signifies the result. Decision Trees are designed to imitate human reasoning abilities when making decisions, making them simple to comprehend.

Linear Regression

Linear Regression is used for numerical research, which provides a causal association between dependent and independent variables. Any of the calculation methods include Bayesian linear regression, the Least square estimation, Mixed models, Quantile regression, etc.

K-Means Clustering

K-means clustering is another general method that is mostly applied in machine learning. Although machine learning is mostly thought of as a very recent technology, the basics have been around for way longer than anyone may have anticipated. It provides a simple way to cluster data into various groups and a quick way to discover the categories of groups in an unlabeled dataset.

Artificial Neural Network

It is a supervised learning model which is focused on arithmetic formulas, input method, and output method for the SCE process. The Hopfield Network and ANN are the best way to predict costs. It has the capability of logic, although it gets a huge quantity of training data and there are zero outcomes for the design [6].

The given Table 1 depicts the comparison between the various methods used in software cost estimation.

5 Accuracy Metrics

Cost and Effort estimation is a time-consuming task because there are several variables that affect these figures, making it impossible to arrive at an accurate estimate.

Table 1 Comparison of methodologies

Methodology	Example	Advantages	Disadvantages
Algorithmic method	COCOMO	Estimating costs is simple and provides a clear picture of the result	Elements of earlier projects are needed
	Function point	To achieve a quick result, methods and techniques are impartial	Duration, reliability, and physical labor are not taken into account
	Putnam model	For performance measurement, size, duration, and resource are simply added up	Levels of Software Development Life Cycle are not included
Non-algorithmic method	Analogy-based estimation	Knowledge and skill are applied for real projects	Attributes are required
	Expert judgment	Domain, architecture, and new technology are the foundation of cost estimation	Knowledge of parallel projects is needed
	Top-down	It only requires the smallest amount of project information	It is really difficult to spot moderate errors
	Bottom-up	Estimation in Errors is fairly consistent	Movements at the component level and in time are never tracked

Organizations employ software measurement systems to keep track of the expense of software production and the consistency of the final product. Several metrics are typically applied in the hopes of improving software creation and development process control.

Accuracy metrics provide a change among the expected effort and genuine effort. Various accuracy metrics are:

5.1 *Mean Magnitude Relative Error*

This is a very well-known criterion applied to estimate software effort and cost (Briand et al. 1999). The Magnitude of Relative Error (MRE) for every single examination is given by:

$$\mathrm{MRE_i} = \frac{|\mathrm{ActualEffort_i} - \mathrm{PredictedEffort_i}|}{\mathrm{ActualEffort_i}}$$

The summing of MRE across N observations completes the MMRE operation.

$$\text{MMRE} = \frac{1}{N}\sum_{i}^{N}\text{MRE}_i$$

5.2 Mean of Magnitude of Error Relative to the Estimate

MMER is an alternative technique that can be applied as a substitute to the MMRE. MER is the abbreviated form of Magnitude of Error Relative to the estimate (Kitchenham, et al.,2001). MER is parallel to MRE but for that the denominator is the projected effort as an alternative to the definite effort. Equations for MER and MMER are:

$$\text{MER}_i = \frac{|\text{ActualEffort}_i - \text{PredictedEffort}_i|}{\text{PredictedEffort}_i}$$

MMER is then accomplished by the summation of MER over N observations

$$\text{MMER} = \frac{1}{N}\sum_{i}^{N}\text{MER}_i$$

MRE and MER are inversely connected, as shown by Eqs. (1) and (3). In other terms, an attempt to enhance one system will have a negative impact on the other, since the denominator of the MRE is the real effort, while the denominator of the MER is the expected effort. Nonetheless, it is important that both MMRE and MMER be used for assessment. For example, if the MMRE is high and the MMER is tiny, this indicates that the projects' average real effort is less than the average expected effort. Broad MMER values, on the other hand, mean that the average expected effort is smaller than the average real effort.

5.3 Prediction Performance

Prediction (PRED) calculates the average proportion of estimates which were within X% of the actual values. This is denoted as PRED(X) and the formula is given below.

$$PRED(x) = \frac{k}{n}$$

where x is the extreme MMRE (or MMER), n is the total figure of projects, and k is the figure projects in the set of n projects whose MMRE (or MMER) <= x.

For example, PRED (25) gives the percentage of projects that were estimated with MMRE (or MMER) less than or equal to 0.25.

5.4 Mean Absolute Error

Mean Absolute Error (MAE) is the average of the absolute errors among the actual and the projected effort as demonstrated in Eq. (6).

$$\mathrm{MAE} = \frac{1}{N}\sum_{i=1}^{N} |E_a - E_p|$$

where *Ea* is the actual effort and *Ep* is the predicted effort.

5.5 Root Mean Square Error

It can be estimated as the square root of the mean square error and can be defined as:

$$\mathrm{RMSE} = \sqrt{\frac{1}{N}\sum_{i=1}^{N} (\mathrm{ActualEffort} - \mathrm{PredictedEffort})^2}$$

5.6 Median Magnitude of Relative Error

It can be calculated as:

$$MdMRE = Median_i(MRE_i)$$

5.7 Mean Balance Error

It can be calculated as:

$$\text{MBER} = N^{-1}\Sigma_i \frac{|\text{ActualEffort}_i - \text{PredictedEffort}_i|}{\min(\text{ActualEffort}_i, \text{PredictedEffort}_i)}$$

6 Software Metrics

Computer analytics is a common method applied to evaluate the development of the software. Good measurements can lead to the construction of models suitable for forecasting software products and processes. Metrics and models are applied to calculate output and result efficiency by calculating the expense and timing of the product. The widely utilized scale and overall flexibility are used to build metric-based applications. It is split into three categories [26].

6.1 Process Metrics

Process metrics, also recognized as management metrics, are applied to assess the characteristics of the software development process. Cost metrics, effort metrics, advancement metrics, and reclaim metrics are all examples of process metrics. Process metrics assist in estimating the scale of the final structure and deciding if a project is on track.

6.2 Product Metrics

Product metrics, also recognized as quality metrics, are applied to assess the software's characteristics. Non-reliability metrics, functionality metrics, efficiency metrics, usability metrics, cost metrics, size metrics, complexity metrics, and style metrics are all examples of product metrics. Product metrics leads towards improving the efficiency of various device components and comparisons.

6.3 *Size of Metrics*

Line of Code

It is the oldest and most simple metrics for determining the size of a computer coded data. It is commonly used to calculate and compare programmers' productivity.

- Productivity is calculated as LOC/man-month.
- LOC is any line of program text that does not have a message or a blank line, regardless of the number of statements or sections of statements in the program text.

Token Count

A computer database is described as a set of tokens that can be categorized as operators or operands in this metric. These simple symbols can be used to describe all software science metrics. Tokens are the names given to these symbols. The essential actions are:

$n1 =$ count of unique operators.
$n2 =$ count of unique operands.
$N1 =$ count of total occurrences of operators.
$N2 =$ count of total occurrence of operands.

In terms of the total tokens used, the size of the program can be expressed as

$$N = N1 + N2$$

Function Count

- A larger device called a module can be used to help measure the scale of a large unit software product. A module is a reusable code segment that can be compiled separately.
- Consider a software product that requires n modules. The size of a module is usually accepted to be about 50–60 lines of code. As a result, the scale of this product of software is estimated to be about $n \times 60$ lines of code.

7 Conclusion and Future Scope

This paper offers all the critical analysis necessary for the cost assessment of software, which specifies and determines the reliability of the software. Several analysis works have shown that the COCOMO dataset is applied to differentiate between the quality and the cost of the program. Study also provides the knowledge of software cost estimation measurements and software matrices for calculating the output more efficiently. Appropriate incorporation of techniques is the essential prerequisite for improved efficiency and expense. The comparative analysis is focused on the efficiency of SCE approaches.

AI models for SCE are used in current technologies to increase estimates precision even if the software requirements are ambiguous, inadequate, and undefined. It is often experienced that only one processor model or methodology for software estimation cannot be used at a time. Therefore, more than one solution needs to be taken. Cost estimation techniques are convincing to the human estimator; otherwise, human intuition must be valued. The criteria and inputs used by the different models implemented must be checked and modified accordingly. Evidence of suitability will only be achieved if these recommended AI models are incorporated into the estimation tool and developed over time through an association by enhanced data in input constraints.

References

1. N.E. Fenton, S.L. Pfleeger, Software metrics: a rigorous and practical approach. PWS Publishing Co (1997)
2. V.R. Montequín, J. Villanueva Balsera, C. Alba González, G. Martínez Huerta, Software project cost estimation using AI techniques, in *Proceedings of the 5th WSEAS/IASME International Conference on Systems Theory and Scientific Computation*, pp. 289–293 (2005)
3. J.M. Minguet, J.F. Hernández, La calidad del software y sumedida (2003)
4. R. Tripathi, P.K. Rai, Comparative study of software cost estimation technique. Int. J. Adv. Res. Comput. Sci. Softw. Eng. **6**(1) (2016)
5. D. Manikavelan, R. Ponnusamy, Software quality analysis based on cost and error using fuzzy combined COCOMO model. J. Amb. Intell. Human. Comput. 1–11 (2020)
6. P. Phannachitta, On an optimal analogy-based software effort estimation. Inf. Softw. Technol. **125**, 106330 (2020)
7. Q. Xiaotie, M. Fang, Summarization of software cost estimation. Proc. Eng. **15**, 3027–3031 (2011)
8. S.A. Deshmukh, S.W. Ahmad, Using classification data mining techniques for software cost estimation. 2229–5518 (2016)
9. V.R.Arulmozhi, B. Vijaya Nirmala, N. Deepa, A software cost estimation approach using Least Square-Support Vector Machine Technique. 2395–695X (2015)
10. Z.A. Khalifelu, F.S. Gharehchopogh, Comparison and evaluation of data mining techniques with algorithmic models in software cost estimation. Proc. Technol. **1**, 65–71 (2012)
11. S. Shivangi, U. Kumar, Review of various software cost estimation techniques. Int. J. Comput. Appl. **141**(11), 31–34 (2016)
12. R. Poonam, S. Jain, Enhanced software effort estimation using multi layered feed forward artificial neural network technique. Proc. Comput. Sci. **89**, 307–312 (2016)
13. S. Densumite, P. Muenchaisri, Software size estimation using activity point, in *IOP Conference Series: Materials Science and Engineering*, vol. 185, no. 1, p. 012013. IOP Publishing (2017)
14. O.P. Sangwan, Software effort estimation using machine learning techniques, in *2017 7th International Conference on Cloud Computing, Data Science & Engineering-Confluence*, pp. 92–98. IEEE (2017)
15. P.A. Whigham, C.A. Owen, S.G. Macdonell, A baseline model for software effort estimation. ACM Trans. Softw. Eng. Methodol. (TOSEM) **24**(3), 1–11 (2015)

16. F. Sarro, A. Petrozziello, M. Harman, Multi-objective software effort estimation, in *2016 IEEE/ACM 38th International Conference on Software Engineering (ICSE)*, pp. 619–630. IEEE (2016)
17. Y. Masoudi-Sobhanzadeh, H. Motieghader, A. Masoudi-Nejad, Feature select: a software for feature selection based on machine learning approaches. BMC Bioinform. **20**(1), 170 (2019)
18. P. Pospieszny, B. Czarnacka-Chrobot, A. Kobylinski, An effective approach for software project effort and duration estimation with machine learning algorithms. J. Syst. Softw. **137**, 184–196 (2018)
19. M.M. Al Asheeri, M. Hammad, Machine learning models for software cost estimation, in 2019 *International Conference on Innovation and Intelligence for Informatics, Computing, and Technologies (3ICT)*, pp. 1–6. IEEE (2019)
20. B. Baskeles, B. Turhan, A. Bener, Software effort estimation using machine learning methods, in *2007 22nd International Symposium on Computer and Information Sciences*, pp. 1–6. IEEE (2007)
21. M.Z. Tunio, H. Luo, C. Wang, F. Zhao, A.R. Gilal, W. Shao, Task assignment model for crowdsourcing software development: TAM. J. Inf. Process. Syst. **14**(3) (2018)
22. M. Lawal, P. Genevès, N. Layaïda, A cost estimation technique for recursive relational algebra, in *Proceedings of the 29th ACM International Conference on Information & Knowledge Management*, pp. 3297–3300 (2020)
23. S.Z. Iqbal, M. Idrees, A.B. Sana, N. Khan, Comparative analysis of common software cost estimation modelling techniques. **33** (2017)
24. M. Ullah, R. Ali, M. Ahmad, T. Khan, F. UlMulk, Software cost estimation–a comparative study of COCOMO-II and Bailey-Basili Models, in *2019 International Conference on Advances in the Emerging Computing Technologies (AECT)*, pp. 1–5. IEEE (2020)
25. L.R. Nerkar, P.M. Yawalkar, Software cost estimation using algorithmic model and non-algorithmic model a review. Int. J. Comput. Appl. **2**, 4–7 (2014)
26. G. Singh, D. Singh, V. Singh, A study of software metrics. IJCEM Int. J. Comput. Eng. Manage. **11**, 22–27 (2011)

A Comprehensive Review of Routing Techniques for IoT-Based Network

Bishwajeet Kumar and Sukhkirandeep Kaur

Abstract The trend toward the adoption of IoT has been increasing from the past few years because of its diverse applications in various domains. Billions of smart objects will communicate among themselves for acquiring information in the next generation for IoT-based network. Along these lines, routing insight assumes a significant part in guaranteeing communication between the nodes. The primary purpose of this study is to recognize the gaps in network utilization optimization, resulting in a longer stability of IoT (Internet of Thing)-based network. Till now, the majority of the distributed work zeroed in mostly on energy effectiveness, node dependability, quality of service (QoS), and network lifespan with WSN (wireless sensor network) viewpoint. The principle point of this work is to address the difficulties and different issues for IoT-based network IoT along with comparative study of different routing protocols for IoT-based network considering different parameters such as energy consumption, end-to-end delay, scalability, and packet delivery ratio (PDR).

Keywords IoT · Routing protocol · Bandwidth · Processing capability · PDR · End-to-end delay

1 Introduction

Over the last number of years, WSN technology has become popular and demands a new Internet of Things (IoT) utilization. Such a concept of creating new omnipresent and smart IoT applications has marked the start of a new era [1]. IoT consists of two words, "Web" and "Things". Web represents Internet and Thing represents different types of smart sensors. IoT secures the omnipresent communication by using the wireless media among different types of smart sensors, actuators, cell phone, etc.

B. Kumar (✉) · S. Kaur
Lovely Professional University, Phagwara, Punjab, India

S. Kaur
e-mail: sukhkirandeep.2328@lpu.co.in

A. K. Luhach et al. (eds.), *Second International Conference on Sustainable Technologies for Computational Intelligence*, Advances in Intelligent Systems and Computing 1235,
https://doi.org/10.1007/978-981-16-4641-6_12

In providing omnipresent computing capable of connecting both physical and virtual worlds, IoT plays a key role [2]. The correlation between WSNs as well as the Internet of things has a tremendous impact on quality of life and also leads to massive benefits. IoT also gains significant recognition from different universities, to help develop newborn technologies and utilization based on smart institutions, smart healthcare system, IoT-based modern irrigation and agriculture system, and modern cities. Omni-present concept in IoT provides significant ways for increasing the effectiveness of food production chain management [3]. For the healthcare framework, patient should convey clinical sensor to follow indispensable boundaries, for example, internal heat level, pulse, ECG (electrocardiogram), and breathing framework. Medical care centers will also be able to use automated remote monitoring to assess a medical symptom [4]. The Smart Santander project proposes an experimental community-based test facility to support traditional smart city applications and services in terms of overall implementation. Depending on the applications, the facility contains more than 20,000 IoT modules, divided into layouts of tens or hundreds of nodes, such as environmental monitoring and smart parking [5]. For operations that affect them, data must be sent to the Base Station (BS) for various IoT applications. The transmission of data from application to base station (Bs) can be achieved at the cost of proper data transmission mechanism or through routing protocol that are one important component of IoT to increase the stability of the network, minimization of end-to-end delay and scalability.

Nonetheless, the attributes of WSNs/IoT represent a few difficulties to the advancement of fruitful correspondence conventions because of inadequate assets and the trickiness of low-power remote connections that regularly fizzle regarding nature of administration (QoS) necessities [6].

The remainder of the paper is facilitated as follows. Challenges in routing for IoT-based networks is defined in Sect. 2. Section 3 describes different routing protocols for IoT-based networks. Section 4 describes trends in IoT and Sect. 5 discusses various energy-efficient routing protocols for IoT-based networks, and Sect. 6 concludes with the discussion.

2 Challenges in Routing for IoT-Based Network

The Internet of Things is a standard climate where routing is a muddled occupation for various factors. A portion of these contemplations are referenced underneath [7].

2.1 Network

The source hub where from that information gets delivered and the objective hub for which the information is foreordained may exist on the homogeneous or heterogeneous organizations. Along these lines, moving information just as control bundles across the organizations is again a difficult errand.

2.2 Connectivity

The accessibility between any two devices in the IoT organization might actually be consistent. This depends upon the organization and the climate. Viable trade of data within the sight of inadequacies in the organization is a bottleneck.

2.3 Limited Resource

Resource constraint, which involves energy supply, computing capacity, memory capabilities, wireless communication range, and bandwidth of wireless communication, is one of the most critical challenges for IoT. In several cases, this restriction can impact routing. Due to the limitation in range of wireless communication, routing is performed with the help multi-hop way, i.e., multiple relay nodes are required for transferring the data packet form source to destination.

2.4 Congestion Control

Regulation of congestion is an issue that usually exists across all network types. It happens because of the sudden rise in traffic, a complicated phenomenon. The consequences of congestion are packet loss and unnecessary delays. It should be done at any node to protest the congestion load balancing. Congestion minimizes a network's lifetime. It is therefore critical that the routing protocol attempts to resolve congestion.

2.5 Deployment of Node

It may be a fundamental question to be discussed in the IoT. By correctly settling the node deployment in sensor networks, the difficulty of the problem in IoT can be solved. Communication, routing, etc., for example. In an IoT-based network, nodes can be perform either by setup or manually. On account of manual rendition to send the information way will be foreordained and on account of arrangement, the hub will be coordinated arbitrarily with the help of ad hoc routing setup.

3 Routing Protocols in IoT

In IoT, gadgets predominantly cooperate with one another for information handling, store and investigation from source to target gadgets. In order to transfer data between devices securely, reliable protocol must promote low-energy consumption and scalability [8]. Routing is characterized as the way toward moving information bundles starting with one then onto the next gadget, which helps to keep the mobile channel's route among devices intact. Routing must always select the best or maximum throughput to the destination, and it must use protocols to get from source to destination [9]. Either an intradomain network or an interdomain network may be used to communicate.

In IoT, routing protocols are three types depending upon wireless communication which are given below.

1. Organization of Network.
2. Discovery of Route.
3. Operational Protocol.

Figure 1 shows the logical characterization of steering convention for IoT-based association.

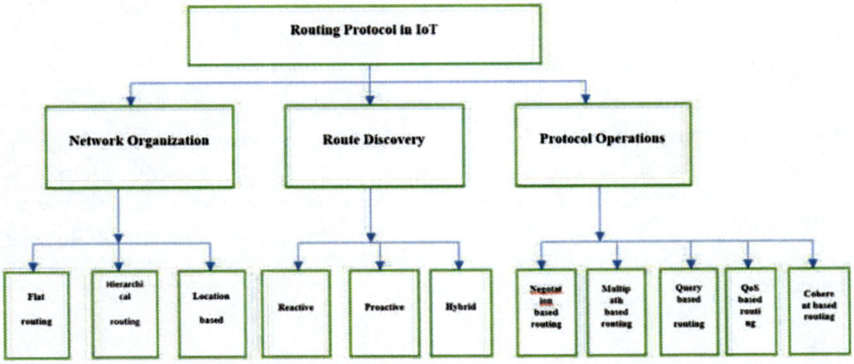

Fig. 1 Taxonomy of protocol for IoT routing

3.1 Establishment of Network

In IoT, foundation of organization assumes a vital part in the utilization of the directing convention. A concise outline of the qualities of an organization is given by this organization association.

The various protocols involved in a network structure are defined as follows.

A. **Flat-Based Routing**
Flat routing is additionally perceived as horizontal-based routing. This kind of convention is basic in an organization with a level or flat construction. Every node is treated evenly in this network, and consists of the same features. There is no need for any steps to be taken here to coordinate the network and its traffic. A schedule based on contention is flat-based routing [10]. Such kind of network is used in IoT to provide an effective solution for many independent issues that arise due to their minimum operational complexity and high performance.

- **Sensor Protocol for Information via Negotiation (SPIN)**: On account of this calculation [11], current information will, in general, be part into metainformation and shipped off adjoining hubs to dodge information from being sent needlessly. It sends the data and asks without any destruction for a legitimate data. Via energy consumption, SPIN increases the network lifetime.
- **Directed Diffusion (DD)**: In the case of this algorithm [12], query is sent continuously from Base Station (BS) to the neighboring nodes. In the wake of getting the inquiry from BS, the hub containing the ideal information sends it right back to BS. By selecting the optimal return path, energy consumption is achieved here.
- **Rumor Routing (RR)**: RR [13] is a variety of directed diffusion. In RR, it will think about every mathematical occasion and inquiries. It is the events used to flood the events that are less than queries than the operation. RR may use agents that inundate the network events. Agents fly to another event from one event. Agents store data in a table of events. The RR algorithm helps to minimize contact costs and the consumption of resources.

B. **Hierarchical-Based Routing**
In this directing, the organization geography is part into numerous progression layers, for example, bunches and group heads, to diminish the size of the steering table dependent on their energy levels [14]. When the node contains the highest energy, it is used to transmit the data in hierarchical terms, and whatever node contains the lowest energy is used to detect the data.

C. **Location-Based Routing**
Position of the senor hub is found in this directing component. When the hubs are close, the sign force of a hub is utilized to decide the situation of the hub [15]. Information shared between adjacent nodes is used to notify the relative coordinates and distance separated by a node. The transmission course will be characterized in this sensor region, in view of the locale and area of the adjoining

Table 1 Correlation between hierarchical, flat, and location-based routing

Hierarchical structure	Flat structure	Location-based routing
1. Reservations-based scheduling approach used by hierarchical structure 2. Collision can be prevented easily 3. The task cycle can be shortened in hierarchical terms based on sleeping of the nodes 4. Routing is not ideal for routing 5. The guarantee of hierarchical justice 6. Aggregation of hierarchical data is carried out by the cluster leader Local hierarchical synchronization and global synchronization are essential	1. Centered on contention in flat scheduling 2. Flat collisions have occurred 3. When it is idle, flat nodes will sleep and the duty cycle is a variable 4. Routing is great 5. There is no guarantee of flat justice 6. Aggregation of flat data is performed by neighbors 7. It is not necessary for flat synchronization	1. In this case, scheduling is based on location 2. Collisions can be detected 3. The nodes will sleep in the place while it is idle 4. In location based, the routing is optimal 5. Fairness can be assured in such kind of routing 6. The aggregation of data is done by clustering in location-based routing 7. In this scenario, synchronization is required

hubs. The ground floor depicts progressive level and area-based conventions are distinctive as a rule.

As given in Table 1, we like to assess the various leveled and level-based directing.

3.2 Discovery of Route

Disclosure of courses is only the safeguarding of steering information from birthplace to target. It joins a receptive, mixture, and useful convention (mix of responsive and proactive conventions).

A. **Reactive Protocol**: No data about the hubs and the courses that are made is found in the receptive steering convention. In dynamic networks that accept continuous changes in topology, this kind of reactive routing is most probably used. Each time dynamic topologies are supported by IoT, this receptive directing extraordinarily affects it. These responsive directing conventions can be partitioned into AODV, AOMDV, and TORA

- **AODV**: AODV [16] is a straightforward steering strategy and is perceived for its effortlessness and proficiency. AODV empowers the making of courses to reasonable objections, yet AODV disallows these courses from being overseen by the hubs when there is no dynamic correspondence. The loop-free problem will be solved in AODV by adding the target gathering numbers. AODV comprises three unmistakable sorts of messages. By route order, route exploration will be done. Path answers will determine the final paths. Connect break error messages are alerted to path errors on an active network route. A steering table in the AODV ought to be kept everything under control to hold all the data about the ways, regardless of whether they are brief courses.

- **TORA**: The temporarily ordered routing algorithm [17] is a protocol for on-demand routing. TORA is intended to limit the message of power. The algorithm, called connection reversal, is used by TORA. TORA entails three activities, first building a way from source to sink hub, second, remove the route when it is invalid between two hubs, Finally, it will keep the path open. If the path fails, an additional route will be allocated to approach the destination. Whenever established at the objective, TORA endeavors to make a directed acyclic graph (DAG).

B. **Proactive**: In static networks, proactive protocols are often used where the topology cannot be modified more than once. This constructive protocol preserves data that can be called a routing table in a tabular format. "Table-driven protocols" are constructive protocols, since routing is based on a routing table. Again, this constructive protocol has numerous types, such as low-energy adaptive clustering hierarchy (LEACH) and routing protocol for low-power lossy network (RPL).

- **OLSR**: The optimized link state protocol (OLSR) [18] is a replacement to the connection state calculation. It requires exchanging messages regularly to establish a proper network topology at each and every node. In order to minimize the packet size and retransmissions, OLSR was improved by multipoint relay flooding and messaging. With a variety of hops, OLSR offers ideal routes.
- **GPSR**: Voracious perimeter stateless routing (GPSR) makes it workable for a hub to send messages to its closest neighbor, who should likewise be nearer to the objective the data is proposed to make a trip to. In a solitary bounce, a hub needs to review the situation of a neighbor. Eager sending is a calculation used to compute the ways of hubs and move the data to the sink of the hub. Directing choices are made progressively in the GPSR. GPSR comprises a vector of the size, a vector of the way, and a connection state directing calculation.
- In order to summarize the constructive routing protocol, the GPSR protocol consists of good performance compared to OLSR, and low overhead.

C. **Hybrid Protocol**: It has similar highlights as useful and responsive conventions. Once more, cross-breed directing is part into different steering conventions, like optimized link state routing, zone routing protocol, etc.

- **ZRP**: For hybrids, ZRP is a protocol. The very fast transmission of data and the overhead minimization is the main use of ZRP. It is not possible to spread ZRP across the entire network. The distances between the nodes will be established by the radius of the field. Two kinds of intra and inter-zone routing protocol areas are composed of ZRP. Inside the routing zones, the intra-zone works and the inter-zone between routing zones works. The best aspect of constructive routing or reactive routing is most possibly chosen.

- **SOC-M2M**: M2M sensor nodes fill in as IoT building blocks. The essential target of utilizing this convention is to decrease the utilization of energy and increase the existence of the organization. SOC-M2M is a self-arranging gathering, i.e., it is consequently coordinated as bunches and chosen as the top of the group. Each hub will share their offset power with the neighbor hubs in this strategy.
  An SOC-M2M efficient energy routing protocol is compared to ZRP in order to summarize the hybrid routing protocol.

3.3 Protocol Operations

It offers a brief overview of the key organizational features of routing protocols, such as patterns of communication, hierarchy, methods of distribution, and computation. This protocol operation is shown below.

A. **Negotiation and Query-Based Routing**
   This routing would deny the surplus data that existed between the source and destination. The resource negotiation decisions will rely on availability.
   Two stages of route exploration are done by query-based routing: the request stage and the reply stage. Details are transferred from origin to destination by the query process. This query method generates a query packet and transmits it with a reply to the nearest node if any node receives the query.

B. **Multipath, Coherent-Based, and QoS-Based Routing**
   In the WSN, multipath routing is most likely used to primarily boost network performance by offering load balancing capabilities. This can provide a range of benefits, such as fault tolerance, improved bandwidth, or enhanced protection-based conventions, which permit sensor hubs to adjust energy utilization prior to conveying the information to the sink hub with certain pre-decided QoS measurements, like dormancy, energy, unwavering quality, transmission capacity, and so on.
   Data such as time stamping and duplicate suppression is transmitted to aggregators after processing during transparent routing. Coherent processing is commonly chosen in order to conduct energy-efficient routing. An example of coherent data processing is some winning algorithms.

4 Trends in IoT

The Internet of Things (IoT) is perhaps the most unmistakable tech patterns to have arisen lately. In basic terms, it alludes to the way that while "web" at first alluded to the wide-scale systems administration of PCs, today, gadgets of each size and shape—from vehicles to kitchen apparatuses to modern hardware—are associated and share data carefully, on a worldwide scale [19].

Likewise, with each part of our lives, the worldwide COVID pandemic has without a doubt influenced the manner in which this pattern is creating and affecting our lives. In reality, as we know it, where contact between people is, for the time being, more restricted, contact between gadgets, devices, and toys can assist us with staying associated.

Along these lines, here's my look forward to 2021 and a portion of the manners in which we can hope to see this megatrend working out, and assuming an inexorably enormous part by the way we live, work, and play.

4.1 Top Trends in IoT

- **Focus on Security**: The IoT tech market will be considered to be a center around security as mind boggling well-being challenges crop up. These intricacies originate from the assorted and dispersed nature of the innovation. The organization of associated gadgets stays defenseless against assaults. Anyway, what number of gadgets are associated with the web in 2021? The quantity of Internet-associated gadgets has arrived at the 10 billion imprints [20]. Gadget and IoT network hacking will get ordinary. It is up to organize administrators to prevent gatecrashers from doing their business.
- Broadly acknowledged security concerns will help IoT suppliers improve sell capacity. This will be by featuring the safety efforts they have set up. Those advertising start to finish IoT arrangements are destined to appreciate this selling point.
- **Big Data, Analytics, and Machine Learning**: Indeed, even now, conveyed IoT examination and information are discovering applications in IoT organizations. This strategy allows frameworks to trigger cautions or activity sans moving volumes of information to arrange centers. This outcomes in improved execution as organizations work at low inertness. The combination of information streams with AI and AI motors is another pattern. Potential applications incorporate savvy homes and lift upkeep, to give some examples. Discover more by perusing the most current enormous information insights.

  Incorporated examination is presently being implanted into arrangements as suppliers look to accelerate information investigation. Such investigation is straightforwardly taken care of into AI applications. This plan upholds IoT gadgets, cycles, and foundation transformation and advancement. IoT information will be sold as a product, which will be focused on generally at machine makers. Selling IoT information will be utilized to demonstrate item practicality. Late AI measurements mirror these turns of events.

- **The Emergence of Smart Cities**: IoT selection will bring about the foundation of keen urban areas. This early, some US urban communities are utilizing IoT to associate utilities, stopping meters, and traffic signals. Shrewd city projects are presently in the pipeline, an advancement that could just spread around the world. An evidence of this is the increment in shrewd city tech spending, which can reach $158 billion by 2022.
- **Cloud Vendors to Zero in on IoT**: Cloud-based programming merchants will rival IoT stage engineers to become prime objections for IoT. They will be zero in on finding new applications for the innovation and influence their joining abilities. Organizations among sellers and IoT suppliers will multiply to extend use cases and improve existing ones. Google will most likely stand out in looking for such associations.

  Apart from this, IoT is used in customer service, manufacturing, and health care, which are some other feature trending in IoT.

5 Energy-Efficient Routing Protocols for IOT-Based Network

Machine-to-machine correspondence is a term in that identifies with innovations that permit both wired/remote frameworks to send messages starting with one gadget and then onto the next gadget without human mediation, either in a homogeneous or heterogeneous organization. As remote sensors go about as fundamental structure squares of M2M correspondence, this geography is presently an unpredictable nature. Machine gadgets measure the occasions that are enlisted by means of hubs and sent through the legitimate passage. Battery power can be used effectively in M2M communication because certain systems must work longer, and communication and processing energies have the energy drained by sensor nodes. These two techniques help to minimize energy utilization and increase the life expectancy of the network and integrate a sleep technique in a hierarchical organization through a proper topology.

In [20], in different administrative domains, the emerging demands of innovative services are increasing which leads to the development of advanced technologies at heterogenous platform, but this can be devised by IoT. To support seamless services for human, IoT focuses on integrating two terms, network and human. Routing plays crucial role in IoT-based network. The author of this article proposes a routing algorithm that is based on the concept of SFG, i.e., sunflower-based gray wolf optimization algorithm for improving the lifespan of IoT-based networks. In this algorithm, select the best from the group of available paths for routing. The selection procedure depends on the lifespan, rest energy, and delay. The mechanism of SFG algorithm is designed by combining SFO and GWO (grey wolf optimizer) for selecting the best path for routing. In this way, the proposed algorithm SFG outperformed other methods by considering parameters such as network stability, energy conservation, and network delay.

In [21], network stability is the major concern for IoT-based network. Author of this article proposed an energy-efficient link stable routing (EELSR) protocol to minimize the consumption of energy by the node and to enhance the lifespan of the network. To overcome the accuracy of the suggested algorithm, author of this paper puts an analytical model of link stability and optimal route selection algorithms are designed with considering remaining energy of a route, distance of the route, and stability of the route. The proposed algorithm is compared with AODV and REL by considering parameters such as packet delivery ratio and route setup delay. In this way, the proposed algorithm EELSR outperformed other state-of-the-art staring technique.

In [22], IoT required lot of research due to their unique feature compared to traditional wireless sensor networks (WSN). In next generation, it may be possible for the number of interconnected device on the earth to overtake the actual number of people on the earth. So, in this way, security becomes one of the major concerns for IoT-based networks. As a result, the author's goal throughout this article is to present a novel lightweight structure-based data aggregation routing (LSDAR) protocol for IoT-based networks to protect data from malicious disruptions. Firstly, network nodes are decomposed into independent clusters with varying radiuses to prevent energy holes around the base station (BS). Secondly, efficient and loop-free algorithms are constructed deepened upon A-star heuristic algorithm. Experimental result of this paper is more effective in compression of state of the art in terms of energy consumption, network lifetime, packet drop ratio, and end-to-end delay.

In [23], author of this article proposed an EPC-RPL (energy-efficient parent determination for RPL). It comprises dynamic stream and subterranean insect province improvement calculations. The streamlining of the subterranean insect state considers the boundaries; HC, RER, and ETX will pay for the title parent for data transmission. The powerful stream calculation utilizes the crown system, what separates whole organization into number of rings dependent on the inclusion region. It chooses the best hubs among the organization hubs in every crown utilizing inclusion region of hubs and number of neighbors. Along these lines, it smothers the measure of control messages to set up the course in the organization. It accomplishes the information unwavering quality and diminishes the deferral and builds the organization lifetime. Nonetheless, it takes more union effort to build the quantity of hubs in IoT.

In [24], author of this article proposed an EDADA-RPL (energy and delay-aware data aggregation in routing protocol) for IoT-based network. It comprises the parent determination measure and the information total interaction. The main cycle considers the measurements, that is, RER for enhancing the parent in DODAG. The subsequent one called the information conglomeration measure, which utilizes a compacted detecting calculation to eliminate repetitive information from the sensor hubs. DODAG's parent hubs collect data from their kid hubs and send it to their parent hub. Finally, the information parcels arrive at the sink hub from the member. The

EDADA-RPL convention gives predominant execution analyzing RPL and LA-RPL. Notwithstanding, information misfortunes happen in the EDADA-RPL information recuperation measure.

In [25], author of this article proposed an RPL-CGA (chaotic genetic algorithm) in LLN. This algorithm helps in choosing the best parent for information transmission in the preferred parent list and also this proposed algorithm significantly improves the performance of the network (end-to-end delay, average success rate, etc.)

In [26], author of this article proposed a firefly tuned steering convention for IoT-based network. In this article, author discussed a multichannel portion and directing utilizing FA in remote video sensor organizations. Every firefly's situation in FA is named as the channel assignment grouping and estimation of cost by target work planned in MATLAB is determined for every specialist. The reenactment is led utilizing MATLAB. The adequacy of FA furnishes better execution when contrasted and hereditary calculation.

In [27], author of this paper proposed a LA-OF (learning automata-based objective function) in RPL. This algorithm keeps an eye on the issue of dynamic and lossy environment in LLN.

In [28], to extend the life expectancy of an IoT organization, the author of this article recommended EEOPS-RPL (energy-efficient optimal parent selection) in RPL using firefly enhancement calculation. Every hub in the organization is viewed as a firefly in EEOPS-RPL, which computes the current firefly region, the fascination of the fireflies, irregular capacity, speed, and the worldwide best qualities in the network. Energy that isn't being used and that is supposed to be transmitted to choose the ideal parent in the objective organized coordinated non-cyclic chart for information transmission, tally is allure boundaries and distance is a growth boundary. When compared to EPC-RPL and E-RPL, the EEOPS-RPL increases the bundle processing proportion and the organization's life expectancy by 2% to 5% and 5% to 10%, respectively.

In [29], author of this article proposed an LOADng [lightweight on-demand ad hoc distance-vector routing protocol for next generation]. In this routing protocol, as per requirement, route requests are generated to find the path from source to destination node.

In [30], the creator proposed another convention called EMA-RPL [energy-efficient and mobility-aware routing protocol], the objective of this convention is to lessen the MN energy utilization while giving a consistent network, staying away from information misfortune and diminishing the handover delay. EMA-RPL enables better supporting of accessibility of flexible center points and pivoting. Consequence of this convention is superior to RPL and portability mindful variation [MRPL] as far as handover delay, information misfortune rate, singling cost, and energy utilization.

Table 2 illustrates the various current algorithms and their parameters.

Table 2 Comparative study of different routing protocols for IoT-based network

Routing protocol	Parameters			
	energy consumption	End-to-end delay	PDR	Scalability
SFG	Avg	Avg	X	X
EELSR	Low	Avg	Low	X
LSDAR	Low	Avg	Low	X
EPC-RPL	Low	Avg	X	X
EDADA-RPL	X	X	Low	X
RPL-CGA	X	X	High	X
Firefly-based routing	Low	Low	Avg	X
LA-RPL	Low	Low	X	X
EEOPS-RPL	Low	Avg	Avg	X
LOADng-IoT	Low	Avg	Avg	Yes
EMA-RPL	Low	Low	Low	Yes

6 Conclusion

The primary goal of this review study is to examine the difficulties, issues, and distinctive steering convention for IoT-based organizations. The basic plan of all these directing conventions is to limit energy utilization by the devices or increase the strength of the organization. Flat-based routing is appropriate for tiny networks, whereas rumor routing takes advantage of the scalability and low overhead of the flat routing protocol. The hierarchical routing protocol can deal with a variety of situations in which greater number of sensor nodes with better output interact with each other. When it comes to hierarchical routing, data aggregation aims to save resources by reducing message use by decreasing the number of messages between nodes of sensors. Considering these reasons, hierarchical routing is considered to be the easiest. This paper also discussed different routing protocols by considering parameters such as energy consumption, end-to-end delay, PDR, and scalability for IoT-based networks like SGG, EELSR, LSDAR, EPC-RPL, etc. We found that LOADng-IoT and EMA-RPL are the best protocols because both of them support scalability along with minimum energy consumption, low end-to-end delay, and low PDR.

References

1. Y. Yu, R. Govindan, D. Estrin, *Geographical and energy aware routing: a recursive data dissemination protocol for wireless sensor networks*, pp. 211–221. Springer.
2. V.B. Carvalho, Including context in a routing algorithm for the Internet of Things, Doctoral Dissertation. Faculdade de Ciências e Tecnologia, Lisboa, Portugal

3. M. Farooq, G.A. Di Caro, *Routing protocols for next-generation networks inspired by collective behaviors of insect societies: an overview*, pp. 101–160. Springer (2008)
4. G.C. Onwubolu, B.V. Babu, *New Optimization Techniques in Engineering*, 1st edn. (Springer-Verlag, Berlin Heidelberg, 2013), pp. 1–2
5. Y. Tian, R. Hou, An improved AOMDV routing protocol for internet of things, in *Computational Intelligence and Software Engineering (CiSE)*, pp. 1–4 (2010)
6. C.E. Perkins, P. Bhagwat, Highly dynamic destination-sequenced distance-vector routing (DSDV) for mobile computers. ACM SIGCOMM Comput. Commun. Rev. **24**(4), 234–244 (1994)
7. O. Younis, S. Fahmy, Distributed clustering in ad-hoc sensor networks: a hybrid, energy-efficient approach, in *INFOCOM 2004. Twenty-third Annual Joint Conference of the IEEE Computer and Communications Societies*, vol. 1 (2004)
8. W.B. Heinzelman, A.P. Chandrakasan, H. Balakrishnan, An application-specific protocol architecture for wireless microsensor networks. IEEE Trans. Wireless Commun. **1**(4), 660–670 (2002)
9. H.S. Kim, H. Kim, J. Paek, S. Bahk, Load balancing under heavy traffic in RPL routing protocol for low power and lossy networks. IEEE Trans. Mob. Comput. 964–979 (2016)
10. C. Kharkongor, T. Chithralekha, R. Varghese, A SDN controller with energy efficient routing in the Internet of Things (IoT). Proc. Comput. Sci. **89**, 218–227 (2016)
11. A. Mayzaud, A. Sehgal, R. Badonnel, I. Chrisment, J. Schönwälder, Using the RPL protocol for supporting passive monitoring in the Internet of Things, in *Network Operations and Management Symposium (NOMS)*, pp. 366–374 (2016)
12. S. Sharma, D. Puthal, S. Tazeen, M. Prasad, A.Y. Zomaya, MSGR: a mode-switched grid-based sustainable routing protocol for wireless sensor networks, in IEEE Access, vol. 99, pp. 1–1. S.R. Akbar, W. Kurniawan, M.H. Ichsan, I. Arwani, M.T. Handono, Pervasive device and service discovery protocol in XBee sensor network, in *Advanced Computer Science and Information Systems (ICACSIS)*, pp. 79–84 (2016)
13. S.R. Akbar, W. Kurniawan, M.H.H. Ichsan, I. Arwani, M.T. Handono, Pervasive device and service discovery protocol in XBee sensor network, in *2016 International Conference on Advanced Computer Science and Information Systems (ICACSIS)*, pp. 79–84. IEEE (2016)
14. K. Machado, D. Rosário, E. Cerqueira, A.A. Loureiro, A. Neto, J.N. de Souza, A routing protocol based on energy and link quality for internet of things applications. Sensors **13**(2), 1942–64 (2013)
15. B. Karp, H.T. Kung, GPSR: greedy perimeter stateless routing for wireless networks, in *Proceedings of the 6th Annual International Conference on Mobile Computing and Networking*, pp. 243–254 (2000)
16. Z.J. Haas, M.R. Pearlman, The performance of query control schemes for the zone routing protocol. IEEE/ACM Trans. Netw. (TON) **9**(4), 427–438 (2001)
17. P. Sofiane, Bioinspired Routing Algorithm and Efficient Communication within IoT. IEEE Network September/October 2017. https://doi.org/10.1109/MNET2017.1600282
18. K. Kumar, S. Kumar, Energy efficient link stable routing in internet of things. Int. J. Inf. Technol. **10**, 465–479 (2018). https://doi.org/10.1007/s41870-018-0141-0
19. K. Haseeb, N. Islam, T. Saba, A. Rehman, Z. Mehmood, LSDAR: a light-weight structure-based data aggregation routing protocol with secure internet of things integrated next-generation sensor networks. Sustain. Cities Soc. **54**, 101995. https://doi.org/10.1016/j.scs.2019.101995
20. R.P. Pingale, S.N. Shinde, Multi-objective sunflower based grey wolf optimization algorithm for multipath routing in IoT network. Wirel. Pers. Commun. **117**, 1909–1930 (2021). https://doi.org/10.1007/s11277-020-07951-6
21. K. Kumar, S. Kumar, Energy efficient link stable routing in internet of things. Int. J. Inf. Technol. **10**(4), 465–479 (2018)
22. K. Haseeb, N. Islam, A.R. Tanzila Saba, Z. Mehmood, LSDAR: a light-weight structure-based data aggregation routing protocol with secure internet of things integrated next-generation sensor networks. Sustain. Cities Soc. **54**, 101995 (2020)

23. S.S.L. Preeth, R. Dhanalakshmi, R. Kumar, S. Si, Efficient parent selection for RPL using ACO and coverage based dynamic trickle techniques. J. Amb. Intell. Human. Comput. **10**, 1–15 (2019)
24. S. Sennan, S. Balasubramaniyam, A.K. Luhach, S. Ramasubbareddy, N. Chilamkurti, Y. Nam, Energy and delay aware data aggregation in routing protocol for Internet of Things. Sensors **19**(24), 5486 (2019)
25. Y. Cao, M. Wu, A novel RPL algorithm based on chaotic genetic algorithm. Sensors **18**(11), 3647 (2018)
26. A. Singh, A. Goyal, Firefly tuned minimum power routing in wireless video sensor networks for IOT. **3**(3), 78–83 (2016)
27. A. Saleem, M.K. Afzal, M. Ateeq, S.W. Kim, Y.B. Zikria, Intelligent learning automata-based objective function in RPL for IoT. Sustain. Cities Soc. **59**, 102234 (2020)
28. S. Sennan, R. Somula, A. Luhach, G. Deverajan, W. Alnumay, N. Zaman, U. Ghosh, P. Sharma, Energy efficient optimal parent selection-based routing protocol for Internet of Things using firefly optimization algorithm. Trans. Emerg. Telecomm. Technol. (2020). https://doi.org/10.1002/ett.4171
29. M.T. Clausen, J. Yi, U. Herberg, Lightweight on-demand Ad hoc distance-vector routing—next generation (LOADng): protocol, extension, and applicability. Comput. Netw. **126**, 125–140 (2017). ISSN 1389-1286
30. M. Bouaziz, A. Rachedi, A. Belghith, M. Berbineau, S. Al-Ahmadi, EMA-RPL: energy and mobility aware routing for the internet of mobile things. Fut. Gener. Comput. Syst. **97**, 247–258 (2019). ISSN 0167-739X. https://doi.org/10.1016/j.future.2019.02.042

Static Hand Sign Recognition Using Wavelet Transform and Convolutional Neural Network

Rohit Beniwal, Bhavya Nag, Avneesh Saraswat, and Parth Gulati

Abstract Hand gestures play a crucial role in helping the deaf and hearing impairment community to communicate smoothly. Consequently, a robust hand sign recognition system will help the first responders to better communicate and understand those people, who cannot use the spoken languages. Therefore, in this research paper, we propose a novel approach for static hand sign recognition using convolutional neural network as the classifier and discrete wavelet transform as the feature extractor. We then implement our proposed approach using Python programming language. Moreover, we use the MNIST dataset, which is one of the publicly available datasets available at Kaggle website to further implement our research approach. Furthermore, we compared our approach with Hu moments, Zernike moments (Degree 8), and Krawtchouk moments feature extraction techniques along with SVM Linear, Radial, Random Forest, and K-Nearest Neighbor classifiers. As a result, our approach yields the 99.1% accuracy that outperforms the accuracies of above-discussed feature extraction methods and classifiers.

Keywords Sign language · American sign language · Gestures · Orthogonal moments · Convolutional neural network · Discrete wavelet transform

1 Introduction

Human–Computer Interaction (HCI) is a multidisciplinary field of study that focuses on the interaction between humans and computers. Originally, at the advent of personal computers, HCI was mainly performed using mouse and keyboards to enable easy use of word processors, accounting aids, gaming units, etc. Today, the scope of HCI has increased to a much larger extent such that it includes robotics, smartphones, automatic lighting, sensors in home appliances, in car navigation, etc. [1].

A major part of HCI is gesture recognition. These gestures can originate from any part of the body; however, we mainly focus on hand and emotion gesture recognition

R. Beniwal · B. Nag (✉) · A. Saraswat · P. Gulati
Department of Computer Science and Engineering, Delhi Technological University, Shahbad Daulatpur, Main Bawana Road, Delhi 110042, India

A. K. Luhach et al. (eds.), *Second International Conference on Sustainable Technologies for Computational Intelligence*, Advances in Intelligent Systems and Computing 1235,
https://doi.org/10.1007/978-981-16-4641-6_13

areas. Therefore, in this research paper also, we focus on the area of hand gesture recognition. These hand gestures can be of static or dynamic nature [2]. In static gesture recognition, firstly, the focus is on identifying the object of interest (here, hand) and then predicting its gesture using a single image based on its shape, posture, and other properties. On the other hand, in the dynamic hand gesture recognition, the focus is mainly on distinct hand locations and how it changes its position in real time. It means that the emphasis is primarily on the palm and the arm movement in a series of images or in a video.

In the hand gesture recognition area, significant work has been done using various input devices. These input devices include a wired glove, a depth-aware camera, a stereo camera, an image processing controller, and a single camera. A wired glove is a device, which uses an inertial or a magnetic tracking device to send data about the hand gestures, postures, and positions to a computer. On the other hand, a depth-aware camera operates on the information as generated by the short-range camera to depict the features in the three dimensions. Contrarily, the stereo camera method generally uses the output of two cameras to predict an object. Elseways, an image processing controller utilizes a software to capture the motion of the palm and the arm when a gesture is executed [3]. However, a single camera is often convenient, especially in cases when none of the above input or the environment methods are conducive [4–6].

Hand gestures play a crucial role in helping the deaf and hearing impairment community to communicate smoothly. These hand gestures form a major part of the sign languages which are yielded using hands and facial expressions. The most commonly used sign language in the deaf and the hearing-impaired community is American Sign Language (ASL) [7]. Similar to the English language, ASL too consists of 26 alphabets which are depicted using only one hand gesture (usually, static gestures). A powerful hand sign recognition system that is able to detect ASL gestures becomes essential because it will help the elderly, disabled people, and children to interact with the computers efficiently. It will also spare them the need to deal with the traditional interfaces such as a mouse or a keyboard that requires a certain level of skill to operate. Therefore, research in the hand sign recognition field is a key operative challenge as a good hand sign recognition system will help the deaf and hearing-impaired community to communicate better with the general public. Furthermore, such a system will help the first responders to better communicate and understand the people, who cannot use the spoken languages. Additionally, such a system will help in providing an equal opportunity to access videos and visual transmissions.

Therefore, in this research paper, we propose a novel approach for static hand sign recognition using convolutional neural network as the classifier and discrete wavelet transform as the feature extractor. We then implement our proposed approach using Python programming language. Moreover, we use the MNIST dataset, which is one of the publicly available datasets available at Kaggle website to further implement our research approach. Furthermore, we compared our approach with Hu moments, Zernike moments (Degree 8), and Krawtchouk moments feature extraction techniques along with SVM Linear, Radial, Random Forest, and K-Nearest Neighbor classifiers. As a result, our approach yields the 99.1% accuracy that outperforms the accuracies of above-discussed feature extraction methods and classifiers.

The rest of the paper is organized as follows: Sect. 2 briefly talks about the related work; Sect. 3 outlines the proposed research approach followed by Sect. 4, which discusses its implementation and provides the corresponding result and analysis. Lastly, Sect. 5 concludes the research paper.

2 Related Works

As far as hand sign recognition field is concerned, we found methodologies such as Artificial Neural Networks [8–10], the powerful Convolutional Neural Networks [11], Support Vector Machines (SVM) with Gabor filter [12], elastic graph matching [13], Locally Distributed Linear Embeddings [14], Hidden Markov Models [15–17], Histograms of Oriented Gradient [18, 19], Hu Moments [20], Zernike Moments [21], and many more [2] that have been explored in this area.

Among the above techniques, many of them used image moments to extract the global features and the translation invariants for hand sign recognition. These image moments work directly with pixel regions and, thus, are very beneficial in image analysis as they capture essential information about them. Two of the most commonly used image moments are Hu moments and Zernike moments. Hu moments are invariant to all such as translation, scale, rotation, and reflection [20]. The seven Hu moments are calculated using equations from (1)–(7). However, the seventh Hu moment is not invariant to reflection as it shows change in sign.

$$h_0 = \eta_{20} + \eta_{02} \tag{1}$$

$$h_1 = (\eta_{20} - \eta_{02})^2 + 4\eta_{11}^2 \tag{2}$$

$$h_2 = (\eta_{30} - 3\eta_{12})^2 + (3\eta_{21} - \eta_{03})^2 \tag{3}$$

$$h_3 = (\eta_{30} + \eta_{12})^2 + (\eta_{21} + \eta_{03})^2 \tag{4}$$

$$\begin{aligned} h_4 = {} & (\eta_{30} - 3\eta_{12})(\eta_{30} + \eta_{12})\left[(\eta_{30} + \eta_{12})^2 - 3(\eta_{21} + \eta_{03})^2\right] \\ & + (3\eta_{21} - \eta_{03})\left[3(\eta_{30} + \eta_{12})^2 - (\eta_{21} + \eta_{03})^2\right] \end{aligned} \tag{5}$$

$$h_5 = (\eta_{20} - \eta_{02})[(\eta_{30} + \eta_{12})^2 - (\eta_{21} + \eta_{03})^2 + 4\eta_{11}(\eta_{30} + \eta_{12})(\eta_{21} + \eta_{03})] \tag{6}$$

$$\begin{aligned} h_6 = {} & (3\eta_{21} - \eta_{03})(\eta_{30} + \eta_{12})\left[(\eta_{30} + \eta_{12})^2 - 3(\eta_{21} + \eta_{03})^2\right] \\ & + (\eta_{30} - \eta_{12})(\eta_{21} + \eta_{03})\left[3(\eta_{30} + \eta_{12})^2 - (\eta_{21} + \eta_{03})^2\right] \end{aligned} \tag{7}$$

Additionally, Zernike moments are orthogonal moments, which are invariant to rotation. Due to the orthogonality property of these moments, they can also be used for image reconstruction. Zernike moments outperform other moments in terms of image representation and information redundancy. However, one of the disadvantages of Zernike moments is that the higher order Zernike moments are much more susceptible to noise disturbances [21]. The two-dimensional Zernike moment of order m and n is shown in Eq. (8):

$$Z_{mn} = \frac{m+1}{\pi} \sum_{r} \sum_{\theta} P(r,\theta) V^{*}_{mn}(r,\theta) r \leq 1 \tag{8}$$

Here, $P(r, o)$ is the image pixel intensity and $V^{*}{}_{mn}$ is the complex conjugate for Zernike polynomial V_{mn} [15].

Consequent to usage of Hu and Zernike moments, Otiniano-Rodr´ıguez et al. in [22] presented a technique in which the feature extraction was done through Hu and Zernike moments, and the classification for hand sign recognition was done via an SVM. As a result, they achieved 96.27% accuracy. Similarly, Tolba et al. [9] used a unique configuration of neural network, namely, Learning Vector Quantization (LVQ) for hand gesture recognition and yielded 90.83% accuracy. Likewise, Nguyen et al. [10] proposed a method by choosing the best pieces of information using Principal Component Analysis (PCA) followed by neural network classifier for hand gesture recognition giving 94.3% accuracy. Oyedotun and Khashman [11] also presented a methodology where the shape of the hand was extracted using various image processing methods, and then later on for classification purpose, stacked denoising autoencoders and convolutional neural networks were used. As an outcome, they accomplished 91.33% accuracy. Similarly, Chevtchenko et al. [23] proposed a method called Feature Fusion-based Convolutional Neural Network (FFCNN). In this approach, they used Hu and Zernike moments, Gabor features, and the hand contour information to extract the best features. These features were then fed to a convolutional neural network for classification purpose.

Moreover, a novel collection of orthogonal moments called "Krawtchouk moments" was proposed in 2003 to overcome the discretization and approximation error limitations of other moments. These moments were based on Krawtchouk polynomials [24]. Krawtchouk moments possess a very unique capability as it extracts area-specific features from any region of an image. These moments are also scale, translation, and rotation invariant for the better recognition of an object. The Krawtchouk moment for $(n\,x\,m)$ order with an intensity function $f(x,y)$ where the size of the image is $(M\,X\,N)$ is shown in Eq. (9) where K_n represents the Krawtchouk polynomial (order n):

$$Q_{mn} = \sum_{x=0}^{N-1} \sum_{y=0}^{M-1} K_n(x; p1, N-1) K_m(y; p2, M-1) f(x,y) \tag{9}$$

Likewise, Yu et al. also suggested a model using Krawtchouk moments in [25] for static hand sign recognition and consequently they attained 96.2% accuracy. However, in this research paper, we propose a novel approach for static hand sign recognition using convolutional neural network as the classifier and discrete wavelet transform as the feature extractor that further extend the existing literature work where our results outperform the accuracies of above-discussed feature extraction methods and classifiers.

3 Research Approach

The research approach for static hand sign recognition using wavelet transform and Convolutional Neural Network (CNN) is divided into three phases, namely, data pre-processing phase, feature extraction phase, and classification phase. Figure 1 depicts the research approach.

Fig. 1 The proposed research approach

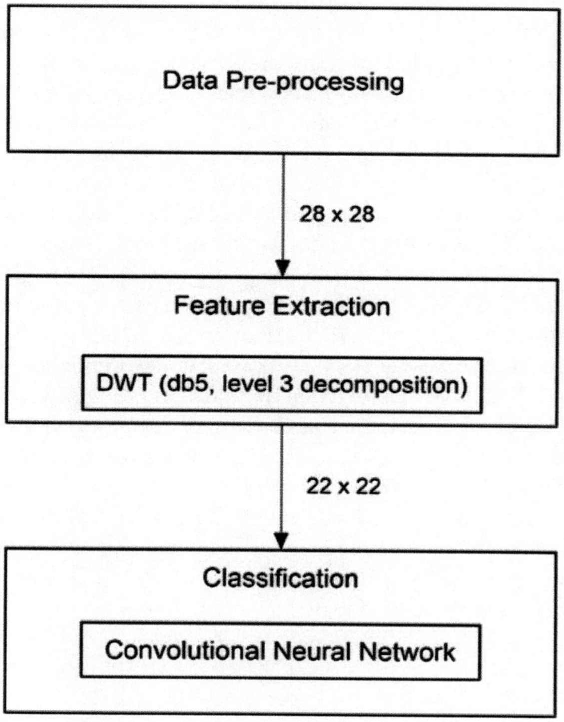

3.1 Data Pre-Processing

In this elementary phase, we will perform data pre-processing in order to convert it into standardized form.

3.2 Feature Extraction

In this second phase, we will perform feature extraction using the discrete wavelet transform method, which will capture information about both the location and the frequency of our hand gestures.

The idea of wavelets was first introduced by Morlet in 1984 as a family of functions used for non-stationary signals [26]. Today, wavelets are abundantly used as a mathematical tool to derive information in image processing, computer graphics, signal processing, in medical imaging, etc. Wavelet transform is similar to Fourier transform, which extracts the information about the frequency from the digital signal. Unlike Fourier transform, wavelet transform provides information about both the frequency and the variations occurring in it over the period of time. It can also analyze a signal at different resolutions. Furthermore, a wavelet transform is very resistant to noise disturbances that occur in any signal or a wave. On the other hand, Discrete Wavelet Transform (DWT) is the discretized variant of the wavelet transform. The DWT captures the scale of the image and is also invertible as well as orthogonal. Here in DWT, image is broken down into four sub-bands of frequency, namely, LL (low-low) quadrant, LH (low-high) quadrant, HH (high-high) quadrant, and HL (high-low) quadrant. The LL block is the low-frequency sub-band, which provides the approximate image of our original image. The LH block derives information about the horizontal characteristics or edges of the original image. The HL block extracts the vertical information while the HH block provides information about the diagonal elements present in the original image. During feature extraction, this breakdown of image in four quadrants helps in reducing its dimensionality. The resulting LL band then can be further decomposed into sub-bands leading to higher levels of decomposition. Figure 2 shows the decomposition of an image into three successive levels.

In our proposed research approach, we will use the db5 wavelet of the Daubechies family [27] with three levels of decomposition. Here, db5 wavelet is an orthogonal wavelet with FIR (finite impulse response) digital filters, which is useful for extracting high-quality stable factors.

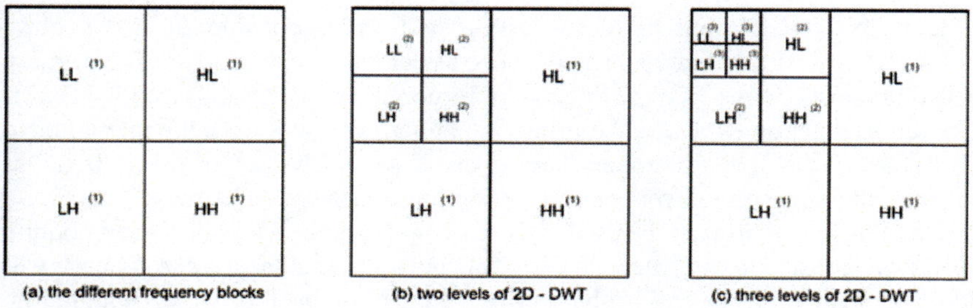

Fig. 2 Decomposition of an image into three successive levels

3.3 Classification

Lastly, in this third phase, the features extracted from the preceding phase will be classified using convolutional neural network, as shown in Fig. 3.

Over the years, numerous methods have been proposed for classifying the image datasets. However, Convolutional Neural Networks (CNN) are the most noteworthy and advantageous classification technique for an image dataset [28]. It is because the successive layers of CNN are able to extract and optimize various features of an image such as edges, lines, circles, and textures. The first layer is the convolution layer in which the scalar product of the input image is taken with the chosen filter. This results in a two-dimensional feature map, which then passes through an activation layer consisting of non-linear activation functions for learning. The next layer is called the pooling layer, where the features are down-sampled. This pooling helps in summarizing the most activated presence of a feature without losing important information. Here, for classification, we can form a complex architecture for our CNN by adding multiple layers performing convolution. The pooling layers will then reduce the dimensionality of our feature map. Lastly, a completely interconnected layer, which is a neural net, is used for finally classifying the images.

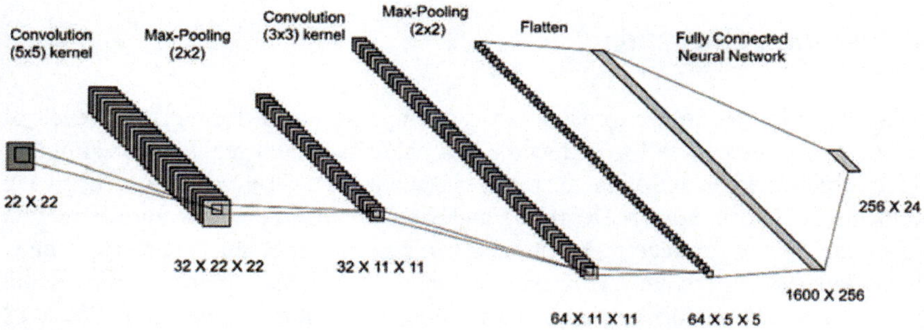

Fig. 3 Our CNN architecture

Our CNN model consists of two layers performing convolution and pooling followed by two fully interconnected layers, as demonstrated in Fig. 3. The primary convolution layer consists of 32 filters with kernels of size (5, 5) and Rectified Linear Activation function (ReLU). The primary pooling layer will have the max filter of (2, 2) size. We will add a dropout here to decrease the chances of overfitting. Our subsequent convolution layer consists of 64 filters, each with a kernel of size (3, 3), and the ReLU function will be used as the transfer function. The subsequent pooling layers will consist of max filters of size (2, 2) and a stride of length 2. Again, we will add dropout after this layer. Finally, we will flatten the output that we will obtain from the above layers and send it to the completely interconnected layer. The completely interconnected neural net layer will then classify them into 24 class labels.

4 Implementation, Results, and Analysis

To implement the research approach, first of all, we picked suitable data known as MNIST, which is publicly available at Kaggle website [29]. This MNIST dataset consists of 34,627 cases of the 24 letters of the alphabet. There are 784 pixels per sample. In this dataset, letters "J" and "Z" are excluded as they require movement for their formation in American Sign Language. We then used the Python programming language to implement the three phases of the proposed research approach. The details for implementation of three phases of research approach are as follows.

4.1 Data Pre-Processing

As the MNSIT dataset is already available in required standardized format, we did not perform any pre-processing.

4.2 Feature Extraction

In this second phase known as feature extraction phase, we performed three levels of decomposition using DWT to extract the optimal features and generated 484 number of features. Moreover, we extracted features from the MNIST image dataset using Hu moments, Zernike moments (Degree 8), and Krawtchouk moments feature extraction techniques. Feature extractors, Hu moments, and Zernike moments yielded 7 features and 25 features, respectively. However, feature extractors Krawtchouk moments and discrete wavelet transform yielded a large number of features, specifically ($28 \times 28 \times 3$) and ($22 \times 22 \times 1$) features, respectively.

Table 1 Accuracies of feature extractors with different classifiers

Feature extractor	Feature classifier			
	SVM linear	SVM radial	Random forest	KNN
Hu moments	14.1	21.5	28.8	26.6
Zernike moments	35.2	52.9	56.17	50.55
Krawtchouk moments	49.2	49.7	62.8	53.8
DWT	79.3	86.2	84.8	75.0

Table 2 Accuracies of different feature extractors with CNN

Method	Accuracy
CNN	92.01
Krawtchouk moments+CNN	96.45
Our proposed model (DWT+CNN)	99.1

4.3 Classification

Lastly, in this third phase, first of all, we created a training and testing dataset. Our training set consisted of 16,967 samples, the validation set has 10,388 samples, and the testing set consisted of 7,272 samples. Further, we used SVM Linear, Radial, Random Forest, and K-Nearest Neighbor classifiers to classify the feature map into the 24 class labels representing the 24 letters of the alphabet. Table 1 shows the achieved accuracies using above-discussed feature extraction methods and classifiers.

Due to the large number of features extracted from Krawtchouk moments and DWT, we selected CNN as the classifier, which works well in classifying a dataset with a large number of features. Furthermore, we compared the results with a basic CNN model, which works both as a feature extractor and the classifier. We can observe from the results derived from our proposed model in Table 2 that our proposed model, which uses DWT as the feature extractor and CNN as the classifier, provides outstanding results.

Here, in the above table, the accuracies of various feature extraction techniques along with CNN classifier are presented. It also shows the accuracy of a basic CNN technique where CNN acts both as a feature extractor and classifier.

It can be observed from the results of Tables 1 and 2 that our research approach, consisting of DWT and CNN, outperforms the results of all other techniques such as basic CNN and Krawtchouk moments+CNN.

Furthermore, we depict the confusion matrix of our research approach as shown in Fig. 4.

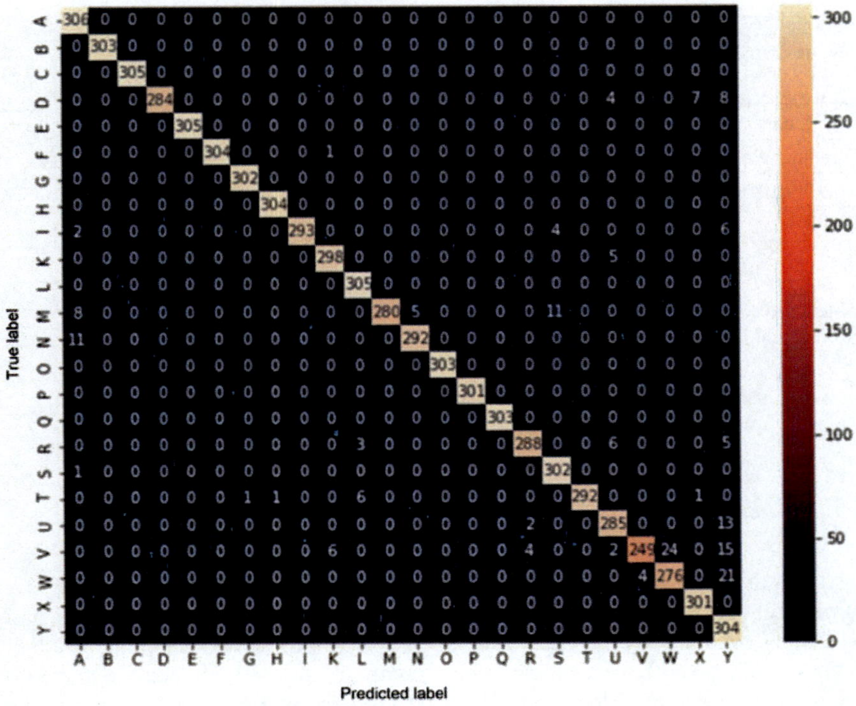

Fig. 4 Confusion matrix of our proposed model

5 Conclusion

In this research paper, we proposed a novel approach for static hand sign recognition using CNN as the classifier and DWT (db5, Level 3 Decomposition) as the feature extractor. We then implemented our proposed approach using Python programming language. Moreover, we used MNIST dataset, which is a publicly available dataset available at Kaggle website, to further implement our research approach. Furthermore, we compared our approach with Hu moments, Zernike moments (Degree 8), and Krawtchouk moments feature extraction techniques along with SVM Linear, Radial, Random Forest, and K-Nearest Neighbor classifiers. As a result, our approach yields the 99.1% accuracy that outperforms the accuracies of above-discussed feature extraction methods and classifiers.

References

1. The Resource for the Global Coatings Industry - Coatings World, https://www.coatingsworld.com.
2. P. Pisharady, M. Saerbeck, Recent methods and databases in vision-based hand gesture recognition: a review. Comput. Vis. Image Underst. **141**, 152–165 (2015)
3. M. Van den Bergh, D. Carton, R. De Nijs, N. Mitsou, C. Landsiedel, K. Kuehnlenz, D. Wollherr, L. Van Gool, M. Buss, Real-time 3D hand gesture interaction with a robot for understanding directions from humans, in *2011 RO-MAN* (2011)
4. M. Weiss Cohen, D. Regazzoni, Hand rehabilitation assessment system using leap motion controller. AI & Soc. **35**, 581–594 (2019)
5. 2015 Index IEEE Transactions on Circuits and Systems for Video Technology Vol. 25. IEEE Trans. Circ. Syst. Video Technol. **25**, 2121–2143 (2015)
6. S. Riofrío, D. Pozo, J. Rosero, J. Vásquez, Gesture recognition using dynamic time warping and kinect: a practical approach, in *International Conference on Information Systems and Computer Science (INCISCOS)*, pp. 302–308 (2017)
7. Sign Language Research, Uses and Practices (2013)
8. N. Ming-Hsuan Yang, M. Tabb, Extraction of 2d motion trajectories and its application to hand gesture recognition. IEEE Trans. Pattern Anal. Mach. Intell. **24**, 1061–1074 (2002)
9. A.S. Tolba, M.A. Elsoud, O.A. Elnaser, LVQ for hand gesture recognition based on DCT and projection features. J. Electr. Eng. **60**, 204–208 (2009)
10. T. Nguyen, H. Huynh, J. Meunier, Static hand gesture recognition using principal component analysis combined with artificial neural network. J. Autom. Control Eng. 40–45 (2015)
11. O. Oyedotun, A. Khashman, Deep learning in vision-based static hand gesture recognition. Neural Comput. Appl. **28**, 3941–3951 (2016)
12. D. Huang, W. Hu, S. Chang, Gabor filter-based hand-pose angle estimation for hand gesture recognition under varying illumination. Expert Syst. Appl. **38**, 6031–6042 (2011)
13. J. Triesch, C. von der Malsburg, A system for person-independent hand posture recognition against complex backgrounds. IEEE Trans. Pattern Anal. Mach. Intell. **23**, 1449–1453 (2001)
14. S. Ge, Y. Yang, T. Lee, Hand gesture recognition and tracking based on distributed locally linear embedding. Image Vis. Comput. **26**, 1607–1620 (2008)
15. G. Fang, W. Gao, D. Zhao, Large vocabulary sign language recognition based on fuzzy decision trees. IEEE Trans. Syst. Man Cybern. Part A: Syst. Humans. **34**, 305–314 (2004)
16. J. Zieren, K.-F. Kraiss, Robust person-independent visual sign language recognition, in *Iberian Conference on Pattern Recognition and Image Analysis (IbPRIA)*, pp. 1–8 (2005)
17. C. Vogler, H. Sun, D. Metaxas, A framework for motion recognition with applications to American sign language and gait recognition, in *Proceedings Workshop on Human Motion*
18. N. Dalal, B. Triggs, Histograms of oriented gradients for human detection, in *IEEE Computer Society Conference on Computer Vision and Pattern Recognition* (CVPR'05), vol. 1, pp. 886–893 (2005). https://doi.org/10.1109/CVPR.2005.177
19. K. Symeonidis, Hand gesture recognition using neural networks (2000)
20. H. Ming-Kuei, Visual pattern recognition by moment invariants. IEEE Trans Inf Theory **8**, 179–187 (1962)
21. C. Teh, R. Chin, On image analysis by the methods of moments. IEEE Trans. Pattern Anal. Mach. Intell. **10**, 496–513 (1988)
22. Hu and Zernike Moments for Sign Language Recognition, in *International Conference on Image Processing, Computer Vision, and Pattern Recogni-tion (IPCV)* (2012)
23. S. Chevtchenko, R. Vale, V. Macario, F. Cordeiro, A convolutional neural network with feature fusion for real-time hand posture recognition. Appl. Soft Comput. **73**, 748–766 (2018)
24. Y. Pew-Thian, R. Paramesran, O. Seng-Huat, Image analysis by Krawtchouk moments. IEEE Trans. Image Process. **12**, 1367–1377 (2003)
25. S. Liu, Y. Liu, J. Yu, Z. Wang, A static hand gesture recognition algorithm based on Krawtchouk moments. Commun. Comput. Inf. Sci. 321–330 (2014)

26. A. Grossmann, J. Morlet, Decomposition of Hardy functions into square integrable wavelets of constant shape. SIAM J. Math. Anal. **15**, 723–736 (1984)
27. I. Daubechies, C. Heil, Ten lectures on wavelets. Comput. Phys. **6**, 697 (1992)
28. M.D. Zeiler, R. Fergus, Visualizing and understanding convolutional neural networks (2013). arXiv:1311.2901
29. Sign Language MNIST Version 1 (2017). https://www.kaggle.com/datamunge/sign-language-mnist

Enhanced A5 Algorithm for Secure Audio Encryption

Akshay Joshi and Arun Mishra

Abstract This paper endeavours to bring out the fundamental mechanism of A5 encryption algorithm used in GSM cellular network, its lacunae and solutions to mitigate them. The A5 encryption algorithm is discussed with special emphasis on the strongest of all versions, the A5/1. The lacunae with respect to security of A5/1 algorithm are then brought out and methodology for mitigating them is discussed. A5/1 algorithm is modified and tested to confirm the lacunae and elucidation for their liquidation is proposed. The maximum period of existing A5/1 algorithm is 2^{54} which is low and is susceptible to correlation attacks, whereas the maximum period of proposed model, i.e. Enhanced A5/1 is approximately 2^{128} on which cyberattacks are difficult to execute.

Keywords A5 · Correlation attacks · Period · Cipher · Clocking · Encryption · Shift registers

1 Introduction

It is aimed at securing personal data from attacks by adversaries by cryptographic methods. The safety of personal data conveyed over the Global Mobile Communication System (GSM) is almost insignificant, though. An opponent with normal hardware can decrypt GMS packages in real time, endangering the confidentiality of every opponent using an ordinary computer [1]. The cellular network is principally designed for the use of GSM-based technology, which is profitable for a wide range of assaults [2]. Although many more implementations are available, A5/0, A5/1 and A5/2 are the typical forms of the A5 algorithm. The geographical location usually defines which of the three versions of the network are implemented. The A5/1 form is the strongest of all three in the USA as well as in European countries. In certain

A. Joshi (✉) · A. Mishra
Department of CSE, Defence Institute of Advanced Technology, Pune, Maharashtra, India

A. Mishra
e-mail: arunmishra@diat.ac.in

A. K. Luhach et al. (eds.), *Second International Conference on Sustainable Technologies for Computational Intelligence*, Advances in Intelligent Systems and Computing 1235,
https://doi.org/10.1007/978-981-16-4641-6_14

export markets such as Asia A5/2, which is an intentional fading of the A5/1 algorithm, is used. The A5/0 has been used in nations subject to US sanctions and in a few third-world countries as an unencrypted version of the algorithm. In this work, we highlight the A5/1 algorithm, the strongest of the three, and highlight its properties and shortcomings. A detailed investigation of the versions of the Algorithm A5 in relation to the A5/0, A5/1 and A5/2 methods starts in this examination. It can only be evaluated on the basis of its design constraints and associated vulnerabilities with extensive knowledge of the design and functioning of the present A5 algorithm. The most prominent attack against stream ciphers has been determined to be correlation attacks and A5 is no exception to this. The A5 algorithm is intended to be strengthened such that it strikes an outstanding balance between maximal period (linear complexity) and immunity to correlation. The achievement was that multiple LFSR was deployed in a unique combination and so a maximum of 2^{128} was obtained as compared to an existing A5/1 maximum of 2^{54}.

2 A5 Encryption Algorithm

In the original phase, GSM was developed for normal telephones, and hence its security mechanism could implement just minimum hardware prerequisites. This technology is now used for telecommunications by an amazing 6 billion people [3]. A5 is the symmetric stream cipher for the encryption of the GSM network air transmissions. The method takes inputs of 228 bits of plain text to give output to 228 bits of chip text. Every 228-bit block contains a 'frame' in which data transmitted by the first 114 bit from Unit X to Unit Y and data received in Unit X from Unit Y is last indicated by 114 bits. Each frame has duration of 4.615 ms, making possible to broadcast 28 frames each second. Different A5 variations have been designed over time, although all feature the same basic premise. With a session key Kc (symmetric), a frame counter and an algorithm A5 creates a key stream of 228 pseudorandom bits (PRAND). Then the key stream is XORed with a 228-bit plain text block to generate a 228-bit cipher text block. The data flow of A5 algorithm is illustrated in Fig. 1.

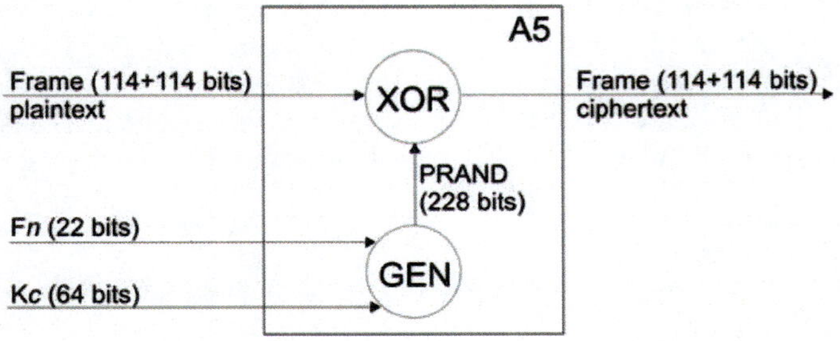

Fig. 1 A5 data flow [4]

Of all A5 variants, A5/1 is the most robust and highly deployed. It is used in the United States and the countries of Europe. Generation uses three linear feedback shift registers of pseudorandom bits in A5/1 (LFSRs). A shift record with its input is a linear function of its former state, known as LFSR. Certain bits define the state of the register and the linear feedback function is XOR of all taped bits. When a register is clocked, the tapped bits are XORed with a minimum result (LSB). It removes the most important bit (MSB) in the register and discards its value. Clocking of its registers is the main functioning of this algorithm. Five successive steps [5] are listed below for the creation of the stream cipher.

LFSRs Initialisation. 3 LFSRs generate the key stream of A5/1 (PRAND). Before the LFSRs can generate random numbers, they have to be initialised. The length of all LFSRs and indices varies during the clocking process. In this phase, the values of all LFSR bits are set to null. Figure 2 shows the specifications of each LFSR.

The tapped bits show the register indices that are used when timed. These are XORed values and returned to the register. Figure 3 shows the initialised LFSR.

LFSR number	Length in bits	Clocking bit	Tapped bits
1	19	8	13,16,17,18
2	22	10	20,21
3	23	10	7,20,21,22

Fig. 2 LFSR information table

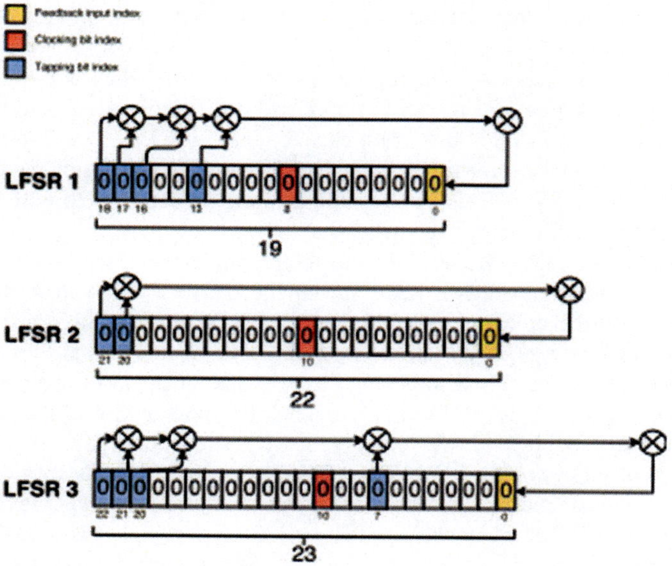

Fig. 3 Initial LFSR

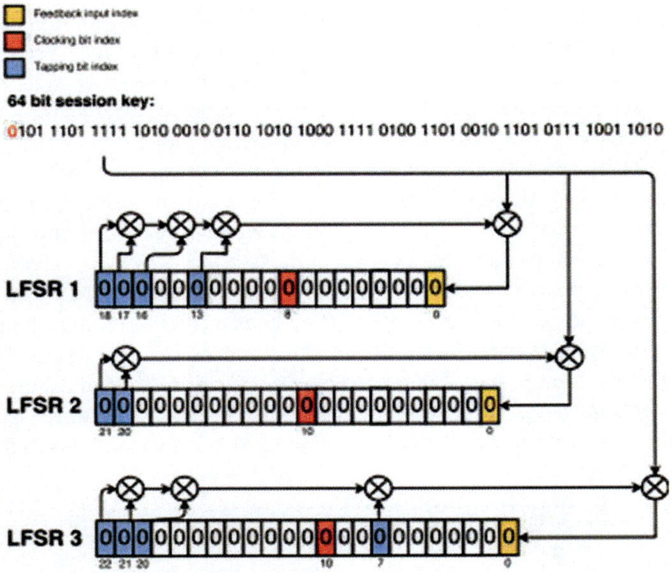

Fig. 4 Session key clocking

Session key Kc clocking. In this phase, 64 times, once for every bit of the Kc session is timed the initialised registers. The 64-bit Kc session bits with registry feedback are simultaneously XORed. The operation result is kept in the LSB of the register. This is shown in Fig. 4.

Frame number Fn clocking. The registries have non-zero values at the end of the last phase. Figure 4 shows the results of step II and step II. This phase is followed by the registers 22 times, Frame Counter Fn once for each bit. The 22-bit frame counter Fn bits are XORed at the same time as register feedback. The operation result is kept in the LSB of the register. This step is similar to the preceding step with only the 22-bit frame counter Fn being used instead of 64-bit Kc.

Majority vote clocking. We have finished collecting external inputs at this point. But before key stream generation, the LFSRs still need to be warmed up. At this stage, through asymmetrical clocking, we clock the records 100 times. In this procedure, the role of clocking bits lies. Asymmetrical watching uses the rule of majority, whereby the bits are chosen. The LFSRs' clocking bits are used for the most bits (LFSR 1-8, LFSR 2-10, LFSR 3-10). A register is clocked if the clocking bit of the register is equal to the majority bit. The A5/1 clocking logic is shown in Fig. 5.

Generation of Key stream (*PRAND*). At this stage, the registries are warmed up and ready to generate the PRAND key stream that is utilised to encrypt the plain text. In this stage, the LFSRs are clocked for the production of pseudorandom key stream 228 times using asymmetrical clocking. For final calculations throughout each cycle, the most significant bit (MSB) of each LFSR is used. Each LFSR's MSBs are XORed,

CB of $R1$**:** $R1[8]$	0	0	0	0	1	1	1	1
CB of $R2$**:** $R2[10]$	0	0	1	1	0	0	1	1
CB of $R3$**:** $R3[10]$	0	1	0	1	0	1	0	1
R1 clocked?	✓	✓	✓	–	–	✓	✓	✓
R2 clocked?	✓	✓	–	✓	✓	–	✓	✓
R3 clocked?	✓	–	✓	✓	✓	✓	–	✓

Fig. 5 A5/1 clocking logic

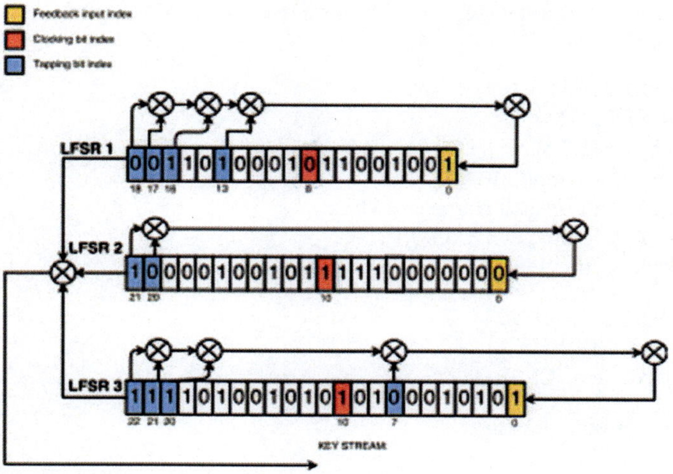

Fig. 6 Key stream generation

and the result is put in the key stream. As one bit of key stream is produced at each cycle, we generate a 228-cycle key stream per 228 bit. In this stage, Fig. 6 shows the process.

On analysis of the A5 stream cipher, it is found that the algorithm has certain lacunae which makes it susceptible to attacks by adversaries. These limitations/vulnerabilities of A5 are summarised below:

- The algorithm (A5/1, A5/2) was not designed with modern cryptographic knowledge. Its maximal exhaustive search complexity is (only) 2^{64} [6].
- In whatever versions of the A5/1 algorithm deployed, ten minor bits from the Kc 64-bit session key are always set to zero. When zeroing 10 bits, the search complexity is purposely decreased by around 16 percent in deployed systems.
- Encryption is applied after error correction (digital audio).
- The key length is short, and feedback polynomials are scarcely spaced making exhaustive search attacks/brute force attacks feasible [7].

3 The Modified A5/1

On completion of analysis of the existing A5/1 design, a trial was undertaken to modify the A5/1 algorithm by changing the values of feedback polynomials, session key, frame number, and clocking bits. The aim of this trial was to check the impact of changing feedback polynomial on maximum period. Figure 7 gives the LFSR information table of modified A5/1. The length of LFSRs is its tapped bits and its clocking bits are changed here.

The modified A5/1 was implemented using the steps used in implementation of A5/1 algorithm. The implementation of modified A5/1 is given below:

Initialisation.

- Length of LFSR1 = 5
- Length of LFSR2 = 7
- Length of LFSR3 = 9
- No of clocks for random: 50
- Message length in each frame = 46 bits
- Seed: 011,101,100,111,110,000,101
- Primitive Polynomials

    $X^5 + x^2 + 1$:- LFSR1
    $X^7 + x + 1$:- LFSR 2
    $X^9 + x^4 + 1$:- LFSR 3

- Bits for majority clocking

    LFSR1:- 3
    LFSR2:- 4
    LFSR3:- 5
    18 Bit Frame Number 000000000001111101

Seed loading. During this phase, the initial seed is loaded in the modified LFSRs combination as discussed in the previous section. The process of loading of initial seed is same as loading of session key Kc which is done in original A5/1 algorithm. The only difference here is the length and bits of the initial seed and length of the LFSRs.

LFSR Number	Length in bits	Clocking bit	Tapped bits
1	5	3	2,4
2	7	4	5,6
3	9	5	8,4

Fig. 7 Modified LFSR information table

Frame loading. This phase is similar to the last phase except that LFSRs are not in zero state and frame number is loaded instead of the seed.

Random clocking. At this stage, the LFSRs are clocked for 50 times as part of its warming up. The clocking logic used at this stage is same as A5/1 clocking logic. No output is taken from the LFSRs at this stage of random clocking. Figure 10 depicts this phase.

Clocking message length times for generation of PRAND. Now, the LFSRs are clocked message length times, i.e. 46 times to generate the required PRAND sequence.

A few bits of key stream generated with modified A5/1 are ***10000011110010001010110100011000110000000001111***0000001001110111011110010
111011000100110000010011111111001010000111010011010011111011010000000
1001001110101000010010100111000101010110010010101010010000111110011100
010011101111101001001001010010000000110111100011011011111001001111100001
0011011001001001100001101100001101001011010101110010010000111001111101
010110101101100000101011010100110011101110001011101011100100110110001
1011110001101111000101101100100101000111001011011110001010000000101000
0100001011001110111010111011111111101101100011011101011010011110111110
0100000000100011011011011010000110111010010111010111101100010010111100
1001101110110011101101100000100110010111100010111011111001100000101000
1111111011110100000111011010010100010000100010000110010011000111100001
0110010000001001111001000110000100011010101000011011001000100001001000
011000111111010010001110001010010101010010000110010001111101001010100
1110101101010101001100101101001000011001011010100010101111000100101111
0110011100101101100110011000011110001010011001111010100100011001111110
1010011101111101110010011110110101111

4 Observations

On completion of the above-mentioned experiment, following is submitted:

- The stream is repeated only once at 7012290th position in the entire generated key streams achieving a maximum period of close to 2^{22}. Thus, it is inferred that reduction in degrees of feedback polynomials has also led to reduction in maximum period.
- In order to design an LFSR-based enhanced A5 algorithm which has maximum period, it is imperative to have a design that uses dense polynomials as feedback polynomials. If we use dense polynomial as feedback polynomial, then we can achieve higher immunity against attacks even with shorter LFSRs.
- One of the main parameters for analysing random number generators of the LFSR is the linear complexity. The shortest LFSR that gives the same output sequence is defined as the length. A finite-state machine sequence over a finite field has

a limited linear complexity. The importance of linear complexity is important because after only analysis of 2n bit of the key stream, a fundamental tool called the Berlekamp–Massey algorithm [8] can reproduce an LFSR. Reproduction of this LFSR basically entails the breaking of the stream chip. It is therefore essential to have a higher degree of linear complexity.

- Some developers attempt, in some asymmetrical fashion, to attain larger linear complexity by mixing output from several output sequences. The concern here is that one of the internal output sequences (usually the output of individual LFSR) can be correlated and attacked by a linear algebra with the aggregated key stream. This is known as the attacks of correlation. Therefore, an optimal balance between correlation immunity and linear complexity is required.

5 Proposed Enhanced A5

Based on the observations from previous experiment the Enhanced A5 is designed by using five LFSRs for output generation. This model used four batches of LFSRs. Each batch consists of 5 LFSRs with different degrees of feedback polynomials. It is to be noted that all the feedback polynomials are primitive polynomials as it is imperative for achieving maximum period and are selected from pool of standard primitive polynomials [9]. A high maximum period compared to original A5/1 (approx. 2^{128}) is achieved by using this model. Using four different batches increases the randomness of pseudorandom number generation making correlation attacks difficult to execute. Salient features of the Enhanced A5 algorithm are discussed in subsequent paragraphs.

A total of 23 LFSRs are used in this model. Three LFSRs are common to the LFSRs used in original A5/1 algorithm. Remaining 20 LFSRs are selected with different degree primitive polynomials as their characteristic polynomials. The first three LFSRs act as trigger for selection of any one batch of LFSRs. Remaining 20 LFSRs are clubbed into four batches, namely, Batch A, Batch B, Batch C and Batch D consisting of five LFSRs each. One of these batches as selected by selector trigger is used for generation of pseudorandom sequence. The details of LFSRs are listed below.

Three LFSRs which are common to original A5/1 are

- $X^{19} + x^{18} + x^{17} + x^{14} + 1$
- $X^{22} + X^{21} + 1$
- $X^{23} + X^{22} + X^{21} + X^{8} + 1$

Batch A

- $X^{17} + X^{6} + 1$
- $X^{21} + X^{2} + 1$
- $X^{23} + X^{3} + 1$
- $X^{25} + X^{3} + 1$

- $X^{37} + X^6 + X^4 + X + 1$

Batch B

- $X^{17} + X^3 + 1$
- $X^{31} + X^6 + 1$
- $X^{33} + X^{16} + X^4 + x + 1$
- $X^{35} + X^2 + 1$
- $X^{19} + X^5 + X^2 + X + 1$

Batch C

- $X^{11} + X^2 + 1$
- $X^{15} + X + 1$
- $X^{25} + X^3 + 1$
- $X^{37} + X^6 + X^4 + 1$
- $X^{45} + X^4 + X^3 + X + 1$

Batch D

- $X^7 + X + 1$
- $X^9 + X^4 + 1$
- $X^{37} + X^5 + X^4 + X^3 + X^2 + X + 1$
- $X^{41} + X^3 + 1$
- $X^{43} + X^6 + X^4 + X^3 + 1$

All the polynomials of all the LFSRs are primitive polynomials. The seed value for initialisation of LSFRs is taken at random and the decision rule deployed for clocking is majority of the middle bit of the respective LFSRs. As discussed in previous section, initially, three LFSRs are used as trigger for selection of one of the four batches of LFSRs vis-a-vis A, B, C and D. The first three LFSRs will generate a stream of bits of which only the first two bits are taken as selector trigger. The selector trigger logic is tabulated in Fig. 8.

Figure 9 shows the selection of LFSR batches based on the selection trigger. Whichever batch of LFSRs is selected on the basis of selector trigger is then clocked as per the decision rule (majority of the middle bit) to generate a 128-bit block cipher. After this, we again use the initial set of LFSRs for triggering one of the four batches of LFSRs and the process is repeated for continuous generation of pseudorandom sequence as per the user's requirement.

Fig. 8 Logic table for LFSR batch selection

First Two bits	Batch Selected
00	Batch A
01	Batch B
10	Batch C
11	Batch D

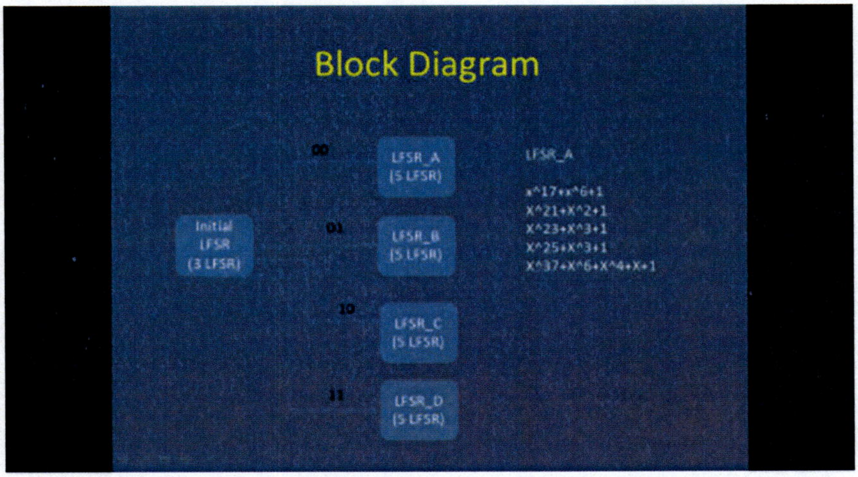

Fig. 9 Block diagram for LFSR batch selection

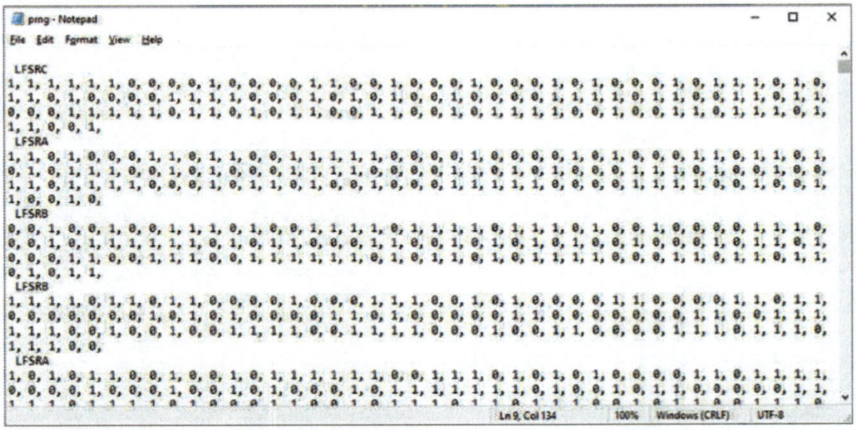

Fig. 10. 128-bit block cipher—Output

Output of the algorithm is shown in Fig. 10 which highlights that the 128-bit blocks are generated randomly from one of the four LFSR batches each batch containing five primitive polynomials.

A test was carried out to check the repetition in the output of the algorithm. 40 initial bits were matched out of 128-bit blocks. However, no matches were found in 2.5 million counts. An online tool called Berlekamp–Massey algorithm tool was used to identify if any other polynomial can generate similar pattern. The Berlekamp–Massey algorithms take 2n inputs where n is number of the output bits of the proposed

Algorithm	Sum of maximum degrees of feedback polynomials	Maximum period of generated output
Original A5/1	64	2^54
Enhanced A5/1	123 (Batch A)	2^122
	135 (Batch B)	2^134
	133 (Batch C)	2^132
	137 (Batch D)	2^136

Fig. 11 Maximum period comparison

algorithm and produces output as a polynomial which is capable of generating similar output patterns. The proposed algorithm was checked using the online generator and results showed no such patterns. The maximum period of Enhanced A5/1 model is compared with the original A5/1 algorithm and is tabulated in Fig. 11.

6 Conclusion

On analysis of the existing A5/1 algorithm and its flaws, Modified A5/1 algorithm and Enhanced A5 algorithm, it is implied that through use of a random seed and primitive polynomials as LFSR feedback polynomials, it is possible to generate a non-repeating PRAND which can be utilised for secure voice encryption. Choosing a combination of sparse and dense polynomials of different degrees is the most suitable trade-off between complexity and processing time. It is also noted that the maximum period is increased manifolds (up to 2^{128} as compared to 2^{64} achieved on original A5/1) by using higher degree feedback polynomials and increased number of LFSRs. Use of different batches of LFSRs has been particularly advantageous for increasing the randomness of the generated key stream and providing immunity against correlation attacks. It is concluded that on giving emphasis on proper selection of A5/1 design parameters vis-à-vis selection of primitive polynomials, selection of key/ initial seed and clocking logic, it is possible to achieve an Enhanced A5 stream cipher which does not repeat itself at a shorter period and is not vulnerable to cyberattacks.

References

1. A. Shamir, A. Biryukov, D. Wagner, Real time cryptanalysis on a5/1 on a pc. https://cryptome.org/a51-bsw.htm
2. SANS Institute InfoSec reading Room. The gsm standard (an overview of its security). https://www.sans.org/reading-room/whitepapers/telephone/gsm-standard-an-overview-security-317
3. Internet Archive. Gsm: Global system for mobile communication. https://web.archive.org/web/0140208025938http://www.4gamericas.org/index.cfm?fuseaction=page&sectionid=242
4. T. Stockinger, GSM network and its privacy - the A5 stream cipher. November 2005
5. O.D. Jensen, K.A. Andersen, A5 Encryption in GSM. June 2017

6. D. Wagner, M. Briceno, I. Goldberg, A pedagogical implementation of a5/1. May 1999
7. M. Roe, R. Anderson, A5. http://jya.com/crack-a5.htm, 1994
8. E. Casey, *Berlekamp-Massey Algorithm* (University of Minnesota, REU Summer, 2000)
9. Bruce Schneier. Applied Cyptography, 2nd edn. (1996)

Stock Market Prediction Techniques: A Review Paper

Kirti Sharma and Rajni Bhalla

Abstract Due to the non-linear and highly volatile nature of the Stock market, it has become a very challenging task for researchers to make accurate predictions. Improving the efficiency of predictions has become the main goal of many researchers. From the traditional approach of working with historical dossiers to using the latest machine learning and deep learning techniques, researchers are busy finding out the best possible ways of accurate predictions. Many new models are suggested that can make good estimations of stock prices. Investors are interested in knowing both the immediate next-day prices and as well as future share prices in the long run. This paper inspects the algorithms and techniques that are useful for making accurate predictions.

Keywords Stock market · Fundamental analysis · Technical indicators · Machine learning · SVM · Random forests · Decision trees · ANN · KNN

1 Introduction

The Stock Market has become an integral part of modern society. Today, the modern economy greatly depends on the stock exchange. It is a place where capitalists can deal with stocks. This capital can be used in various investment projects which can prove beneficial. It can help the masses and organizations in dealing with their monetary troubles. Financial markets constitute an important part of the total infrastructure for every society that has passed the stage of largely domestic economies. The stock market is part of the financial markets. Businesses can raise capital for the extension of their business by selling shares, when people invest in shares, which otherwise could have been kept idle; this money benefits many economic sectors which strengthens the economy.

Making accurate predictions will benefit the traders and they can gain profits. But getting the predictions right is not an easy task, as there are many components that

K. Sharma (✉) · R. Bhalla
School of Computer Application, Lovely Professional University, Jalandhar, India

A. K. Luhach et al. (eds.), *Second International Conference on Sustainable Technologies for Computational Intelligence*, Advances in Intelligent Systems and Computing 1235,
https://doi.org/10.1007/978-981-16-4641-6_15

affect the stock prices, i.e., bureaucracy, Country's monetary status, the mindset of masses/social conception, environmental, and many more.

The method for making the best possible predictions is by examining the past dossier. To do this prediction, many techniques are used, from fundamental analysis to the technical indicators like SMA, EMA, MACD, Volume, RSI, various ratios, Machine learning, etc. The best possible selection of methods and techniques may provide the optimal solution to the Stock Market.

2 Literature Review

Sai et al. [23] have used "Support Vector Machine" to forecast stock prices for large and small companies and have used three markets and employed prices for daily as well as up-to-the-minute frequencies. As per the authors, this model gives more accurate numerical results as per his set benchmarks. He has undertaken four features; price volatility, price momentum, sector volatility, and sector momentum.

Alexander et al. [1] have used "Linear Regression" and "Random forests and SVM", and as per their analysis, both of these methods fail. As per them, "regression" is for sure the better method and gives better accuracy and can be used to make money in a real scenario.

Hegazy et al. [7] have proposed a machine learning model "LS-SVM-PSO" that integrates the Particle Swarm Optimization (PSO) algorithm and LS-SVM (least square support vector machine) for stock price prediction. LS-SVM-PSO is built upon historical data along with relative strength index, money flow index, exponential moving average, stochastic oscillator, and moving average convergence/divergence. The author concludes that their model proposed better prediction when compared to the artificial neural network with Levenberg–Marquardt (LM).

Shunrong Shen and Tongda Zhang [28] have proposed a new algorithm. As per the authors, the performance of a stock market predictor heavily depends on the correlation between the data used for training and the current input for prediction. This algorithm was applied along with various regression algorithms. This new algorithm's numerical results prediction accuracy is 75% approx.

Iacomin [9] has used different "machine learning algorithms" from the easiest of them like SVM to the complex ones like PCASVM. The author has suggested a new algorithm that shows better results for recent data. He has shown SVM along with PCA gives better results in terms of profit.

Nayak et al. [15] have built two different models, one that makes predictions on a daily basis and the other on the monthly basis. Supervised ML techniques are considered. The authors have taken historical price data along with sentiments. As per them, Decision Boosted Tree had outperformed SVM and Logistic Regression.

Mankar et al. [14] have used sentiment analysis on the Twitter data and stock's closing values. After comparing various algorithms, the author concludes that the Support Vector Machine proved to be the most efficient and beneficial in making predictions.

Raut et al. [22] have implemented machine learning with supervised classifiers to get the best accuracy. They have tested their results using binary classification which is done using the SVM classifier. They have found evidence that input data has the same importance as ML Techniques. As per the authors, SVM can induce extra features and can make better forecasting.

Pimpalkar et al. [20] have developed a new prediction software. For classification along with regression, they have used SVM and RNN. As per the authors, their software can accurately predict the next day's market price, as they have considered multiple elements such as pattern recognition, data of the history of the stock market, and Sentimental analysis. They have also used a graph to show past trends to help investors. Their study shows that SVR is the most efficient amidst all.

Zhou et al. [33] have built a forecasting model and have named it as GAN-FD. Deep Learning forms the basis of this model. Authors have used adversarial networks with thirteen simple technical inputs and have avoided complicated data. They have done rolling segmentation to check the effects of the features update cycle and came to the conclusion that when the model is small in size then the update cycle can give better forecasting results.

Parmar et al. [19] have taken data from Yahoo Finance and considered two models, one is regression-based while the other is based on LSTM. Through the means of experiments, authors have come to the conclusion that while both the models gave promising results, still the LSTM proved to be the better one.

Hiransha et al. [8] have used four types of deep learning networks CNN, LSTM, MLP, and RNN on two stock markets NSE and NYSE. From the experiments, the authors have shown that Models based on Deep learning are way better than ARIMA. And out of the four undertaken techniques, CNN due to its ability to consider sudden changes in the system has proved to be the best.

Sharma and Kaushik [27] in their paper have done a quantitative study of the Stock market. Their research concluded that LSTM neural network gives better accuracy in comparison to other techniques like PCA, Decision Tree Classifier, SCM, WB-CNN, CNN, Hidden Markov model, NV Bayes, and ANN.

Jeevan et al. [10] have recommended machine learning to do predictions in the stock market. They have used the RNN-based model and also used backpropagation to collect and to avoid overlapped data along with the long short-term memory method. As per them, the proposed model has shown greatly improved results as compared to the other available models.

Khan et al. [12] have proposed a new framework. Dataset consists of data from Twitter, Yahoo Finance, and Business Insider. Their study shows that removing spam from Twitter data and feature selection is a good effective measure. Authors have concluded that the performance of Random forests is always better in every case that they have considered. They have also created one hybrid algorithm by combining Random Forests, Extra Tree, and Gradient Boosting Classifier, which has shown promising results.

Nti et al. [17] have created a new combination of sentiments-oriented data and thereafter they have removed any replicated sentiments. They have used MLP-ANN (Multi-Layer Perceptron) to do the Stock price estimation on the data from the Ghana

Stock Exchange. The authors believe that the Stock Market price estimation improves if data from the Web is used as input. Authors suggest that the integration of basic and technical ideas can significantly increase the chances of better prediction in Developing nations like Ghana.

Shanthi et al. [26] have used historical data to make predictions. The authors have achieved approximately eighty percent exactness. As per them, their model could be improved if news data and sentimental analysis are considered.

Xia et al. [32] have worked with the Genetic Algorithm. Real data from the stock market is taken. They have given an extension to the standard GA Algorithm by including mutation, selection, and crossover re-definition and code–encode methods. Their evaluation and multiple iterations have shown improved results both in the form of correctness and consistency.

Si et al. [29] have proposed a prediction model made using vector autoregression. They have worked on the S&P 100 and applied DPM (Dirichlet Processed Mixture) model, an extended version on LDA (Latent Dirichlet Allocation). As per their findings, "Twitter's topic-based sentiments" provide greater accuracy over other non-topic-based techniques. This study is only for short-term dynamics-based prediction.

Dang and Duong [6] in their paper have analyzed the data from the Vietnam stock exchange. They have undertaken time-series data along with text mining techniques to do forecasting. They have used the support vector machine as their machine learning methodology. They have concluded that financial news and the share market are related. As per their results, their research has significantly shown improvements. Though the performance can substantially go up if the news can be taken from trustworthy mediums. In this model, each day is limited to only one transaction (buy/sell).

An et al. [2] have investigated the effects of financial reports' release on the stock market. They have worked with four reports: "Total asset, total equity, total revenue, and net income". They have worked with LWRP (Locally weighted Linear regression), SVM, and NB. With this model, they have achieved seventy percent accurate results, although they have taken a very small amount of data, which may lead to high biased and overfitting data problems.

Kalra and Prasad [11] have come with a daily forecasting technique that uses the interrelation of financial news and the share market. In their model, they have first found the polarity in news data, and then they have combined past dossier, variance with sentiments that were converted in numeral values, and made forecasting. Their dataset contained two months of data from four selected banks. As per their finding, KNN gave them better results than any other undertaken algorithms. Dataset is too small, i.e., just two months of data have been used in the study.

Atkins et al. [4] have said that the news data gives improved forecasting of volatility in comparison to the stocks' trend. They have built ML models of LDA and NB for the data from news and forecasting of price trends, respectively. They have worked with the stock index of NASDAQ and DJIA, and equity of Goldman Sach and J.P. Morgan for one year dataset.

Pagolu et al. [18] have studied the changes in stock prices in relation to public sentiments via Twitter data. Authors have used SVM and sentimental analysis to do extraction and analysis of the relationship among Twitter data and the company's stock data. Their major outcome is the analyzer that can classify the sentiments in Twitter data. Authors have considered only Twitter data for analyzing people's sentiment which may be biased because not all the people who trade in stocks share their opinions on Twitter.

Rajput and Bobde [21] have analyzed various techniques for forecasting stock volatility using "Sentimental analysis and data mining". They have found that the forecasting of stocks is very difficult work, and multiple attributes should be undertaken to predict the market with accuracy.

Sai Redddy [24] in his paper has used the support vector machine to make predictions for large and small-cap companies from three dissimilar stock markets. As per him, no issue of overfitting occurred with SVM. He has used SVM with Radial Basis Function Kernel. The only SVM is explored that too with only historical stock data.

Selvamuthu et al. [25] exploited multiple versions of ANN for making predictions. As per them, the accuracy largely relies on the training algorithm. A comparison of the efficiency of the three NN algorithms is done by forecasting over the "tick by tick" and fifteen minutes dataset. All the undertaken algorithms gave 99.9% results when the tick dataset is used. Only data for the "tick by tick" and fifteen minutes dataset is used.

Singh et al. [30] have reviewed various ML algorithms that are commonly applied to make stock forecasting. As per them, not a single model alone can predict the stock market with higher accuracy. Although a hybrid model is created with a focus on maintaining accuracy and with simple computation. Fusion techniques do better predictions. Some models work better with historical data and others with sentiment data. Combining these can result in better accuracy.

Kokila [13] has worked with Random Forest and SVM to make stock forecasting. The dataset is collected from Kaggle. The feature extraction technique is applied to the dataset. They have considered 11 attributes in their study. Their proposed work has shown an accuracy of 91.9% which is more than the other models taken for comparison. But they have worked with limited parameters to make predictions.

Khan et al. [12] have suggested a framework that undertakes news and social media data and makes forecasting for the upcoming ten days. Authors have found that there is a positive effect of feature selection and spam reduction on Twitter data when making predictions. Random Forest gives promising results. Forecasting accuracy in NN can be improved by the means of deep learning. A limited number of stocks' data for a limited period is selected. The dataset could be used to get better accuracy.

Nti et al. [16] in their study have used Ensemble techniques such as boosting, bagging, blending, and super learners (stacking). Using Decision Trees (DT), Support Vector Machine (SVM), and Neural Network (NN), twenty-five different ensembled regressors and classifiers are constructed. SVM ensemble suffered some deficiencies related to input features and dataset sizes. The classical SVM algorithm assumes that

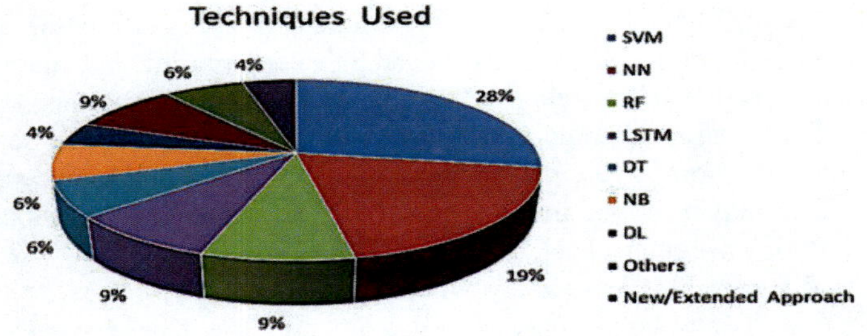

Fig. 1 Summary of the machine learning techniques used in discussed papers

all the features of samples give the same contribution to the target value, which is not always accurate in real problems (Fig. 1).

3 Prediction Methods

3.1 *Fundamental Analysis*

It is a method of finding out the actual value of the stocks. (*Understanding Fundamental and Technical Analysis* [31]) Here, everything is examined about the profile of the company, from the evaluation of its assets, expenses, income, future plans, and the form of the product to the study of the industry including the competition in the market. This helps the investors to make a profit by buying undervalued shares and then selling them when the stocks reach their intrinsic value. These factors help in finding out the actual average value of the company since its commencement. This is not done for short-term estimates rather it is for long-term analysis.

3.2 *Technical Analysis*

It is the study of past data of the market which includes cost and volume to make predictions about future market trends. Technical analysis is about chart patterns and statistical data. Various mathematical formulae are to be applied to do the analysis. As per the analysis, history repeats itself, and this applies to the stock market too. Stock charts, price, volume, and trends are analyzed to make future price predictions. The motive of using TA is using the data in hand to make predictions that give the result that investors look for. Here, we are discussing a few of the important technical indicators that form the basis of future predictions:

Volume—It tells the sum of shares traded on a single day/time. It shows trend directions. When price and volume go up, it means uptrend, which shows the existence of a trend or its reversal. It's the indicator of market health; high volume indicates a strong market. But when prices fluctuate at a lower rate, it shows a kind of fundamental change.

SMA—Simple Moving Average—It uses data from a predefined timestamp and then calculates the average price of the given data. The average calculated for SMA is different from the simple average as SMA discards the old data as soon as a new dataset is provided. Here equal weightage is given to all the data. If Price > SMA, the market is said to be bullish. But if Price < SMA, the market is called bearish.

EMA—Exponential Moving Average—This is the same as SMA except that new data here is assigned greater weightage than old data. EMA is said to give better results whenever there is a big change (Trends/reversal trends) in the market because of their quick responsiveness to the latest price changes.

MACD—Moving Average Convergence Divergence—Appel [3] is an intelligent innovation that has converted the limitations of MA into a practical trading trick. It is for asset movement. It depicts the relation among two moving averages. Two lines are used the first is the moving average convergence divergence line and the second is the signal line. Based on default values, we can create MACD and signal lines.

Default Settings: Fast = 12, Slow = 26, Smooth = 9

$$\text{MACD line} = 12\text{ day EMA} - 26\text{ day EMA} \tag{1}$$

$$\text{Signal Line} = 9\text{ day EMA of MACD Line} \tag{2}$$

Taking MACD and oscillator together can provide an overall view of the current market. Though MACD is very good at clearing out price action but its signal lag is a drawback.

RSI—(Relative Strength Index): J. Wells developed RSI which uses the average gain and loss to measure the velocity of price change. This method uses a single ratio known as RS (a combination of average profit and loss) which represents time momentum.

$$\text{RSI} = 100 - [100/(1 + (\text{AUC}/\text{ADC}))] \tag{3}$$

where

AUC Average of Upward Price Change.
ADC Average of Downward Price Change.

3.3 Four Basic Components of Valuation of Stock

Beattie [5] tells that making investments in stocks largely depends on four key elements that can give a good estimate of the stock price. These are as follows:

Price-to-Book Ratio: The P/B Ratio tells the value of a company. It takes into consideration the book value of a company if it is sold on that day. It includes the valuation of all the physical fixed assets, i.e., anything that can be sold.

Price-to-Earnings Ratio: P/E Ratio tells the time a stock will take to give back the amount an investor has invested in if there is no shift in the business.

Price-to-Earnings Growth: PEG Ratio not only considers incomes and costs but rather also uses the company's past growth record. One simply has to divide the P/E ratio with the company's income growth rate per year to get the PEG ratio.

Dividend Yield: It is the part of a company's profit that's distributed to its stockholders at regular intervals. By this, investors get a kind of interest in their investments. By this even if stock prices go down, investors still can earn some amount.

Individually, these four may not give the desired results, but when taken together investors can get a better estimate of stocks' prices. Although these four may be outweighed by many other specially tailored metrics, these factors always prove helpful in making stock decisions.

3.4 Machine Learning Methods

It grew out of work in artificial intelligence. It makes a machine capable of learning on its own and improving itself with further learning with its own experiences, without the requirement of explicit programming. It can be subdivided further into supervised, unsupervised, and reinforcement learning. Depending on the type of problem, one can choose from the available methods of machine learning.

Support Vector Machine—It is a supervised ML technique, which can be used in both regression and classification problems. SVM looks at extremes of the datasets and the decision boundaries (also known as hyperplane) are drawn near the extreme points in the dataset (Fig. 2). SVM can efficiently work for high-dimensional datasets. A data point is called a vector. Each data sample is plotted on an n-dimensional space, where n is the number of features.

A coordinate is the feature's value. Thereafter classification is done with a hyperplane to divide the space in the best possible way. The aim is to maximize the margin between the nearest point and the hyperplane. If patterns are non-linear, then the transformation of the dataset is done to map the underlying dataset into a new high-dimensional space using kernel function, for example, RBF Kernal, Sigmoid Kernal, Feature Space, etc. Finding an optimal hyperplane is an optimization problem that can be solved by optimization techniques.

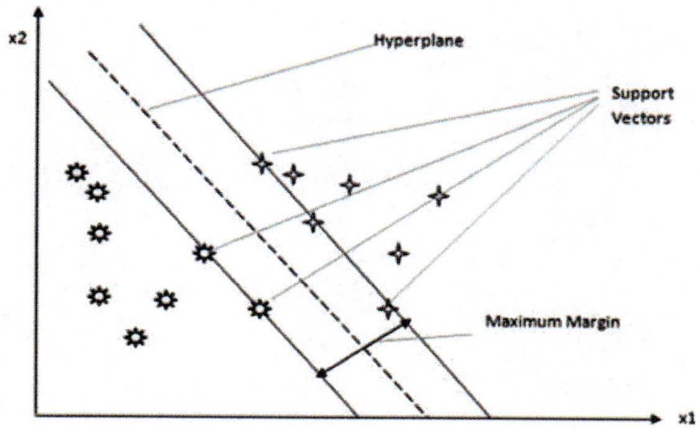

Fig. 2 Support vector machine

Decision Trees—It's a non-parametric supervised learning method that is used both for regression as well as classification problems. It divides the dataset into small subsets. A DT decides by splitting the given tree nodes into subtree nodes until only similar types of elements are left. Dividing the nodes is the main idea. DT does several recursive decision tree tasks before it reaches the end results and when plotted, it looks like a huge tree thus named as a decision tree (Fig. 3). For this, the challenge is to find the attribute for the decision node at every level. Two Attribute Selection Measures that are Information Gain and Gini Index are generally used for the same. One disadvantage of using a decision tree could be overfitting.

Artificial Neural Network—Deep Learning is a very powerful stream of machine learning. NN is influenced by the working of the human brain. ANN does the mathematical calculation on a series of inputs to generate one output.

It follows a layered approach where there are three layers: the input layer which is to provide inputs, the hidden layer (Fig. 4) for processing purposes which uses activation functions, and the output layer to generate output. It can be used both for supervised or unsupervised learning. In comparison with other methods, it gives better prediction results. ANN can learn a non-linear function, it does it through an activation function. ANN can only process inputs in the forward direction.

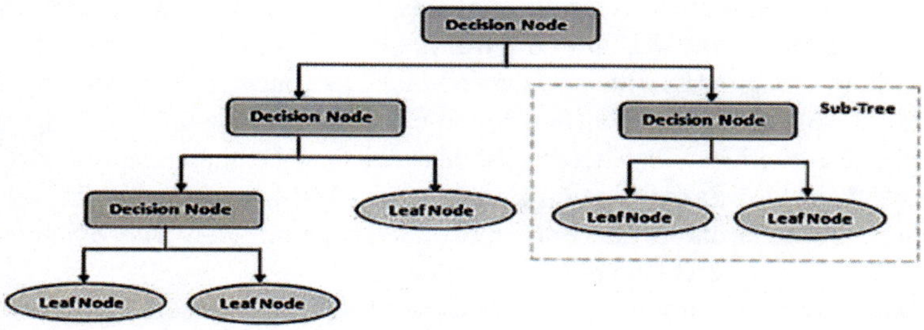

Fig. 3 Decision tree

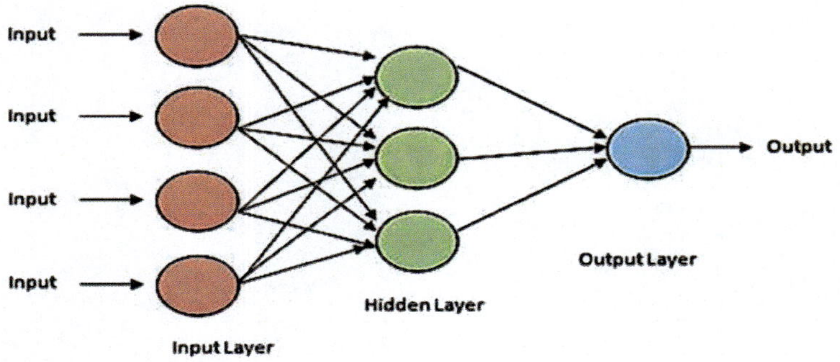

Fig. 4 Artificial neural network

Naïve Bayes: NB is a type of supervised learning technique. It is mainly used for classification although it can also be used for regression problems. It is a simple yet very effective classifier that is generally used for textual data where dimensions are high. NV considers each feature as an individual feature and gives equal weightage to all its feature. Maximum A Posteriori known as MAP is calculated:

$$\text{MAP(M)} = \max \text{P((M/E))} \tag{4}$$

$$\text{MAP(M)} = \max \text{P((M/E)} * \text{(P(M)/P(E))} \tag{5}$$

$$\text{MAP(M)} = \max \text{P((E/M)} * \text{P(M))} \tag{6}$$

Here P(E) will not have any effect on the answer as it is evidence probability. NB considers that all the features are unrelated and thus the presence or absence of a feature doesn't affect other features. Naïve Bayes can be categorized in Gaussian, Multinomial, and Bernoulli NB. One can choose from these as per their dataset. The disadvantage of using NB is that it lacks in the relationship in features which is sometimes very important in datasets.

K-Nearest Neighbors: This is a non-parametric supervised learning algorithm and can be used both for regression and classification problems. It uses feature alikeness to find out where the new data will lie. Here k is the number that identifies the same kind of neighbors of the new data point. KNN considers k nearest neighbors to find out the place of new data (Fig. 5). To calculate the distance, there are many options like the Euclidean distance, the Manhattan distance, the Hamming distance, and the Minkowski distance. The disadvantage of KNN is that finding the value of k sometimes becomes cumbersome; moreover, it's comparatively a costly algorithm as the distance has to be calculated for all the data points.

Random Forest: It is a supervised learning technique and is useful in both regression and classification problems. It is good at restricting the overfitting problem. RF is made up of multiple decision trees. Every DT has the same nodes with distinct data

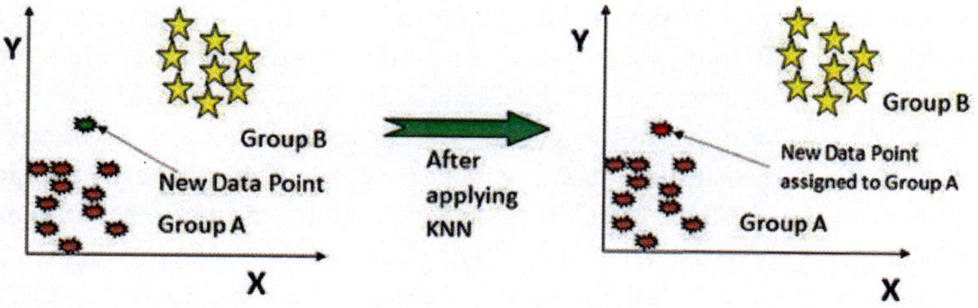

Fig. 5 K-nearest neighbor

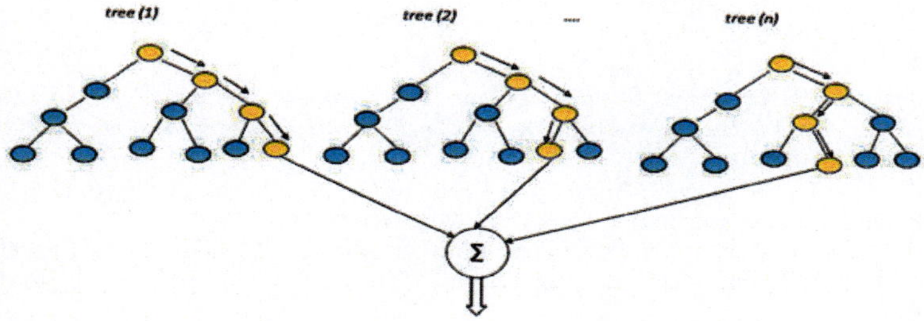

Fig. 6 Random forest

that makes distinct leaves. All the decisions are then joined together to get the result which is their average (Fig. 6). RF provides an extra feature of randomness. The random subgroup of features is considered for dividing a node. It looks for the elite feature in a random subset rather than finding that during splitting the node. Experiments show that RF gives better results for classification problems in comparison with the regression problem.

Hybrid approach: Integrating two or more techniques to make a new prediction technique is called a hybrid technique. It should be kept in mind that the new proposed system gives better accuracy and performance with a less complex mechanism. These models can overcome the problems that other basic models lack in, as they are free to use any combination of techniques and processes. Hybrid models are created specifically for an undertaken problem, so their prediction accuracy level may vary as the problem or dataset changes.

4 Conclusion

This paper summarizes various approaches and methods that are used for making predictions about the stock market. Fundamental analysis helps in giving a good estimate of long-term analysis whereas technical analysis provides a way to analyze chart patterns and statistical data to make future predictions. We have also examined the P/E ratio, P/B ratio, PEG ratio, and dividend which are key elements for stock market prediction. Some of the machine learning techniques that are used by researchers are also discussed in this paper. After doing a detailed study of the techniques that are discussed in this paper, we can conclude that with the inclusion of more features, which can represent the outer world emotions and other influencers with the integration of statistical model along with machine learning algorithms, a hybrid system can provide a better solution to the stock market prediction problem.

References

1. E. Alexander, E. Kawaler, D. Szafir, *Stock Market Analysis and Prediction* (2009) pp. 1–15
2. R. An, C. Qian, W. Zheng, W. Zheng, R. An, *Stock Market Trends Prediction after Earning Release* (n.d.) 06117810, 814
3. G. Appel, *Technical Analysis: Power Tools for Active Investors*. Financial Times/Prentice Hall (2005), https://books.google.co.in/books?id=RFYIAAAACAAJ
4. A. Atkins, M. Niranjan, E. Gerding, Financial news predicts stock market volatility better than close price. J. Finance Data Sci. **4**(2), 120–137 (2018). https://doi.org/10.1016/j.jfds.2018.02.002
5. A. Beattie *The 4 Basic Elements of Stock Value* (2016). Investopedia.Com. http://www.investopedia.com/articles/fundamental-analysis/09/elements-stock-value.asp
6. M. Dang, D. Duong, Improvement methods for stock market prediction using financial news articles, in *NICS 2016 - Proceedings of 2016 3rd National Foundation for Science and Technology Development Conference on Information and Computer Science, May*, (2016), pp. 125–129. https://doi.org/10.1109/NICS.2016.7725636
7. O. Hegazy, O.S. Soliman, M.A. Salam. A machine learning model for stock market prediction. Int. J. Comput. Sci. Telecommun. **4** (2013) (2014). http://arxiv.org/abs/1402.7351
8. M. Hiransha, E.A. Gopalakrishnan, V.K. Menon, K.P. Soman, NSE stock market prediction using deep-learning models. Procedia Comput. Sci. **132**(Iccids), 1351–1362 (2018). https://doi.org/10.1016/j.procs.2018.05.050
9. R. Iacomin (2015). Stock market prediction, in *2015 19th International Conference on System Theory, Control and Computing, ICSTCC 2015 - Joint Conference SINTES 19, SACCS 15, SIMSIS 19*, pp. 200–205. https://doi.org/10.1109/ICSTCC.2015.7321293
10. B. Jeevan, E. Naresh, B.P. Vijaya Kumar, P. Kambli. Share price prediction using machine learning technique, in *2018 IEEE 3rd International Conference on Circuits, Control, Communication and Computing, I4C 2018*, 1, pp. 1–4 (2018). https://doi.org/10.1109/CIMCA.2018.8739647
11. S. Kalra, J.S. Prasad, Efficacy of news sentiment for stock market prediction, in *International Conference on Machine Learning, Big Data, Cloud and Parallel Computing (Com-IT-Con)* (2019), pp. 491–496. https://doi.org/10.1109/COMITCon.2019.8862265
12. W. Khan, M.A. Ghazanfar, M.A. Azam, A. Karami, K.H. Alyoubi, A.S. Alfakeeh, Stock market prediction using machine learning classifiers and social media, news. J. Ambient Intell. Hum. Comput.0123456789 (2020). https://doi.org/10.1007/s12652-020-01839-w
13. S. Kokila, Implementation of extended deep neural networks for stock market prediction. Int. J. Res. Appl. Sci. Eng. Technol. **8**(2), 29–34 (2020). https://doi.org/10.22214/ijraset.2020.2006

14. T. Mankar, T. Hotchandani, M. Madhwani, A. Chidrawar, C.S. Lifna, Stock market prediction based on social sentiments using machine learning, in *2018 International Conference on Smart City and Emerging Technology, ICSCET 2018*, (2018) pp. 1–3. https://doi.org/10.1109/ICSCET.2018.8537242
15. A. Nayak, M.M.M. Pai, R.M. Pai, Prediction Models for Indian Stock Market. Procedia Comput. Sci. **89**, 441–449 (2016). https://doi.org/10.1016/j.procs.2016.06.096
16. I.K. Nti, A.F. Adekoya, B.A. Weyori, A comprehensive evaluation of ensemble learning for stock-market prediction. J. Big Data **7**(1) (2020). https://doi.org/10.1186/s40537-020-00299-5
17. I.K. Nti, A.F. Adekoya, B.A. Weyori, Predicting stock market price movement using sentiment analysis: evidence from Ghana. Appl. Comput. Syst. **25**(1), 33–42 (2020). https://doi.org/10.2478/acss-2020-0004
18. V.S. Pagolu, K.N. Reddy, G. Panda, B. Majhi, Sentiment analysis of Twitter data for predicting stock market movements, in *International Conference on Signal Processing, Communication, Power and Embedded System, SCOPES 2016 – Proceedings* (2017), pp. 1345–1350. https://doi.org/10.1109/SCOPES.2016.7955659
19. I. Parmar, N. Agarwal, S. Saxena, R. Arora, S. Gupta, H. Dhiman, L. Chouhan, Stock Market Prediction Using Machine Learning, in *ICSCCC 2018 - 1st International Conference on Secure Cyber Computing and Communications* (2018), pp. 574–576. https://doi.org/10.1109/ICSCCC.2018.8703332
20. P.S.P. Pimpalkar, J. Karia, M. Khan, S. Anand, Mukherje, Stock market prediction using machine learning. Int. J. Adv. Eng. Res. Dev. Spec. Issue Recent Trends Data Eng. **4**(5), 6–8 (2017)
21. V. Rajput, S. Bobde, Stock market forecasting techniques: literature survey. Int. J. Comput. Sci. Mobile Comput. **5**(6), 500–506 (2016). www.ijcsmc.com
22. R.S. Deepak, S.I. Uday, D. Malathi, Machine learning approach in stock market. Int. J. Pure Appl. Math. **115**(8), 71–77 (2017)
23. K. Sai, R. Vanukuru, V. Kranthi, S. Reddy, Stock market prediction using machine learning stock market prediction using machine learning view project development of microcontroller based automated public transport ticketing system view project stock market prediction using machine learning. Int. Res. J. Eng. Technol. 1032 (2008). https://doi.org/10.13140/RG.2.2.12300.77448
24. V.K. Sai Redddy, Stock market prediction using machine learning. Int. Res. J. Eng. Technol. **05**(10) (2018). https://doi.org/10.1109/ICAC49085.2019.9103381
25. D. Selvamuthu, V. Kumar, A. Mishra, Indian stock market prediction using artificial neural networks on tick data. Financ. Innovation **5**(1) (2019). https://doi.org/10.1186/s40854-019-0131-7
26. D.S. Shanthi, T. Aarthi, A.K. Bhuvanesh, R.A.C. Prabha, Pattern Recogn. Stock Market **9**(3), 106–111 (2020)
27. S. Sharma, B. Kaushik, Quantitative analysis of stock market prediction for accurate investment decisions in future. J. Artif. Intell. **11**(1), 48–54 (2018). https://doi.org/10.3923/jai.2018.48.54
28. H.J. Shunrong Shen, Tongda Zhang, Stock market forecasting using machine learning algorithms. Successful User Experience: Strategies and Roadmaps, 1–344 (2015). https://doi.org/10.1016/c2013-0-19353-1
29. J. Si, A. Mukherjee, B. Liu, Q. Li, H. Li, X. Deng, Exploiting topic based twitter sentiment for stock prediction, in *ACL 2013 - 51st Annual Meeting of the Association for Computational Linguistics, Proceedings of the Conference*, 2(2011) (2013), pp. 24–29
30. S. Singh, T.K. Madan, J. Kumar, A.K. Singh, Stock market forecasting using machine learning: today and tomorrow, in *2nd International Conference on Intelligent Computing, Instrumentation and Control Technologies (ICICICT)* (2019). https://doi.org/10.1109/ASYU48272.2019.8946372
31. Understanding Fundamental and Technical Analysis. (n.d.). Retrieved August 29 2020, from https://groww.in/blog/fundamental-vs-technical-analysis/#2Factors_important_for_the_analysis

32. T. Xia, Q. Sun, A. Zhou, S. Wang, S. Xiong, S. Gao, J. Li, Q. Yuan, Improving the performance of stock trend prediction by applying GA to feature selection, in *Proceedings - 8th IEEE International Symposium on Cloud and Services Computing, SC2 2018*, (2018), pp. 122–126. https://doi.org/10.1109/SC2.2018.00025
33. X. Zhou, Z. Pan, G. Hu, S. Tang, C. Zhao, Stock market prediction on high-frequency data using generative adversarial nets. Math. Prob. Eng. **2018** (2018)

Survey of Various Techniques for Voice Encryption

Sameeksha Prasad and Arun Mishra

Abstract It is well established that wireless communications face significant challenges in terms of maintaining high security, owing to the ease with which attackers can compromise communication channels. As a result, there are ongoing developments toward developing a suitable approach to ensuring high security for wireless communication. Voice signals are highly vulnerable to security threats and are often influenced by network interferences such as reflection. Various methods for encrypting voice signals have already been suggested. Bit shift register schemes are commonly used in voice encryption signals. The LFSRs and FCSRs are two common techniques for voice signal encryption that are based on the bit shift register. The advantage of the LFSR is that it is simple, while that of FCSR is that it is nonlinear. A variety of algorithms have been used with LFSR and FCSR to increase the unpredictability and randomness of the LFSRs and FCSRs.

Keywords FCSR · LFSR · Encryption · Voice data

1 Introduction

The coding theory and cryptography make an extensive implementation of sequences in finite fields. Fast generators of sequences [1–3] with fine statistical properties are useful in a variety of applications, including generating mainstreams for stream ciphers, spreading sequences for CDMA communications systems, generating pseudo-random numbers so that the quasi-Monte Carlo can be integrated, generating sequences for radar ranging and error correction codes, and generating pseudo-random numbers so that the quasi-Monte Carlo [4] can be integrated. There are two parts comprised in a feedback shift register [4]. The initial is the shift register that can be described as a sequence of bits. The bits are considered to determine the

S. Prasad (✉) · A. Mishra
Department of CSE, Defence Institute of Advanced Technology, Pune, Maharashtra, India

A. Mishra
e-mail: arunmishra@diat.ac.in

A. K. Luhach et al. (eds.), *Second International Conference on Sustainable Technologies for Computational Intelligence*, Advances in Intelligent Systems and Computing 1235,
https://doi.org/10.1007/978-981-16-4641-6_16

length of this section. The n-bit shift register is the register with its n-bit long section. Each and every bit of the shift register is shifted to the right, one bit, every time during the requirement for a bit. A novel leftmost bit's quantification is achieved as a result of the register's other bits. The XOR of bits selected in the register is commonly known as the feedback function [1]. A tap sequence is defined as the list of these bits. The output of the shift register is obtained in form of one bit which is often known as LSB. Linear feedback shift registers (LFSRs) are generally implemented as part of keystream generators in stream ciphers. Particular criteria are taken into account for the parts of keystream generators. These parameters include the main-stream's period, linear complexity, and statistical measurements. The length of the sequence produced as output prior to initializing its repetition is equal to the time of a shift register [5]. If the n-bit LFSR's feedback polynomial is primitive and its initial state is non-zero, the output sequence created by this LFSR has a maximum length of 2n -1. The linear feedback shift register generators are calculated using only one metric: linear complexity or linear span. The length "n" of the shortest LFSR capable of imitating the generator output is known as this. The statistical behavior of the sequence is found with the help of various statistical tests [6–8]. These tests are often carried out to check for random distribution, distribution of ones and zeros in a sequence, linear dependence among substrings which are of fixed length, and the level of compression which is performed on the tested sequence and it determines that whether a sequence is complex enough to be considered random or not.

In various cases, LFSRs or the generators developed using LFSRs are deployed to produce the suitable sequences. The linear feedback shift registers (LFSRs) act significantly to generate the sequences across finite fields. There are an enormous number of theories available on LFSRs and its literature is also wide-ranging [9]. The LFSRs are simple techniques whose deployment is done for producing huge period sequences across the finite fields in an effective manner. In mathematical interpretation, these techniques are regarded as finite state machines with output or linear recurrences over fields. These techniques are revealed as finite state device. Consider q_i and a_j as the components of a finite field F. The contents of linear feedback shift registers (LFSRs) are shifted to the right position for changing its positions. To achieve this, $q_1 a_{k-1} + ... + q_k a_0$ are inserted in the leftmost cell, and the a_0 is the general value which is raised to a_{k-1} to generated LFCRs sequences. With the support of linear recurrence, an infinite sequence $A = a_0, a_1, \ldots a_{k-1}, a_k, a_{k+1} \ldots$ is created from the initial values $a_0, \ldots, a_{k-1}$ using linear feedback shift register (LFSR).

As shown in Fig. 1, a_k is the value of linear feedback shift register which can define the number of bits that get shifted during the process.

$$a_k = q_1 a_{k-1} + \cdots + q_k a_0 \tag{1}$$

To evaluate the linear feedback shift registers (LFSRs), many related algebraic constructs are introduced. The generating function denoted by $\alpha(A, x) = \sum_{i=0}^{\infty} a_i x^i$ is a power series that is related to the output sequence [10, 11]. The connection polynomial is a polynomial represented with $q(x) = \sum_{i=1}^{k} q_i x^i - 1$ which has

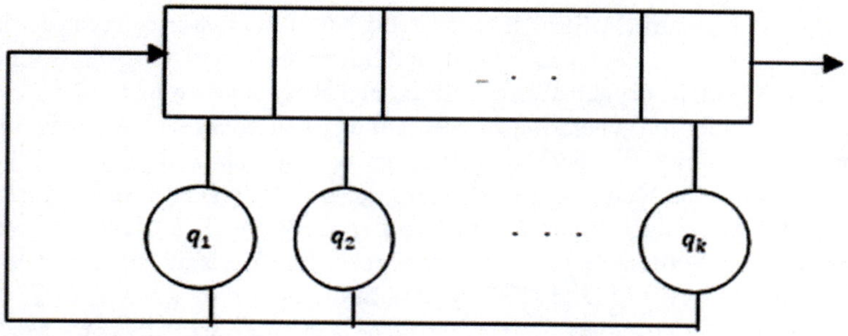

Fig. 1 A linear feedback shift register [9]

relation with the coefficients of the linear recurrence. The Fundamental Theorem of LFSR is used to link the generating function to the relation polynomial. The FCSRs are a new class of sequence generators that was founded in 1994. FCSRs (Feedback with Carry Shift Registers) have much of the same properties as linear feedback shift registers (LFSRs). The FCRSs and the LFSRs have a lot in common. The FCSR, on the other hand, provides the extra memory that is constantly carried from one phase to the next. The simplified view of feedback with carry shift register (FCSR) is seen in Fig. 2. To alter the states of FCSR, a linear combination is determined as follows:

$$\sigma = q_1 a_{k-1} + \cdots + q_k a_0 + m \tag{2}$$

In Fig. 2, M corresponds to memory, while q1 and q2 are connection numbers. The setting is done as $a_k = \sigma$ mod N and $m' = \lfloor \sigma / \mathrm{N} \rfloor$. The major part of novel state is denoted with $(a_k, \ldots a_1)$ where as m' m, on the other hand, is used to show the new value of the additional memory. Under any boundless implementation of feedback

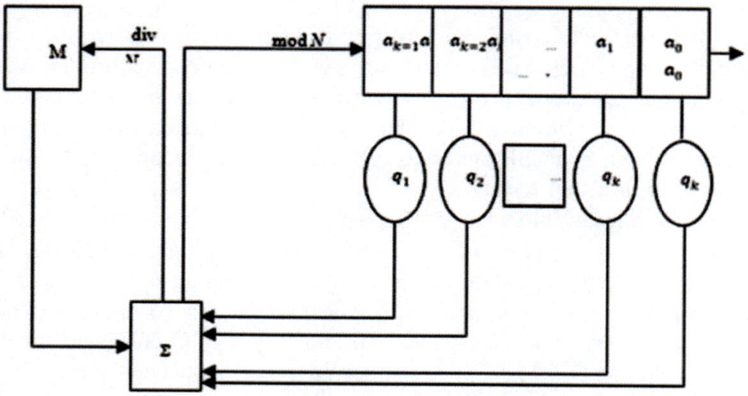

Fig. 2 A feedback with carry shift register [10]

with carry shift registers (FCSRs), it is observed that extra memory is created without restriction. Goresky and Klapper formulated the feedback with carry shift registers (FCSRs) [12, 13]. The cryptanalysis of the summation combiner has a major impact on the interpretation of generating FCSR. Although some researchers have looked into recurrences with carry, with the aim of developing fine random number generators for applications such as quasi-Monte Carlo simulation. They only considered a few special cases, such as Add with Carry recurrences. The MWC recurrences are used to generate these recurrences. Feedback with carry shift registers (FCSRs) is similar to multiply with carry (MWC). With the help $a_i q_i \in \{0, 1, \ldots, N-1\}$ for $i = 0, \ldots, k-1$ andm_{k-1}, there are two sequences explained by MWC as $a_i \in \{0, 1, \ldots, N-1\}, m_i \in Z$ for $i = k, k+1, \ldots$ with the equation which is expressed as

$$a_n + N_{mn} = q_1 a_{n-1} + \cdots + q_k a_{n-k} + m_{n-1} \quad (3)$$

For $i = k, i = k+1, \ldots$.. These a_n and m_n are described in a unique way. The produced sequence is expressed as $A = a_0, a_1, \ldots, a_{k-1}, a_k, a_{k+1}$. [14]. The first algebraic structure that has an association with feedback with carry shift register (FCSR) is known as the connection number which can be expressed as $q = q_k N_k + q_{k-1} N_{k-1} + \cdots + q_1 N - 1$. This is the arithmetic equivalent to linear feedback shift registers (LFSRs).

2 Analysis of Various Research Works for Generation of Pseudo-Random Sequence

Klapper et al. embodied some of the most basic characteristics of FCSRs. To achieve the comparative goal, some basic information about linear feedback shift registers (LFSRs) were first checked [15]. Following that, feedback with carry shift registers (FCSRs) and their relationship with N-adic numbers are discussed. As a result, the periodicity of certain FCSR sequences was investigated, and the exponential representation of such sequences was discovered, as well as the maximum period FCSR sequences. An arithmetic equivalent of cross-correlations was considered, as well as the problem of FCSR synthesis and the efficient execution of FCSRs in a parallel architecture.

This work applies a Galois-based FCSRs framework for voice encryption which can be shown in Fig. 3. There is the need for extra carry cells amid each pair of normal cells. The state transition is measured by computing $\sigma_i = a_i + q_i a_0 + c_i$. The obtained value of a_{i-1} is $\sigma_i modeN$. The new values of c_i correspond to $[\sigma_i / N]$. This device generates a periodic sequence if and only if an FCSR generates it with a connection number $\sum_{i=1}^{k} q_i N^i - 1$. The key advantage is that any state shift of the various cells can be done in parallel. This reduces the amount of time it takes to shift states and is independent of k.

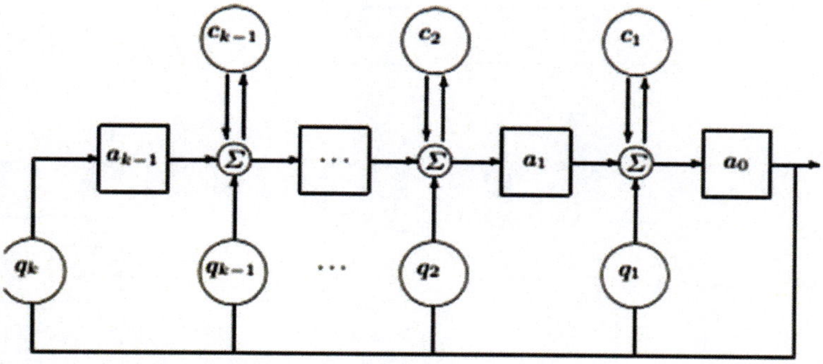

Fig. 3 Galois mode implementation of FCSRs [15]

Han et al. introduced σ−LFSRs that is also known as a novel family of word-oriented linear feedback shift registers (LFSRs) [16]. Some fundamental instructions assist in developing the σ−LFSRs for obtaining high efficiency for them. Unlike the theory of traditional LFSRs, the research was carried out on the basic properties of σ− LFSRs including minimal polynomial and coordinate sequence and the properties of primal σ−LFSRs, etc. An algorithm was presented for searching for the primary σ−LFSRs, and various primary σ−LFSRs were achieved with special properties called HHZ-1 or HHZ-2. Simultaneously, the σ−LFSRs were utilized while setting up the RNGs and stream ciphers. The outcomes demonstrated that the σ − LFSRs that were highly efficient and having good pseudo-random properties had been implemented as the building blocks in the formation of modern stream ciphers.

Rajesh et al. established an architecture (optimized) of edge and pulse-triggered SIPO and SISO registers and focused on evaluating their quantum cost, delay, and garbage output [17]. Subsequently, an optimized design of linear feedback shift register (LFSR) was recommended for which reversible gates were implemented. These gates were utilized to generate the random bit. The military applications had made the execution of optimized design of reversible LFSR due to cryptography. The cryptanalysis became simple as the recommended model was linear. Xilinx ISE 14.7 was employed to conduct the simulations and synthesis procedures.

On the feedback path of an LFSR, a Feynman gate is used for exclusive OR operation, and it can be used with any flip-flops to copy the corresponding output. If we load flip-flops with any sequence other than all zeros, the LFSR is activated, and the pseudo-random sequence of all 0's and 1's is generated. The size of the LFSR pattern count will be equal to "2n-1," where "n" is the total number of flip-flops and "n" is the total number of 0's and 1's.

Kiruthiga et al. suggested a novel architecture of LFSR with the utilization of CDFF and GDI [18]. The suggested architecture provided resultant bits whose sampling was done to perform n-bit random number generation. The complexity of the architecture was found very simple. This technique provided the least power, led to alleviate area, and minimized the usage of the transistor in comparison to the traditional methods. The comparison of suggested architecture was done with other

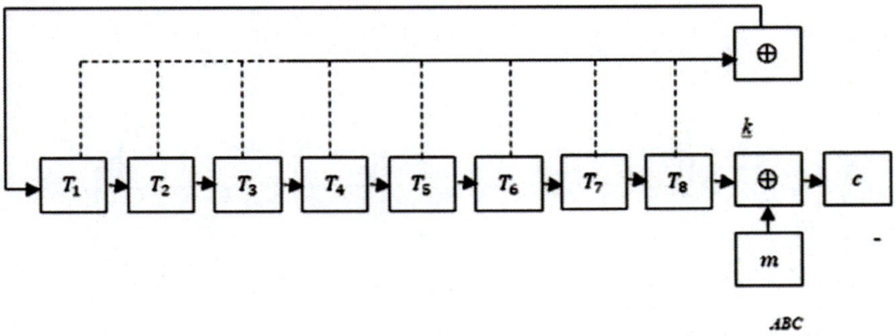

Fig. 4 Cryptosystems based on an 8-bit linear feedback shift register [19]

linear feedback shift register (LFSR) concerning the above metrics. The outcomes demonstrated that the suggested architecture saved average power utilization around 33%. Tanner EDA tool was utilized to generate the net outcomes and deploy the suggested methods.

Borodzhieva et al. focused on implementing FPGA of cryptosystems on the basis of LFSRs [19]. For this purpose, English texts were encrypted and decrypted for the simulation. The educational procedure in the University of Ruse had employed this application for which the assimilation of the material that the students studied was allowed. This approach aimed at presenting the evidence to indicate that the solution on the basis of learning approach along with conventional tools and information was assisted in maximizing the interest of the student in STEM and enhancing their achievements.

This work performs linear feedback shift registers-based FPGA Implementation of Cryptosystems for voice encryption shown in Fig. 4 which represents an 8-bit linear feedback shift register based cryptosystem.

The following are the measures for encrypting Fig. 4: 1. Create a pseudo-random number generator using an 8-bit LFSR and D (or JK) flip-flops, with input taken from the flip-flop outputs with numbers 2 and 5. Determine the pseudo-random binary number series that begins at 10010001 and continues for 8 clocks of abstract automata time. 2. Encrypt the RGB message, which is sent in either 7-bit ASCII or 8-bit EBCDIC code. For this, the pseudo-random number generator from Fig. 4 may be used, which is deployed as the mainstream generator. Analyze the operation of a cryptosystem using an 8-bit LFSR. Use the file LFSR WG xx.bit (xx = 01…60) until the FPGA programming is complete. After filling in the table, describe the procedure to test it on the laboratory board so that the decrypted text can be calculated and a circuit can be built in ISE Project Navigator to generate the bit file.

Sasi et al. presented an LFSR based on a hybrid CMOS-memristor that is fast, small, and low power [20]. The MRL framework, which consisted of 64 CMOS devices and 64 memristors, was used to implement a 4-bit linear feedback shift register (LFSR). In comparison to CMOS-based LFSR circuits, the presented design was more efficient in terms of area. The simulation results showed that the proposed solution reduced power consumption by up to 66 percent, which was significantly less than the CMOS-based approach.

Saha et al. projected a technique for encrypting and decrypting the images through LFSR. A pseudo-random number was generated from the projected technique [21]. The performance of the linear feedback shift registers (LFSR) was randomized, and a stable LFSR was achieved using cryptography. The final encrypted image was made more unpredictable. A two-level procedure was adopted in order to encrypt the image. Initially, the pixels were re-arranged, and subsequently, the substitution was carried out. The outcomes of the experiment exhibited the robustness and security of the projected technique. The scheme proposed by the author also encrypts voice signals.

Zode et al. introduced a random number generator based on LFSR for which system monitor of FPGA was utilized [22]. The random seed generation was carried out using temperature variation that occurred in the chip and variations of the power supply voltage. Xilinx Virtex-5 ML505 board was executed to deploy the introduced system. The results of experiment depicted that the On-chip resource of the introduced system had consumed less power and taken less area. NIST Statistical test suite was employed for testing the randomness outcomes, and all the tests were passed.

Pomeranz et al. discussed that the major objective was to study the approach for the PDFs with the implementation of LFSR as the decompression logic [23]. The confidence for the detection of significant Path Delay Faults was maximized when an extended subset of PDFs was targeted. The presented technique was used to detect an expanded subset of target PDFs using the same collection of linear feedback shift register (LFSR) seeds. The scan-in states were achieved using extra clocking of the LFSR for different novel two-cycle experiments based on identical seeds. The results of the experiments showed that the proposed method was efficient.

The literature survey is conducted on various other work also. The authors proposed various methodologies which are based on shift registers for voice encryption. As illustrated in Table 1, the various techniques of voice encryption are compared in terms of their description and results. The description describes the methodology which is applied for the implementation. The outcomes describe the results of the methodology.

3 Conclusion

The bit shift register schemes are much more effective than traditional encryption schemes for encrypting voice signals. For the encryption of voice signals, the FCSRs and LFSR are two common schemes based on the bit shift register. FCSR is non-linear but complex, whereas LFSR is simple but linear. The LFSR is a rational sequence of GF rings, while the FCSR is focused on rational 2-adic numbers [24]. However, the LFSR's output is linear and simple to crack, while the FCSR's output is quadratic. Since the FCSR's quadratic output is modified nonlinearly, it is more resistant to algebraic and correlation attacks than the LFSR's linear output. On the other hand, the FCSR demonstrates that it is insecure when implemented alone,

Table 1 Comparison table

Author	Year	Description	Outcome
Rajesh et al.	2018	Developed an architecture that centerd on evaluating the garbage performance, delay, and quantum cost of edge and pulse-triggered SISO and SIPO registers	The cryptanalysis becamesimple as the recommended model was linear. Xilinx ISE 14.7 was employed to conduct the simulations and synthesis procedures
Kiruthiga et al.	2020	Suggested a novel architecture of LFSR with the utilization of CDFF and GDI. The suggested architecture provided resultant bits whose sampling was done to perform n-bit random number generation	The outcomes demonstrated that the suggested architecture saved average power utilization around 33%. Tanner EDA tool was utilized to generate the net outcomes and deploy the suggested methods
Borodzhieva et al.	2019	Focused on implementing FPGA of cryptosystems on the basis of LFSRs. For this purpose, English texts were encrypted and decrypted for the simulation	This approach aimed at presenting the evidence to indicate that the solution on the basis of learning approach along with conventional tools and information was assisted in maximizing the interest of the student in STEM and enhancing their achievements
Sasi et al.	2017	A LFSR based on a fast, low-area, low-power hybrid CMOS-memristor was presented. The MRL framework, which consisted of 64 CMOS devices and 64 memristors, was used to implement a 4-bit linear feedback shift register (LFSR)	The outcomes of simulation depicted that the presented approach mitigated the power usage up to 66% that was founded lower as compared to the CMOS-based approach
Saha et al.	2018	Projected a technique for encrypting and decrypting the images through LFSR. A pseudo-random number was generated from the projected technique	A two-level procedure was adopted in order to encrypt the image. Initially, the pixels were re-arranged and subsequently, the substitution was carried out. The outcomes of experiment exhibited the robustness and security of the projected technique. This scheme is used for the voice encryption
Zode et al.	2019	Introduced a random number generator based on LFSR for which system monitor of FPGA was utilized. The random seed generation was carried out using temperature variation that occurred in the chip and variations of power supply voltage	The results of experiment depicted that the On-chip resource of the introduced system had consumed less power and taken less area. NIST Statistical test suite was employed for testing the randomness outcomes, and all the tests were passed

(continued)

Table 1 (continued)

Author	Year	Description	Outcome
Pomeranz	2020	Discussed that the major objective was to study the approach for the decompression logic in PDFs by implementing LFSRs	The scan-in states were achieved using extra clocking of the LFSR for different novel two-cycle experiments based on identical seed. The results of the experiments showed that the proposed method was efficient

and it is more expensive to implement the FCSR than the LFSR [25]. As a result, an additional algorithm must be included in the FCSR implementation to prevent predictable performance. In relation to image and text, the voice signal has a special attribute and structure. Unlike text and picture, the voice can be identified using the jumbled and diffused information from all parts of the voice signal. As a result, combining the improved LFSR or FCSR with the unpredictable encryption scheme will increase security even further.

References

1. D. Pei, Z. Lin, X. Zhang, Construction of transition matrices for ternary ring feedback with carry shift registers. IEEE Trans. Inf. Theory **61**(5) (2015)
2. A. Ali, Feedback with carry shift registers and (in-depth) security of ciphers based on this primitive, in *15th International Bhurban Conference on Applied Sciences and Technology (IBCAST)* (2018)
3. A. Ali, Oppel-2: a new family of FCSR-based stream ciphers, in *International Conference on Emerging Technologies (ICET)* (2014)
4. L. Ke, Z. Lin, Distribution properties and 2-adic complexity of clock-controlled FCSR sequences, in *Ninth International Conference on P2P, Parallel, Grid, Cloud and Internet Computing* (2014)
5. F. Arnault, T.P. Berger, M. Minier, Some results on FCSR automata with applications to the security of FCSR-based pseudorandom generators. IEEE Trans. Inf. Theory **54**(2) (2008)
6. F. Masoodi, S. Alam, M.U. Bokhari, An analysis of linear feedback shift registers in stream ciphers. Int. J. Comput. Appl. **46**(17), 0975–8887 (2012)
7. N. Hayati, Y. Suryanto, K. Ramli, M. Suryanegara, End-to-end voice encryption based on multiple circular chaotic permutation, in *2nd International Conference on Communication Engineering and Technology (ICCET)* (2019)
8. PL. Chithra, R. Aparna, Voice signal encryption scheme using transformation and embedding techniques for enhanced security, in *2nd International Conference on Imaging, Signal Processing and Communication (ICISPC)* (2018)
9. Y. Wang, Z. Tan, Modified RSA encryption algorithm based on chaos and its application in voice encryption system, in *14th International Conference on Computational Intelligence and Security (CIS)* (2018)
10. A.M. Raheema, S.B. Sadkhan, S.M. Abdul Sattar, Performance evaluation of voice encryption techniques based on modified chaotic systems, in*6th International Engineering Conference "Sustainable Technology and Development" (IEC)* (2020)

11. G.R. Bagwe, D.S. Apsingekar, S. Gandhare, S. Pawar, Voice encryption and decryption in telecommunication, in *International Conference on Communication and Signal Processing (ICCSP)* (2016)
12. H. Jahanshahi, A. Yousefpour, F.E. Alsaadi, A new fractional-order hyperchaotic memristor oscillator: dynamic analysis, robust adaptive synchronization, and its application to voice encryption. Appl. Math. Comput. (2020)
13. C. Misra, M.K. Hota, Hybrid technique for voice recognition, encryption and analysis using MATLAB, in *International Conference on Communication and Signal Processing (ICCSP)* (2018)
14. M.A. Riyadi, N. Pandapotan, M.R.A. Khafid, T. Prakoso, FPGA-based 128-bit chaotic encryption method for voice communication, in *International Symposium on Electronics and Smart Devices (ISESD)* (2018)
15. A. Klapper, A survey of feedback with carry shift registers, in *International Conference on Sequences and their Applications* (2004)
16. W. Han, X. Liu, G. Zeng, G. Tan, *Word-Oriented Linear Feedback Shift Registers: $\Sigma-$LFSRs* (Higher Education Press and International Press, Beijing-Boston, 2014)
17. K. Rajesh, G. Umamaheswara Reddy, FPGA implementation of optimized reversible linear feedback shift registers, in, *2nd International Conference on I-SMAC (IoT in Social, Mobile, Analytics and Cloud) (I-SMAC)I-SMAC (IoT in Social, Mobile, Analytics and Cloud) (I-SMAC)* (2018)
18. S. Kiruthiga, M. Shangeeth, S.P. Rajesh Kumar, R. Sowndarya, LFSR using CDFF and GDI, *in 6th International Conference on Advanced Computing and Communication Systems (ICACCS)* (2020)
19. A. Borodzhieva, I. Stoev, V. Mutkov, FPGA implementation of cryptosystems based on linear feedback shift registers for educational purposes, in *29th Annual Conference of the European Association for Education in Electrical and Information Engineering (EAEEIE)* (2019)
20. A. Sasi, A. Amirsoleimani, A. Ahmadi, M. Ahmadi, Hybrid memristor-CMOS based linear feedback shift register design, in *24th IEEE International Conference on Electronics, Circuits and Systems (ICECS)* (2017)
21. S. Saha, R.K. Karsh, M. Amrohi, Encryption and decryption of images using secure linear feedback shift registers, in *International Conference on Communication and Signal Processing (ICCSP)* (2018)
22. P. Zode, P. Zode, R. Deshmukh, FPGA based novel true random number generator using LFSR with dynamic seed, in *IEEE 16th India Council International Conference (INDICON)* (2019)
23. I. Pomeranz, Extra clocking of LFSR seeds for improved path delay fault coverage, in *IEEE Transactions on Very Large Scale Integration (VLSI) Systems* (2020)
24. N. Courtois, W. Meier, Algebraic attacks on stream ciphers with linear feedback, in *Advance in CryptologyEurocrypt 03*, volume 2656 of lectures Notes in Computer Science, ed. by E. Biham (Springer, 2003), pp. 345–359
25. A. Klapper, M. Goresky, Cryptoanalysis based on 2-adic cryptanalysis based on 2-adic rational approximation, in *CRYPTO 1995*, ed. by D. Coppersmith. LNCS, vol. 963 (Springer, Heidelberg, 1995), pp. 262–273

Identifying K-Most Influential Nodes in a Social Network Using K-Hybrid Method of Ranking

Sarthak Koherwal, Divianshu Bansal, and Pulkit Chhabra

Abstract Online social networks contain a lot of information about people, their interests, activities, behaviour patterns, etc., and can be efficient to find the most influential people based on various trades. Studying these networks can help solve a variety of purposes such as rumour prevention, controlled information flow, finding target audience for products and movies, etc. This can help to spread/prevent the flow of information to people very quickly by identifying the most critical or the most influential nodes in the network, from where the information spread is maximum. This paper focuses on creating a hybrid centrality and formulating an algorithm to find these nodes with the most collective influence.

Keywords Hybrid node centrality · SIR (Susceptible-Infected-Recovered) model · Spreading influence maximization

1 Introduction

Social networks are a powerful way of communication and spreading information. The information carried over through a social network can spread very quickly and often gain much more acceptance than conventional methods of information spreading. With the advent of social media and globalization, information has become a very powerful tool and hence must be regulated very carefully. Companies all around the globe have taken note and are trying to design new potential methods that can be used for advertisements and have the potential to appeal to a much wider user base at far lower costs.

Viral marketing is one such popular approach, where a company approaches people with a large social network, to promote and share their products to try to make the products viral through a word of mouth approach. Social networks can also cause a spread of rumours that can cause all sorts of turmoil. Also due to the advent of COVID-19, it can be of paramount importance to help to recognize people with

S. Koherwal · D. Bansal (✉) · P. Chhabra
Department of Computer Science and Engineering, Delhi Technological University, New Delhi 110042, India

A. K. Luhach et al. (eds.), *Second International Conference on Sustainable Technologies for Computational Intelligence*, Advances in Intelligent Systems and Computing 1235,
https://doi.org/10.1007/978-981-16-4641-6_17

the highest influence, i.e., the highest risk of spreading the virus further due to their interactions in their social network. It is hence essential to find a way to be able to quickly spread new information or prevent from doing so at a lower cost and to a wider target audience base. This paper tries to choose a handful of nodes(people) in a social network with an extremely high potential to influence the social network around them, which can help(prevent) some target information to become viral at a very low cost and appeal to a wide potential user base.

2 Literature Review

The total amount of nodes adjacent to a particular node in a graph is known as the Degree Centrality [1] for that node. The degree of a node can also be thought of as the immediate risk of a node getting affected by the information passing through the network. Degree centrality is a local metric, useful when the network is relatively small and doesn't give good results. Closeness Centrality [2] of a node can be defined as the average value of the shortest path between the particular node and all other nodes in the network. Thus, a node in the centre of the network will be closer to all the other nodes and have a higher Closeness Centrality. Betweenness Centrality [3] can be defined as the measure for the count of instances; a node acts as a bridge when a random shortest path is chosen between any two other nodes in the graph. Nodes having high betweenness centrality play an important role in the spreading of information by virtue of their control over information passing between other nodes in the network. If these nodes were not present in the graph, it can result in disruption of communication between vertices in the graph as most of the messages have to pass through them. Eigenvector Centrality [4] measures a node's centrality taking into account the importance of its neighbours. It says that a node having important nodes (neighbours) will be more influential compared to a node with less important adjoining nodes (neighbours). It is calculated by performing a matrix calculation on the adjacency matrix to determine the principal eigenvector.

Percolation Centrality [5] assigns a centrality value to a node while taking into account the contribution of the node in spreading the contagion through the network. K-shell Centrality [6] can be measured by continuously removing nodes having less degree; this leads to the pruning of nodes until all the nodes have been removed. The major pitfall to this approach is that the same Centrality value is assigned to all nodes present in the same shell. To overcome this problem, Neighbourhood Coreness Centrality [6] was developed. It is calculated as the summation value of the K-shell values of the neighbours of a particular node. Thus, high neighbourhood Coreness value signifies that the node is located in the centre of the network as it will be connected to many other nodes in contrast to a node present on the outskirts of the network. It was used to create a new centrality called the Extended Neighbourhood Coreness Centrality [7], given by the sum of Coreness values of all the neighbours of a particular node.

H-index centrality [8] calculates the H-index of the node, in order to determine the most influential spreaders. It is defined for a particular node as the maximum possible value *h*, so that it has at least *h* neighbours with a degree at least *h*. But there can be many nodes with the same value *h*, which can lead to clashes. To solve that, Extended H-index centrality [8] was defined as the sum of H-index value of all the neighbouring nodes.

Mixed Seed Detection (MDD) [9] Centrality defined by Zeng et al. used the K-shell centrality to find the influence of a node by removing all the affected neighbours at each subsequent step. The node which removed the majority nodes via MDD in lesser iterations was used as more effective.

Belfin et al. [10] defined a Superior Seed Detection (SSD) centrality as an upgrade over MDD as it could select top K-nodes and find their influence over the network, instead of a single isolated node. It is worth noting that due to overlaps in the network, MDD can't be used to find the top K-influential nodes as influence is not a scalar arithmetic quantity and two nodes can have overlapping influence. SDD value is calculated as the normalized sum of the page rank centrality, eigenvector and sum of the degree of the node.

Gravity Centrality was proposed by Ma et al. [11] by considering the shortest distance between two nodes as the distance and each node as its mass to find the influential spreaders.

While there exists Neighbourhood Coreness Centrality, the research was focused on a slightly different modification of our own to be able to solve the drawback of K-shell centrality. The approach has been explained in the next section of the paper.

This paper discusses about K-hybrid Centrality which is a mixture of enhanced K-shell (K-batch) value and H-index Centrality. The algorithm used ranks all nodes present in a network based on their K-hybrid value, then a subset of these nodes is chosen and the spreading process is simulated with the help of the SIR model.

The rest of the paper is structured as follows: Section 3 briefly discusses our methodology for calculating of K-hybrid centrality and ranks them, then it explains the process of selection of influential spreaders and explains the algorithm used with the help of a sample network. Section 4 lists the performance metrics and datasets used. Section 5 compares the method used in this paper with other centralities using different performance metrics. Finally, Sect. 6 concludes the paper.

3 Methodology

Selection of the K-most influential nodes by K-hybrid Centrality method is done by the identification of the centrally located and well-connected nodes in a social network, i.e., well-connected nodes present at the core of the network. The idea is to calculate and rank the nodes based on the increasing order of their K-hybrid values and comparing the influence of K-hybrid Centrality with K-shell and ENC centrality. Then it picks the P most influential nodes from such network. The value

P is a parameter of our algorithm, which is calculated by experimental observation to maximize the influence.

3.1 K-Shell Centrality

The K-shell Centrality of a node is defined by the core in which the node lies. K-shell Centrality basically depicts the closeness of a node to the network's core. The higher value of this centrality corresponds to the closeness of the node to the network's core. It also takes care of the global importance of the node when predicting its spreading capability. The K-shell value of a node v is represented by k(v).

$$\text{k(v)} - \text{the k-core in which the node is present} \tag{1}$$

The problem with K-shell Centrality is that many nodes can have the same K-shell value and so there is no conclusive way to decide which node to choose from those having the same K-shell value.

In order to overcome this problem, a small modification of K-shell Centrality is chosen as an input for the K-hybrid rank algorithm. This value is called K-batch value of a node.

3.2 Batch and K-Batch Value

The concept of batch value is loosely based on K-shell value. K-shell algorithm recursively removes nodes with the smallest degree from the network and the Batch value of a node is equal to the no. of recursive calls made for removing this node. The purpose of calculating this value is that the nodes with similar K-shell values may require different numbers of recursive calls before they are removed. The higher this value, the closer they are to the core of the Network. The Batch vlue of node v is represented by b(v).

$$\text{b(v)} - \text{The number of recursive calls required for removing this node} \tag{2}$$

For calculating the K-batch value, the sum of K-shell value and Batch value is taken.

$$\text{kb(v)} = \text{k(v)} + \text{b(v)} \tag{3}$$

3.3 H-Index Centrality

The largest such value h for which the node has at least h neighboursing nodes with a degree at least h gives the H-index centrality of a node. It considers a node to be important if it is well connected to other important nodes up to two levels. H-index gives importance to the neighbourhood of a node while determining its spreading capability. The H-index of a node v is given by

$$H(v) = \max(\min(D(i)) >= h) \tag{4}$$

where i is a neighbour of v and D(i) is its degree.

3.4 K-Hybrid Centrality

K-hybrid Centrality of a node is equal to its H-index value times its K-batch value.

For a node v, its K-hybrid value is defined as

$$kh(v) = kb(v) * H(v) \tag{5}$$

K-batch value basically considers the properties of a node with respect to the whole network while also giving importance to the local topology of the node. K-batch value lays importance on the global characteristics of a node while H-index considers its local properties.

This paper visualizes a small network consisting of 11 nodes in order to demonstrate how our algorithm gives different results when compared to other algorithms like K-shell and Extended neighbourhood Coreness Centrality. Figure 1 shows a diagrammatic representation of the graph.

The table below shows the different centralities considered and their values for each node. The best spreader node for each centrality is shown in bold. From the table,

Fig. 1 Sample Network

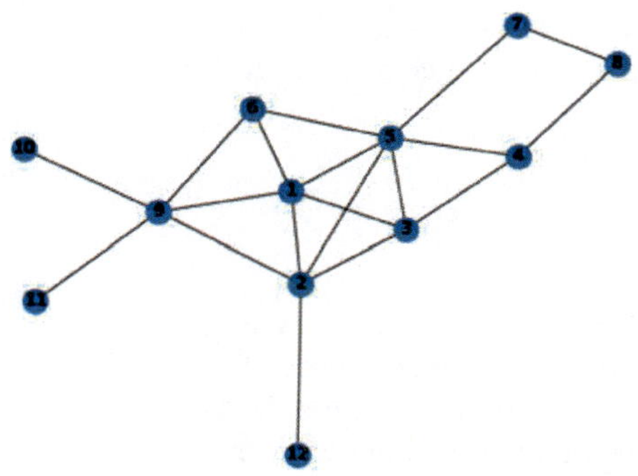

Table 1 Centrality values of each node in our network for different centralities under consideration

Node	K-shell value	ENC	K-hybrid Value
1	**3**	54	**216**
2	**3**	42	126
3	**3**	46	138
4	2	30	60
5	**3**	**55**	165
6	2	38	114
7	2	19	38
8	2	13	26
9	2	31	62
10	1	10	10
11	2	20	40
12	1	10	10

one can see that the ENC selects node 5 best spreader while multiple nodes have the highest K-shell value. However, contrary to ENC, our algorithm will choose node 1 as the most influential spreader due to its central importance and good connectivity to other important nodes (Table 1).

3.5 *Selection of Spreaders*

While selecting the K-most influential nodes, one has to take into account a lot of factors. It needs to be ensured that the nodes selected should not be too close to each other, otherwise their influence will overlap and the spread will be low. The spreaders should not be concentrated in a small area but should be distributed all around the network. Also, it can be helpful if the spreaders are at the centre of the network, that way information can cover a larger amount of nodes instead of circling back. It also helps if the spreaders themselves have a large number of first and second neighbours, thus ensuring that the information spreads quickly and effectively. Based on these ideas, the algorithms take into account most of these factors and hence can make an informed choice on what node to pick.

3.6 *SIR Model*

SIR model is a commonly used model due to its simplicity and clarity. This paper uses the SIR model to evaluate, tabulate and eventually improve upon the performance of our algorithm. It inputs the list of spreaders chosen by our algorithm, the infection probability (alpha) and the recovery probability (beta). It stimulates how the chosen set of spreaders would spread the information inside a social network. As it is a stochastic model, the experiment can be repeated several times and an average can be taken to get a clearer picture. It helps to estimate the pattern of infection(information) spread and the extent of it at any given time.

The social network of nodes is split into three categories at any given time:

Infected: At a particular point of time, these are the nodes that have the ability to spread the information to their previously uninfected and susceptible neighbouring nodes.

Recovered: These nodes were infected at a previous stage and cannot spread the information and also can't be infected again.

Susceptible: These nodes have neighbouring infected nodes but are not infected themselves at a given time. Though the information can be passed to them by these neighbouring nodes immediately.

At t = 0, a set of chosen spreaders are the nodes that are currently infected. At each successive stage, the currently infected nodes pass this information to currently susceptible nodes with a probability (alpha). And after they have passed the information, these nodes can become recovered with a probability beta. Such nodes won't be able to pass the information forward again. This process is continued until there are no more nodes left to pass the information forward anymore. As the nodes are chosen randomly or stochastically, so the algorithm is to run the program 100 times. Then the average value of the results was taken to get a clearer and more definite picture of the scenario.

4 Datasets and Performance Metrics

Infected Scale: We recorded the total amount of nodes that achieved the information (total infected nodes + total recovered nodes at that timestamp) at each given timestamp and compared the results for various algorithms for the same. Then, in order to get a clearer picture, the same is represented in the form of a graph. This method helped to efficiently compare the spreading capabilities of the chosen set of spreaders on the social network and helped to compare various algorithms and centrality measures and give an essence of which algorithm worked better on a particular dataset.

The infected scale at any instant t is given by

$$F(t) = R(t) + I(t)/n \tag{6}$$

where function R represents total recovered nodes till time t, function I represents total infected nodes at timestamp t and the variable n represents total nodes in the network (Table 2).

Table 2 Description of network datasets used

Network	No. of modes	No. of edges	Type	Description
P2P-Gnutella	6301	2077	Directed	Gnutella peer-to-peer network from August 2002
Cond-Mat	23,133	93,497	Undirected	Collaboration network of Arxiv condensed matter category
Dolphins	62	159	Undirected	Social network of association between bottlenose dolphins
Football	115	613	Undirected	A network of American Football games
PGP Network	10,638	24,301	Undirected	Web of trust of PGP network

5 Results

See Fig. 2.

6 Conclusion

This research concluded with an improved centrality measure which used a newly tried out centrality and an existing one in a hybrid manner. This can be used to achieve influence maximization by ranking and choosing the nodes in the network, based on our algorithm. This paper uses established real-world networks (both undirected and directed graphs) to test our algorithm and achieved promising results.

The proposed algorithm outperformed K-shell Centrality and ENC Centrality in all scenarios. This paper used the Infected Scale Performance Metric to compare the algorithm with other measures and found out that it performs in an all-rounded fashion even though it's not too complex computationally.

Hence, it can be concluded that this tool is a useful one for influence maximization when it comes to social networks.

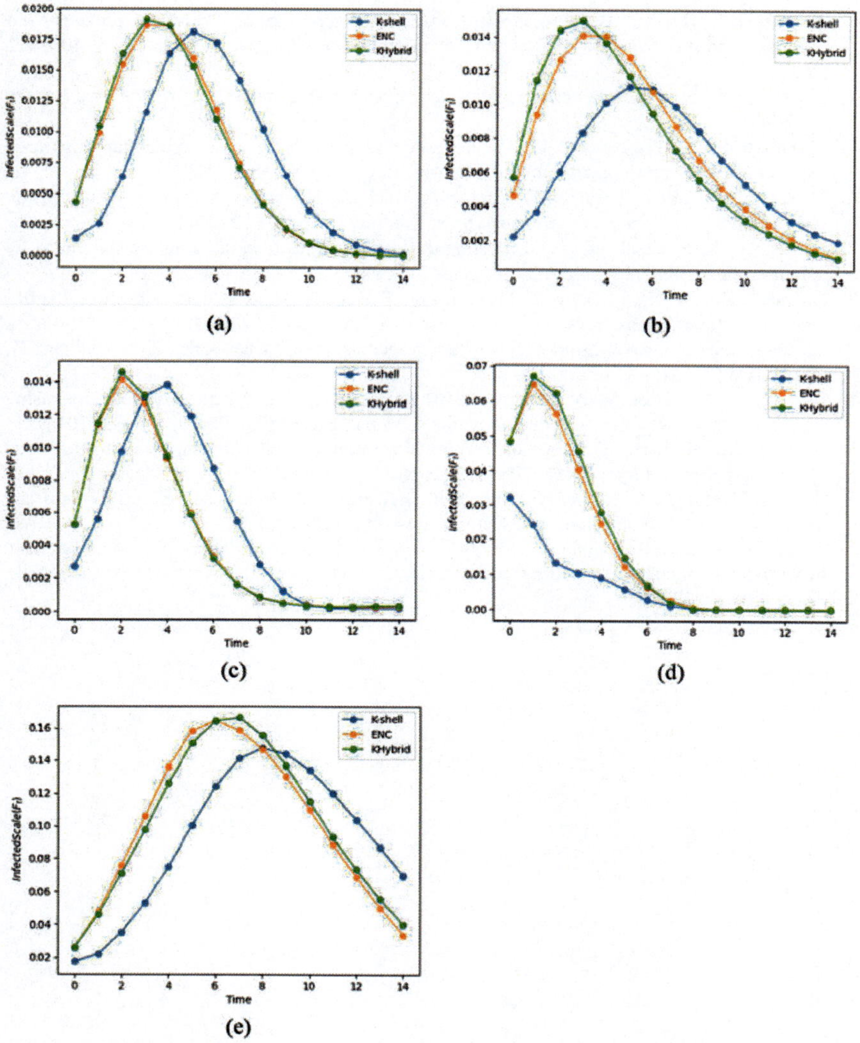

Fig. 2 Infected scale versus time for **a** P2P-Gnutella, **b** COND-MAT, **c** Dolphins, **d** Football, **e** PGP network

References

1. Y. Yustiawan, W. Maharani, A. Gozali, Degree centrality for social network with Opsahl method. Procedia Comput. Sci. **59**, 419–426 (2015)
2. K. Okamoto, W. Chen, X. Li, Ranking of closeness centrality for large-scale social networks. Front. Algorithmics. 186–195 (2008)

3. A. Abbasi, L. Hossain, L. Leydesdorff, Betweenness centrality as a driver of preferen- tial attachment in the evolution of research collaboration networks. J. Informetrics **6**, 403–412 (2012)
4. E. Estrada, J. Rodríguez-Velázquez, Subgraph centrality in complex networks. Phys. Rev. E. **71** (2005)
5. M. Piraveenan, M. Prokopenko, L. Hossain, Percolation centrality: quantifying graph-theoretic impact of nodes during percolation in networks. PLoS One **8**, e53095 (2013)
6. M. Kitsak, L. Gallos, S. Havlin, F. Liljeros, L. Muchnik, H. Stanley, H. Makse, Identification of influential spreaders in complex networks. Nat. Phys. **6**, 888–893 (2010)
7. J. Bae, S. Kim, Identifying and ranking influential spreaders in complex networks by neighborhood coreness. Phys. A **395**, 549–559 (2014)
8. Q. Liu, Y. Zhu, Y. Jia, L. Deng, B. Zhou, J. Zhu, P. Zou, Leveraging local h-index to identify and rank influential spreaders in networks. Phys. A **512**, 379–391 (2018)
9. A. Zeng, C. Zhang, Ranking spreaders by decomposing complex networks. Phys. Lett. A **377**, 1031–1035 (2013)
10. R.V. Belfin, E. Grace Mary Kanaga, P. Bródka, Overlapping community detection using superior seed set selection in social networks. Comput. Electr. Eng. **70** 1074–1083 (2018)
11. L. Ma, C. Ma, H. Zhang, B. Wang, Identifying influential spreaders in complex networks based on gravity formula. Phys. A **451**, 205–212 (2016)
12. K. Singh, S. Singh, A. Luhach, A. Kumar, B. Biswas, Mining of closed high utility itemsets: a survey. Recent Adv. Comput. Sci. Commun. **14**(1), 6–12 (2021)
13. A. Kumari, R. Behera, K. Sahoo, A. Nayyar, A. Kumar Luhach, S. Prakash Sahoo, Supervised link prediction using structured-based feature extraction in social network. Concur. Comput.: Pract. Exp. (2020)

Security Threats in IoT: Vision, Technologies and Research Challenges

Anu Raj and Shiva Prakash

Abstract Internet of things is a group of numerous types of objects "things" that are embedded with sensors, software, and different technologies to communicate and share data with other devices over the internet. It supports a wide variety of applications which include smart homes, smart TV, smart car, smart city, smart agriculture system, smart health monitoring systems, emergency services, digital fingerprint systems, etc. It is a gigantic technology that enables us to connect with anything, anyone at any time and any network. Nowadays, IoT networks are susceptible to various security and privacy threats due to the heavy usage of IoT objects. These threats are becoming a bottleneck in using the IoT smoothly. There is an excessive need of deploying efficient security and privacy solution to ensure integrity, confidentiality, access control, and authenticity in IoT devices. This paper mainly focuses on security threats in IoT along with its vision, technologies, and research challenges. Various emerging technologies such as fog computing, edge computing, blockchain, and machine learning to increase the strength of security in IoT are also discussed. The paper also provides a comparative analysis of various IoT security techniques based on their benefits and IoT solutions. This framework will be useful to new researchers in uncovering the other security threat as well as finding efficient IoT security solutions.

Keywords IoT · Confidentiality · Authentication · Blockchain · Fog computing · Edge computing · Machine learning

1 Introduction

Internet of Things is a collection of various kinds of physical objects or devices that are normally connected over the internet which enable them to communicate with each other. It connects user to user interaction, so that they could communicate with each other while sitting in any corner of the world. In this new era of technology,

A. Raj (✉) · S. Prakash
Department of Information Technology & Computer Application, Madan Mohan Malaviya University of Technology, Gorakhpur, India

A. K. Luhach et al. (eds.), *Second International Conference on Sustainable Technologies for Computational Intelligence*, Advances in Intelligent Systems and Computing 1235,
https://doi.org/10.1007/978-981-16-4641-6_18

devices have become an imperative part of our life. The objective of IoT is to make devices able to take decisions or automatic systems with various types of latest technologies [1]. The advancement of IoT technologies in the world foresees various advances in smart homes, smart irrigation systems, smart health care systems, and other areas [2]. IoT lifts the connectivity and raises the popularity of devices. It supports a wide variety of applications which include smart homes, smart TV, smart car, smart city, smart agriculture system, smart health monitoring systems, emergency services, digital fingerprint systems, etc. Expert Systems, machine learning, and data mining are the main components of IoT applications. In recent years, billions of IoT devices have rapidly increased, with a prediction of over 50 billion physical devices connected to the Internet by 2020.

It is a big revolution in terms of technology that represents the future of computation and data-sharing technologies.

The very first vision of the Internet of Things was to tag physical things that uniquely identify the things using RFID readers. This technique was adopted within diverse business areas which was economically feasible. In today's world, a new concept has emerged to acquire the perception of identifying a worldwide framework of interconnected networks. The recent progress in the field of technologies has extended the vision of the IoT by incorporating other technologies such as wireless sensor networks, machine learning, cloud computing. In the current scenario, IoT is providing a real-time intelligence platform for the collaboration of various types of smart devices via the Internet. This progress is aspired not only by the success of RFID tag reader but also by the accomplishment of wireless networks, such as 5G, WiMAX, LTE, and its wide-area connection capabilities. It is axiomatic with the worldwide adoption of smart devices like smart TVs, smartphones, smart fridges, smart cities, etc. Additionally, the earlier form of IoT mainly emphasizes information services that provide the same through a centralized interface having remote access capabilities. Now recent service providers provide smartphone-based or server-based centralized systems to access sensor information all over the world. The development of IoT is mainly based on dynamic technical innovation in various fields which extends from wireless sensors to nanotechnology [3]. But due to the continuously increasing large number of devices, it may be possible to various security threats to networks as well as devices. IoT devices need to be fully secured and better quality of services for various IoT applications. Many research challenges are covering all the domain which includes various issues of data collection and processing, designing, management and using a multi-national, multi-industry, multi-technology framework, the business-related issues like developing new IoT models, and the organizational, political and social issues of innovative technology which promises to change the vision of our live-in a new direction over worldwide. There are plenty of technical challenges to overcome it requires a thorough evaluation of IoT solutions. The objective of this paper is to identify various security issues and research challenges for the future IoT which may help future researchers in this field. This survey will be helpful to researchers as a relevant source for security enhancement for upcoming IoT applications.

The paper is organized in this way: Sect. 2 of the paper discusses the literature review of the IoT. Section 3 is about various security issues and threats in IoT at all four layers of IoT application. Section 4 is about IoT security using various technologies such as fog computing, edge computing, blockchain, and machine learning. Section 5 of this survey paper provides a brief comparative study and discussion about the techniques which is previously mentioned in Sect. 4. Section 6 is described the various research challenges and issues. The last section of this paper is Sect. 7 which contains the conclusion.

2 Literature Review

The integration of smart devices over the Internet emerges various privacy, security, and various challenges because most of the Internet-related technologies were not supported IoT. This section gives an introduction to the venerability and survey on various security issues in IoT.

Nia and Jha [4] explained various security issues which can be foremost to destroy the IoT devices. The author introduced two new security issues in devices that can be exploited by the intruder which exponentially increases the number of vulnerabilities and unpredicted uses of information in the IoT environment. Anne et al. [5] surveyed mainly on the security issues and various research challenges related to IoT middleware and explained various aspects of existing protocols and their security issues in IoT devices. Bonetto et al. [6] surveyed cases related to secure connections that record the convention stack design. The author proposed a lightweight methodology to make one to one secure communication channel which requires negligible embracement of artificial IoT gadgets. Guizani et al. [7] explained various trust management technologies for IoT devices based on their merits and demerits. The author surveyed about what are the techniques used to secure the various IoT devices and networks. Sicari et al. [8] explained various solutions to privacy and security-related threats in IoT which are confidentiality, authorization, privacy, integrity, etc. The author summarized that most of the privacy and security threats in IoT are open problems that should be identified and disclose to protect the IoT devices from these privacy and security issues. Harris et al. [9] explored Lamina, a flexible framework that setup protection for client gadgets in cooperation with IoT spaces. Lamina furthermore ensures everyone in the IoT space can accumulate sufficient information to clients navigating the space to give Targeted information and organizations while at this point getting the client's identities. Chiang et al. [10] surveyed the research challenges in fog computing, focused on communication of IoT devices and their networking issues. Nowadays, fog computing covers both wireless as well as wired network scenarios, data processing, and sharing are done across both hardware and software, stays not only on edge network but also among end-users. O'Leary et al. [11] studied how the devices or object in IoT creates data; how it is utilized to produce semantic signs, and dissect the significance of utilizing 'Big Data', and the 'Internet of Things'. Riggins et al. [12] proposed how the joining of medical heath monitoring system and versatile application with tele-medication can improve the production and patient's experience. Dimitrov et al. [13] explored the behavioural, managerial, and business challenges of IoT that are essential for each person in the organization, MNC companies, and society. Wang et al. [14] investigated two ideas of verification in IoT initial one is Attribute-Based Encryption, and the other is Cipher content Policy Attribute-Based Encryption. Attribute-Based Encryption provides a public key encryption algorithm that allows a grained get to control, flexible key management, and information allocation. Table 1 describes the contributions of the past related works on IoT security. Though a lot of research works have been done

Table 1 Main contributions of the research work on IoT security

Year	Author	Technologies	Related work
2016	Mosenia and Jha [15]	Cloud computing	Emerged various challenges faced by the application layer of IoT
2017	Yu et al. [16]	Edge computing	Discussion on edge computing to provide security to IoT
2017	Lin et al. [17, 31]	Fog computing	Established relationship between IoT and fog computing
2017	Yang et al. [18]	IoT	Found new vulnerabilities of IoT objects
2017	Chen et al. [19]	LDS security algorithm	Explored security threats of location-dependent services
2017	Ngu et al. [5]	Work on middleware layer	Explained security problems associated with the IoT middleware layer. Obtained new security issues & challenges in IoT
2018	Farris et al. [20]	SDN and NFV security	Proposed work on trust management methods for IoT
2019	Din et al. [7]	Trust management techniques	
2019	Hasina et al. [21]	Trust manage techniques	Discussion on the various emerging security threats and current techniques emphasize accomplishing a foremost trust in the applications
2019	Meneghello et al. [22]	IOT	Compared and established relationship among various IoT technologies to ensure security features such as integrity confidentiality, remote access control, authenticity, and durability, etc.
2020	Sikarwar et al. [23]	Cloud platforms	Various types of IoT based cloud platforms are discussed depending on their applicability

in this field, there are some limitations in every aspect of IoT. This table shows the previous related work their corresponding author and their year of publication and what are security threats were emerging in IoT application.

3 Security Threats in IoT

There are four main layers in IoT architecture which are as follows: (1) sensing layer, (2) network layer, (3) middleware layer, and (4) application layer. Each layer plays a very important role in IoT applications. Every layer uses various techniques such as and different protocols such as Zigbee and Z wave that emerges new issue in the IoT application. In this section, different types of security attacks in IoT are described.

Sleep Deprivation, Booting, Node Capturing, Malicious Code Injection, etc. are the main security threats at the sensing layer [24]. The network threats arise at the network layer when the intruder access the control over IoT communication channel and hack device to cause damage. Website hijacking, DDoS/DoS Attack, Data Transit, Routing Attacks, etc. are security threats in this layer. The software attack happens at the middleware layer when IoT objects got some faults that enable the intruder to gain control over IoT objects and harm the network. Flooding attacks, cloud malware attacks, SQL injection attacks, and Men in the middle threats are the main attacks in this layer. Gateway is a hidden layer in the IoT application that connects a huge number of physical objects, things, and various cloud services [25, 32]. Secure Onboarding attacks, Extra Interfaces threat, Encryption threat, Firmware attacks, etc. are the main security threats. The last but not least layer, the application layer interacts with end-users and delivers services to them. There are various IoT applications which include smart cars, smart cities, smart irrigation systems, smart mirrors, etc. This is the most significant layer, but it also contains some security and privacy issues. Data Thefts, Access Control, Service Interruption, Malicious Code Injection Attacks, etc. are various security threats [26].

It is concluded that there is a requirement for proper security solutions on the Internet of Things because there is a huge amount of IoT objects which only focus on software security threats and do not consider hardware security issues. Figure 1 shows the security threats on different mentioned layers. There is various kind of threats that can exploit IoT objects. The physical attack happens on the sensing layer when the attacker is very near to the IoT devices, False Data Injection threat, eavesdropping and Interference.

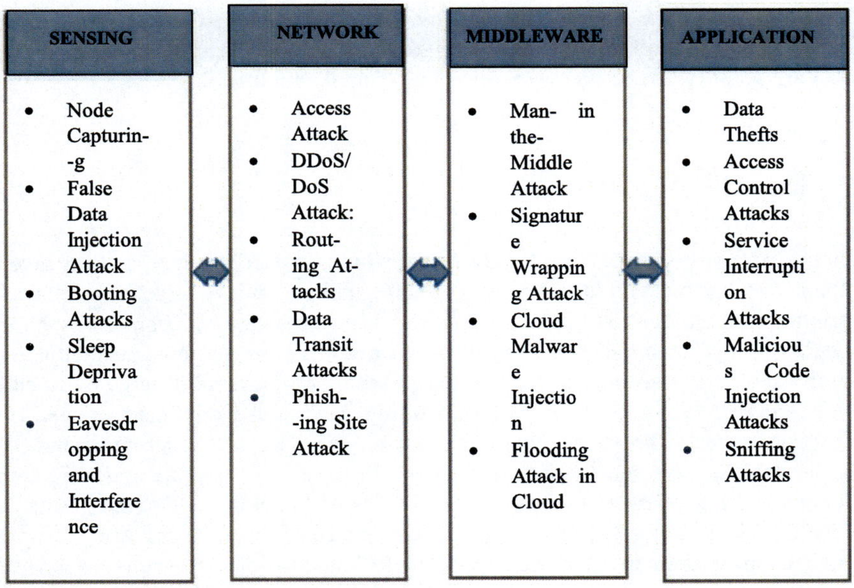

Fig. 1 Various kind of IoT threats [21]

4 IoT Security Using Various Technologies

This section contains IoT security using different current technologies such as blockchain and machine learning. These technological solutions play a very important role in enhancing the security of IoT application. These solutions are composed of four different categories which are as follows:

(1) Blockchain
(2) Fog computing [27]
(3) Machine learning
(4) Edge computing

Figure 2 shows various IoT security solutions using various techniques [28]. All four techniques are described below.

4.1 Blockchain

Blockchain and IoT are significant advancements that will profoundly affect the IT business. These two innovations target improving the general straightforwardness, permeability, level of solace, and trust for the end clients. The internet of things gives

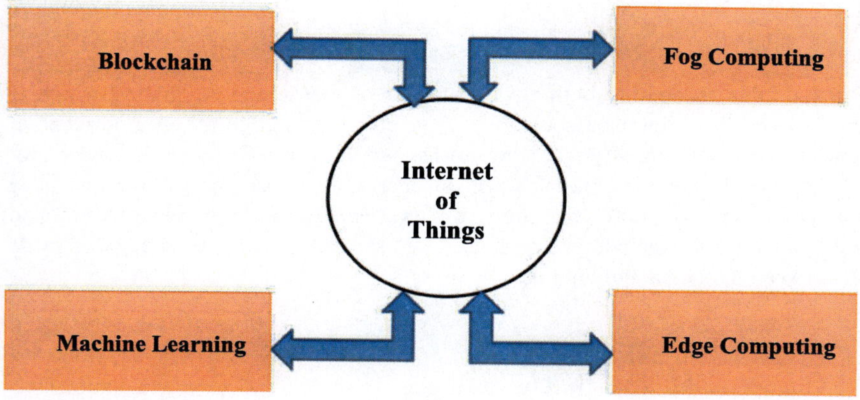

Fig. 2 IoT security using various techniques [21]

continuous information from sensors as well as blockchain gives security to board to the handled information utilizing a conveyed, decentralized, and shared record.

4.2 Fog Computing

It is a system that takes straightforward choices concerning the information that is gathered by the sensors though the intricate choices are taken on the cloud. Fog and distributed computing are two free advances that include numerous applications. Internet of things has furnished clients with various kinds of brilliant gadgets and their applications. Additionally, a cloud gives an extremely viable answer for store and oversee information that can be gotten to from any place and is broadly utilized by numerous associations. IoT is creating a remarkable measure of information, which puts a great deal of strain on the Internet organization. The incorporation of fog computing and IoT presents a time of new freedoms of devices and difficulties for handling, putting away, overseeing, and getting information all the more successfully.

4.3 Machine Learning

The areas of AI have pulled us in critical conditions over past years. A large number of areas use AI for their turn of events, and it is being utilized for IoT security issues as well. This technique can provide a solution for IoT objects against digital threats by developing an alternating way to handle these threats and protect devices from them.

4.4 Edge Computing

Fog and edge computing both are expansions of distributed computing which is mainly used by different associations. Fog, cloud, and edge all are comparable; however, they build up different layers of IoT applications. The cloud is sent at a lot bigger scope that necessities to deal with an enormous measure of information and is arranged at similarly more separation from its clients. To deal with the issues looked at distributed and edge computing is used as a solution where a little edge server is set between the client and the fog.

5 Comparison and Discussion

This section is representing a comparative study of various IoT security techniques based on their benefits and solutions as shown in Table 2. Various research issues and threats are also discussed corresponding to the IoT security techniques and their specific solutions arrangements on which the analysts are working upon and further investigations are going in the associated field. The given table shows the comparisons between various techniques based on their security issues in IoT, advantages, and possible solutions.

The above table summarizes the critical analysis of the different IoT security Techniques which includes blockchain, machine learning, fog computing, and Edge Computing. From the above table, we concluded that every IoT security technique has its method of securing data. More effective and reliable methodologies need to develop that can reach all the devices, as well as pr vent, abandoned utilization of computation power. Various kinds of Machine Learning and blockchain-based methods should be intended to verify the information in the devices to protect from threats.

6 Research Challenges

Internet of Thing provides a new dimension to the Internet and can add to broad monetary benefits but it also faces some difficulties. This section discusses some of the general research challenges in IoT.

Security and Privacy: Authentication, authorization, integrity, and encryption and decryption are considered open challenges in IoT, as a large part of the information being procured and conveyed, contains individual data [29]. All the communication between clients in the blockchain is public, so it may be possible that the private data of clients can be disclosed which can lead to a dangerous threat.

Table 2 Comparison among various IoT security techniques

Technology	Research issues and threats	Benefits	IoT solution
Blockchain	Privacy in IoT devices	Data encryption using the hash key	Utilize permissioned blockchain that can protect the IoT gadgets
	Heavy load on cloud service and services insufficiency	Information collected by IoT object store in blockchain	Data are updated on different nodes on the IoT network, so there is no single point disconnection of the network
	Data manipulation	Prevention from data loss and spoofing attacks	Due to blockchain technology, objects are interlocked them self, if any object refreshes information the framework rejects it
Fog computing	Decentralization	Server-aided computation	Verifiable computation, server-aided exchange, big data analytics
	Real-time services	To offer secure types of services, some authenticity methodology has been propose	Identity recognition, access management, intrusion detection
	Transient storage	To identify and secure private data	Recovery from attacks, data distribution, identifying and protecting sensitive data
Machine learning	Eavesdropping	Provide an efficient security mechanism for eavesdropping	Machine learning methods such as of floating strategy based on Q-learning and Bayesian techniques can be used
	Spoofing	Protection from spoofing attack	Distributed Frank Wolfe techniques, support vector machines, and deep neural network model
	Digital authentication	Provide security to IoT devices and provide trust to the users	Advanced fingerprinting is one of the forthcoming and promising arrangements

(continued)

Table 2 (continued)

Technology	Research issues and threats	Benefits	IoT solution
Edge computing	Security and accessibility	It increases the overall security of data at the user's edge level	Message queuing telemetry transport and message queuing telemetry transport are well known conventions for the edge computing
	Speed & latency	Edge computing overall increase the efficiency and speed of the services	With a decrease in organization inertness, continuous applications will profit by improved reaction time and more prominent in general client experience
	Distributed computing	Protect network from advanced persistent threats	In this mechanism, the computation and processing are provided at the end-user edge

Data Storage and Management: Data storage and management is an open exploration challenge to discover balance, which will be ideal for IoT objects. The accomplishment of AI machines relies upon these elements just as variety in selecting information. IoT applications are making a tremendous measure of information, and it is an important issue for spotless and pre-measure that information precisely.

Latency challenges: Based on the situation, an IoT organization may create a lot of information that needs more opportunity for preparing and calculation that possibly expand the general exhibition and leads to data latency issues in the organization.

Scalability: Devices collaborates inside an open environment. Essential features, for example, connection and service request/response, etc. require to work equally efficiently in both environments, small large-scale framework [30]. The Internet of Things needs a new feature and methodology to acquire an effective strategy for scalability.

Vulnerability: A large number of dangers which include malware and malignant code builds the test of recognizing, verifying, and forestalling them continuously in IoT organizations. A dedicated lightweight cryptographic solution is highly required for internet of things [31–33].

Resilience and Reliability: In industrial surroundings or emergency use cases temporary cases cannot be acknowledged. Consequently, versatility and unwavering quality issues in IoT should be explored from a general framework see and also contain angles like accessibility, strength, and adaptability of the correspondence and equipment [34]. The privacy and security using blockchain technology and lightweight cryptographic solutions for IoT security relies upon its methodology of implementation and the utilization of software and hardware [35]. These are various kinds of research problems that should be identified in near future for better IoT implementation in this world.

7 Conclusion

Internet of Things depends on Internet, sensors, cloud services, and various techniques which enable the communication possible numerous IoT devices by implementing different protocols. Now a day, IoT is everywhere and creating a rapid change in our everyday life. We summarized from the literature survey that there are several major issues such as security of devices during the connection process and security of the network in IoT. In this paper, we have presented various security threats at different layers of an IoT application. We also listed the security issues associated with the sensing layer, network layer, middleware layer, gateways, and application layer. We have also discussed the existing solutions to IoT security threats including blockchain, fog computing, edge computing, and machine learning. This paper concluded that there is a need to do many steps to improve the security of IoT devices such as authentication, secure booting by using a digital certificate, data encryption, and secure the applications to make only authorized users access data of IoT device. There is also a need to find a method to stop some kinds of attacks in IoT such as side-channel which presents a challenge that is not solved by researchers. Designing and implementing security in conventions that are basic for designers to utilize is an absolute necessity for the future of IoT. This paper will be helpful to serve as a valuable resource for security enhancement for upcoming IoT applications.

References

1. M. Mohammadi, A. Al-Fuqaha, M. Guizani, J.S. Oh, Semi-supervised deep reinforcement learning in support of IoT and smart city services: IEEE IoT J. **5**(2), 624–635, IEEE (2017)
2. P. Datta, A survey on IoT architectures, protocols, security and smart city based applications, in *8th ICCCNT* (IEEE, IIT Delhi, 2017)
3. S. Madakam, Internet of Things (IoT): a literature review. J. Comput. Commun. **3**, 164–173, Scientific Research Publishing (2015)
4. A.M. Nia, N.K. Jha, A comprehensive study of security of internet-of-things. Trans. Emerg. Top. Comput. 20–23, IEEE (2014)
5. A.H. Ngu, M. Gutierrez, V. Metsis, S. Nepal, Q.Z. Sheng, IoT middleware: a survey on issues and enabling technologies. Internet Things J. **4**(1), 1–20, IEEE (2017)
6. R. Bonetto, N. Bui, V. Lakkundi, A. Olivereau, A. Serbanati, M.Rossi, Secure communication for smart IoT objects: protocol stacks, use cases, and practical examples: world of Wireless, mobile and multimedia networks, in *IEEE International Symposium* (IEEE, 2012), pp. 1–7
7. I.U. Din, M. Guizani, B.-S. Kim, S. Hassan, M.K. Khan, Trust management techniques for the internet of things: a survey. IEEE Access **7**, 29763–29787 (2019)
8. S. Sicari, A. Rizzardi, L.A. Grieco, A. Coen-Porisini, Security, privacy and trust in Internet of Things: the road ahead, Comput. Networks **76**, 146–164 (2015)

9. A.F. Harris, H. Sundaram, R. Kravets, Security and privacy in public IoT spaces, in *25th International Conference on Computer Communication and Networks* (IEEE, 2016), pp. 1–8
10. M. Chiang, T. Zhang, Fog and IoT: an overview of research opportunities. Internet Things J. **3**(6), 854–864, IEEE (2016)
11. D.E. O'Leary, The Internet of Things and Internet of Signs. Intelligent Systems in Accounting, Finance, and Management, Wiley Online Library, **20**(1), 53–65 (2013)
12. F.J. Riggins, S.F. Wamba, Research directions on the adoption, usage, and impact of the internet of things through the use of big data analytics, in *48th Hawaii International Conference on System Sciences* (IEEE, 2015), pp. 1531–1540
13. D.V. Dimitrov, Medical internet of things and big data in healthcare. Healthcare Inf. Res. **22**(3), 156–163 (2016)
14. X. Wang, J. Zhang, E.M. Schooler, M. Ion, Performance evaluation of attribute-based encryption: toward data privacy in the IoT, in *47 International Conference on Communications* (IEEE, 2014), pp. 725–730
15. A. Mosenia, N.K. Jha, A comprehensive study of the security of internet-of-things. IEEE Transact. Emerg. Top. Comput. **5**(4), 586–602 (2017)
16. W. Yu, F. Liang, X. He, W.G. Hatcher, C. Lu, J. Lin, X. Yang, A survey on the edge computing for the internet of things. IEEE Access **6**, 6900–6919 (2018)
17. A. Raj, S. Prakash, Internet of everything: a survey based on architecture, issues, and challenges, in *5th Uttar Pradesh Section International Conference on Electrical, Electronics and Computer Engineering, MMMUT*, pp. 857–862, IEEE. pp. 2–4, Gorakhpur (2018)
18. Y. Yang, L. Wu, G. Yin, L. Li, H. Zhao, A survey on security and privacy issues in internet-of-things. Internet Things J. **4**(5), 1250–1258, IEEE (2017)
19. L. Chen, S. Thombre, K. Järvinen, E.S. Lohan, A. Alén-Savikko, H. Leppäkoski, M.Z.H. Bhuiyan, S. Bu-Pasha, G.N. Ferrara, S. Honkala, J. Lindqvist, L. Ruotsalainen, P. Korpisaari, H. Kuusniemi, Robustness, security and privacy in location-based services for future IoT: a survey. IEEE Access **5**, 8956–8977 (2017)
20. I. Farris, T. Taleb, Y. Khettab, J. Song, A survey on emerging SDN and NFV security mechanisms for IoT systems. Commun. Surv. Tutorials **21**(1), 812–837, IEEE (2018)
21. V. Hassajia, V. Chamola, V. Saxsena, D. Jain, P. Goyal, B. Sikdar, A survey on IoT security: application areas, security threats, and solution architectures. IEEE Access **10** (2019)
22. F. Meneghello, M. Calore, D. Zucchetto, M. Polese, A. Zanella, IoT: Internet of Threats? A survey of practical security vulnerabilities in real IoT devices. IEEE Internet Things J. (2019)
23. R. Sikarwar, P. Yadav, A. Dubey, A survey on IoT enabled cloud platforms, in *9th International Conference on Communication Systems and Network Technologies* (IEEE, 2020)
24. Y. Sharaf-Dabbagh, W. Saad, On the authentication of devices in the Internet of Things, in *17th International Symposium on A World of Wireless, Mobile and Multimedia Networks* (IEEE, 2016), pp. 1–3
25. C. Fife, Securing the IoT Gateway: securing-the-IoT-gateway/, online; Accessed 2019/02/09
26. G. Verma, S. Prakash, Internet of Things for Healthcare: Research Challenges and Future Prospects: Advances in Communication and Computational Technology (Springer Nature Singapore, 2021)
27. S. Srivastava, S. Prakash, Review on enhanced energy efficient routing protocol of WSN and use of IoT, in *International Conference on Recent Trends in Science & Technology* (Mysuru, India, 2020)
28. K. Christidis, M. Devetsikiotis, Blockchains and smart contracts for the internet of things. IEEE Access **4**, 2292–2303 (2016)
29. G. Verma, S. Prakash, A study towards current trends, issues and challenges in Internet of Things (IoT) based System for intelligent energy management, in *4th International Conference on Information Systems and Computer Networks* (Mathura, India, 2019), pp. 358–365
30. R. Sharma, S. Prakash, P. Roy, Methodology, applications, and challenges of WSN- IoT, in *International Conference on Electrical and Electronics Engineering,India* (Gorakhpur, 2020), pp. 502–507

31. J. Lin, W. Yu, N. Zhang, X. Yang, H. Zhang, W. Zhao, A survey on internet of things: architecture, enabling technologies, security, and privacy, and applications. IEEE Internet Things J. **4**, 1125–1142 (2017)
32. G. Verma, S Prakash, Design and implementation of modified unicode strategy for data security in IoT. Int. J. Adv. Sci. Technol. **29**, 6271–6294 (2020)
33. A.K. Luhach, A.K. Luhach, Analysis of lightweight cryptographic solutions for internet of things. Indian J. Sci. Technol. **9**, 1–7 (2016)
34. I. Batra, A.K. Luhach, N. Pathak, Research and analysis of lightweight cryptographic solutions for internet of things, in *Second International Conference on Information and Communication Technology for Competitive Strategies* (2016), pp. 1–5
35. A.K. Luhach, S. Kumar, Layer based security in Internet of Things: current mechanisms, prospective attacks, and future orientation, in *International Conference on Smart Trends for Information Technology and Computer Communications* (Springer, Singapore, 2016), pp. 896–903

Predict Foreign Currency Exchange Rates Using Machine Learning

Nilesh Patil, Sweedal Masih, Jeneya Rumao, and Veena Gaurea

Abstract Through trade, also known as import and export activities, countries around the globe have become more linked and interconnected in today's business world. The exchange rate is critical in allowing a monetary transaction to take place in a structured and coherent manner during a trading period. The support vector machine (SVM) has proven to be a successful method for estimating the financial markets after conducting research on various forecasting methods. In this research, the prediction of exchange rate of USD/AUD and other 59 countries was done. The forecast and analysis of the US dollar against other foreign currencies has been completed. This research has proven that SVM works best for relatively smaller data with respect to its accuracy.

Keywords Support vector machine (SVM) · Machine learning · Foreign exchange market · Exchange rates · Prediction

1 Introduction

The foreign exchange market is the world's most liquid and the biggest financial market. Exchange rate forecasting is a challenging application of modern time series forecasting that is critical to the success of many companies and financial institutions. The rates are deterministically unstable, noisy, and non-stationary by nature. In such instances, one common assumption is that the historical data includes all of those behaviours. As a consequence, historical data is the primary source of information used in the prediction process [1, 2].There are various number of forecasting techniques available, all of which are focused on various assumptions, mathematical frameworks, and model parameters. However, finding the right methodology for a given forecasting challenge is critical for a better outcome.

N. Patil · S. Masih (✉) · J. Rumao · V. Gaurea
Fr. CRCE, Mumbai, Maharashtra, India

N. Patil
e-mail: nilesh.patil@fragnel.edu.in

A. K. Luhach et al. (eds.), *Second International Conference on Sustainable Technologies for Computational Intelligence*, Advances in Intelligent Systems and Computing 1235,
https://doi.org/10.1007/978-981-16-4641-6_19

Machine learning approaches are useful in a variety of computer science applications, including business decision making, factory chain management, stock forecasting, etc. The main goal of machine learning is to teach computers how to learn to classify data from a given collection of data [3]. Machine learning, including supervised learning techniques, can help users evaluate large amounts of data from a variety of sources and extract valuable knowledge in the corporate world [4]. For example, given the currency exchange rate of data from transactions, supervised learning may predict the fluctuation that occurs over a period of time. The assessment of historical data is the subject of this forecast.

Supervised Classification is one of the machine learning models for efficiently constructing a statistical function f: x–y which links x to a projection y. This means that the provided set of data will have an x, y structure, with x denoting relevant data and y denoting the goal label [5, 6]. The aim of the classifier is to use historical data to forecast desired outcome.

For classification and regression tasks, almost supervised learning methods are used. These techniques are used for the development of Intelligent Systems with the help of AI-based techniques such as artificial neural networks, logic systems, and genetic algorithm [7]. If the goal variable (y) is in form of a class such as nominal or discrete, the task is called classification; if y is constant, the task is called regression.

SVM is a type of supervised machine learning algorithm. The Support Vector Machine (SVM) is a modern and powerful method for gaining knowledge from data and, in specific, for more effectively solving regression and classification tasks [8]. The key benefit of SVM over the NN system is its ability to reduce systemic risk rather than empirical risk minimization.

SVMs (Support Vector Machines) are based on the idea of a "margin," which is the difference between two data classes on either side of a hyperplane. It has been shown that increasing the distance and having the highest possible range between the separated hyperplane and the occurrences on both sides of it can decrease the predicted error rate [9]. As a consequence, SVMs can be used in case of conditional problems such as forecasting a Forex trend based on statistical data which is available as "Low" and "High" and in various time periods.

There are four parts in this paper. The Sect. 1 covers the fundamentals of machine learning as well as the Forex problem. Section 2 starts with a summary of the literature on the supervised learning methodology and how it was used to identify the Forex problem. Section 3 demonstrates the effects of implementing the support vector machine model to learn the patterns of data for the forex market. Section 4 is dedicated to the results and discussions. Section 5 is where the conclusion and future work is included.

2 Related Works

Machine learning's primary use is forecasting. The future predictions can be made using machine learning techniques combined with time series data. In the share market, machine learning techniques are used to forecasts to ck prices. Connor and Korajczyk, Fan, Liao, and Wang, Kelly, Pruitt, and Su have used machine learning techniques for analysis, preprocessing, and transformation [2]. There are various ML applications in forex prediction, in addition to the wide reach of these algorithms in stock return forecasting [10]. They can be expressed as a binary classifier and supervised machine learning models including support vector machines can assist in solving the Forex problem. The hypothesis is viewed as a classification problem issue, with theme thod determining whether or not the high-value rate is likely to occur. Ramakrishnan et al. discovered that Random Forests consistently beat Support Vector Machines and Neural Networks in estimating Malaysian FX when trained with commodity prices [11].

Despite its bad performance in the sign prediction experiment, the Random Forest model proves to be a good model to use in the benefits imulation experiment. It is the RF model, which has a high score in the NMSE and MAE matrices, that compares the actual exchange rate to the value predicted by our models [12, 13]. Although the RF method cannot reliably predict future exchange rate signs, it will consistently generate a predicted value that is reasonably similar to the true value. As a result, when compared to other versions, the RF approach is more adaptable.

The two main financial forecasting methodologies are technical and fundamental assessments. Technical research has received a lot of attention lately, thanks to mounting evidence that markets are less effective than previously believed. As technical data, a moving average with a time delay is used [3]. A moving average has the advantage of mapping out any of the inconsistencies that occur during market days. This model was created using average values from previous weeks to estimate the rate for the next week [3]. The model's forecasting efficiency is assessed using three commonly used statistical metrics: Normalized Mean Square Error, Mean Absolute Error, and Directional Symmetry.

The linear kernel is clearly superior in very few of the results, according to the trend forecast [14]. In terms of directional transition, it was noticed that the radial kernel performs better for JPY, NZD, CHF and the polynomial kernel performs good for the USD and SGD.

The time series forecasting is properly analyzed using ARIMA model. ARIMA is a general model that assumes the time series being forecasted are linear and stationary. One of the most difficult applications of modern time series prediction is currency exchange forecasting, according to [8]. This is because rates are permanently disruptive or non-stationary [15]. Neural Networks have been used to predict the future rates of stocks in the forex sector, as well as to identify electro encephalogram (EEG) [6, 16, 17]. The results showed that using training data methods to predict classification problems can be advantageous.

Several publications have demonstrated that SVMs can be implemented to solve multivariate regression prediction problem, demonstrating that they can be used to solve forecast accuracy or classification errors in a variety of fields [18, 19]. According to [20], there are minimum studies on the use of SVM in forex prediction. Genetic Algorithms and SVM algorithm have recently been combined into foreign exchange trading, according to research published in [2]. For the reasons mentioned above, to forecast the growth or decline of currency values in the financial sector, the SVM based model is used.

3 Methodology

3.1 Supervised Support Vector Machine

The Supervised machine learning technique which is used in this paper uses dataset as input and outputs the patterns recognized to calculate a desired goal. The basic principle of SVM is the search for the best hyperplane that represents as a separator of two classes in the input space [9]. The support vector machine algorithm's goal is to find a hyperplane in an N-dimensional space.

Boser, Guyon, and Vapnik created SVM, which was first introduced in 1992 at the Annual Workshop on Computational Learning Theory. SVM is fundamentally a cohesive combination of decades-old mathematical theories such as margin hyperplane, kernel, and concepts supporting others [9]. However, until 1992, no initiative has been taken to bind these elements together. The margin which has the maximum distance between data points in both classes is called as maximum margin in that plane, is one of several possible hyperplanes for separating the two types of data points. The margin gap can be increased to provide some assistance.

Based on whether the training data set is divided into linear or non-linear hyperplane lines, Support Vector Machines (SVMs) are categorized as linear or non-linear. The formula for a linear line in linear hyperplane is:

Following is a hyperplane which can be written as the set of points x satisfying the function:

$$\mathrm{H}: \mathrm{w} \cdot \mathrm{x} + \mathrm{b} = 0$$

Given is a hyperplane H0 that separates the dataset and satisfies below equation

$$\mathrm{w} \cdot \mathrm{x} + \mathrm{b} = 0$$

Consider two hyperplanes H1 and H2 that also separates the data and have the following equations:

$$w \cdot x + b = \delta \text{ and } w \cdot x + b = -\delta$$

so that H0 is equidistant from H1 and H2.

Figure 4 shows an example of linearly separated data. Measure the hyper plane that separates the two classes using support vectors and find the maximum points to find the best hyper plane between two groups. Margins along the hyperplane are defined as the distance between the hyperplane and the closest data points of each group, also known as the support vector. The best hyperplane is defined in the equation below.

$$f(x) = w^T x + b$$

- w is a weight vector
- x is input vector
- b is bias

3.2 *System Design*

The proposed model employs the SVM algorithm to train on the collected datasets, and the trained data yields a predictive model that defines various features that are extracted during model evaluation.

Block diagram:

According to the block diagram, the process starts with following:

Input: The dataset is from 1950 to 2019, and it is being prepared for training and then forecasting future patterns. For training and research, foreign currency exchange rates from 60 different countries are gathered.

Training: The extraction/collection of raw data, which is the most critical and important step. This data is modelled for feature extraction using identifiers, which are the model's classifiers. Furthermore, the model is trained using a machine learning algorithm based on the data obtained. This is referred to as the Training phase. The model is further evaluated based on the patterns/trends from the collected data after training the data.

Prediction: After applying the appropriate algorithm to the model during testing, a predictive model is generated. The algorithm is chosen based on the dataset size. Following the training of the dataset, future patterns are projected using feature vectors and labels, and the predictive model generates the results as shown in Fig. 1.

Results: The result is a forecast of potential USD dollar prices against the currencies of all 60 countries.

The detailed steps of working of supervised machine learning algorithm that is the Support Vector Machine (SVM) are as follows:

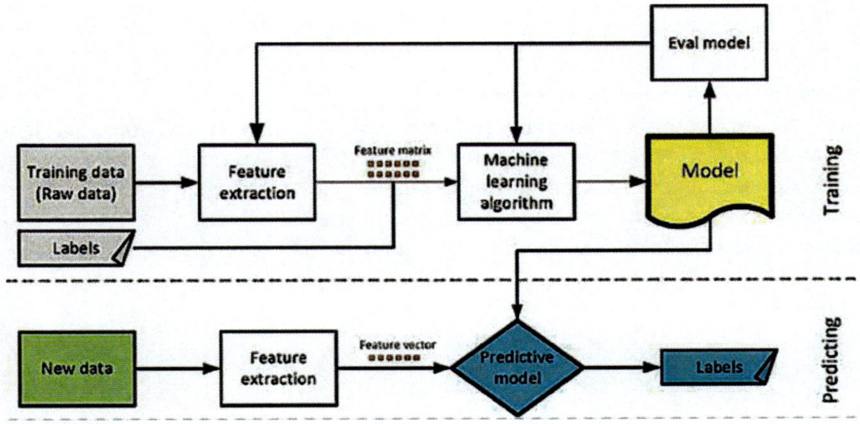

Fig. 1 Block diagram

Step 1: The first step is to choose your features. It is possible that the data set gathered contains features that are not appropriate for the prediction process. As a consequence, these features have been disabled.

Step 2: Choose a currency pair's dataset. In this step, a particular currency pair, such as USD vs. AUD, is selected. Also in this process, data is gathered from various time frames, online sources, and other resources.

Step 3: Using supervised machine learning (SML) models to process data. The dataset is divided into three sections: preparation, testing, and validation. All data is mapped to the [0,1] set.

Step 4: Evaluation models of function vector evaluation metrics are used. The resulting production can have some kind of upward or downward trends.

4 Results and Discussions

4.1 Dataset

The Forex data required for the SVM technique prediction was obtained from Kaggle, an open source community that allows users to find and collect datasets. From 1950 to 2019, the dataset was downloaded for 60 different countries, forecasting the value of the US dollar against other currencies in the economy. This USD/AUD dataset is commonly used in the forex industry. The training data included instances from 12/01/1950 to 23/02/2019.

4.2 Results

Figure 2 represents the Australian dollar's foreign currency exchange rate against the US dollar. The Support Vector Machine (SVM) algorithm is used to predict the forex rate.

Figure 3 shows Google's real-time currency exchange rate prediction for the Australian dollar versus the US dollar. The value predicted in our research using the Support Vector Machine (SVM) algorithm is nearly identical to the forex rate predicted by Google.

When compared, our predicted value using the algorithm (Fig. 2) to Google's real-time predicted value (Fig. 3), it can be seen how accurate the Support Vector Machine (SVM) algorithm is at forecasting foreign currency exchange rates.

The data is trained, and the features are selected based on currency rate inflation. In this paper a common pattern in these data for different countries can be seen, and it has also discovered both an uptrend and a downtrend in some of the data. The key concept behind the SVM algorithm is that it tries to find the best hyperplane based on the labelled data that can be used to identify new data points.

The hyperplane is a straight line in two dimensions. This aided in determining the best fit line for the data. The algorithm tries to find the best hyperplane that separates the classes using these support vectors. By representing the best features of the data in 2d or 3d graphs, visualization makes it easier to study and analyze them. The

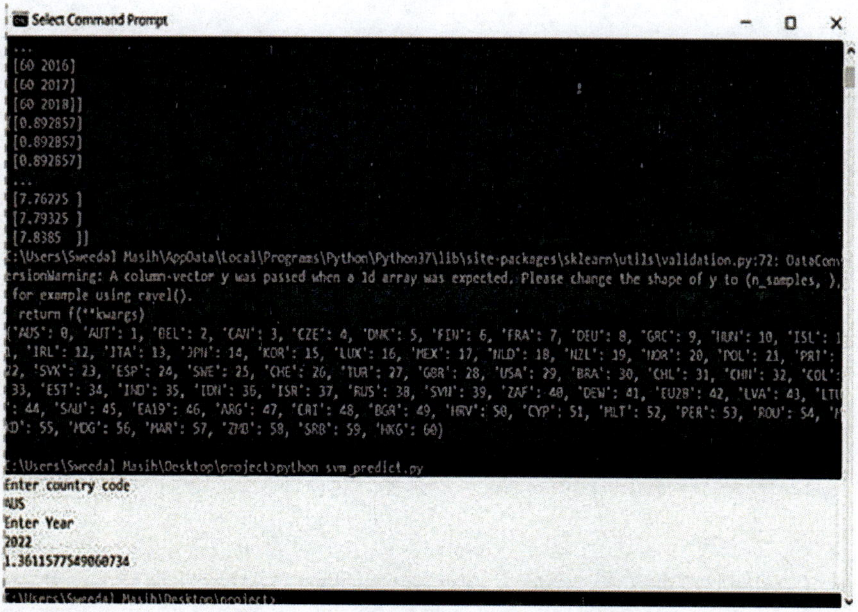

Fig. 2 Currency rates of Australian dollar against US dollar

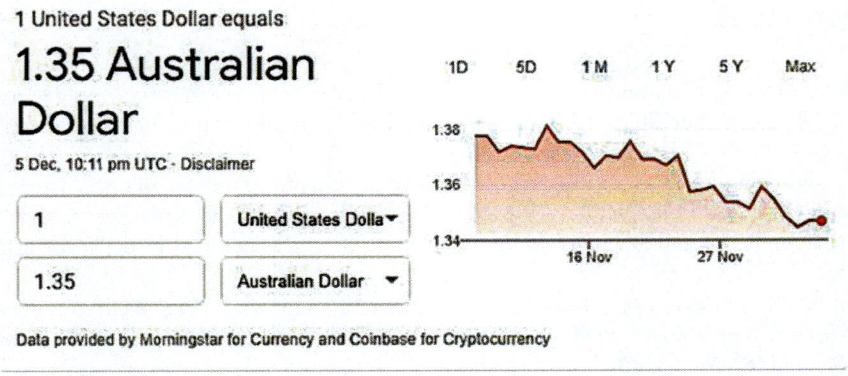

Fig. 3 Google's real-time USD/AUD exchange rate prediction

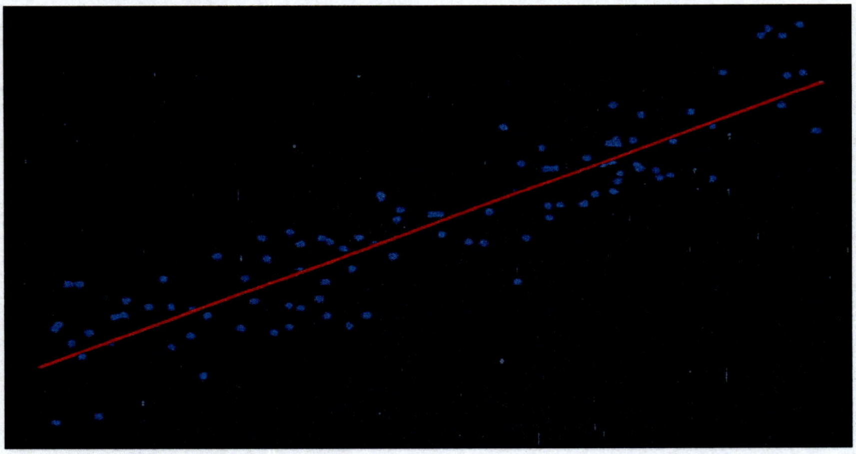

Fig. 4 Graph showing the best line fit

graph below depicts the data. Below is the graph plotted that shows the best fit line for currency (Fig. 4).

4.3 Discussions

The above results are based on the SVM regression model, which produces the best line that connects the data points to form a hyperplane. The best line, intuitively, is the one that is farthest away from both data points (has the largest margin). To arrive at the best solution, it is a must to optimize the margin in both directions (if multiple classes are there, then it should be maximized considering each of the classes).

Tabula, a model visualizer tool, was used to build the graph. The methods that are used to visualize data are known as model visualizers. It is necessary to imagine the training process during model training as this helps to understand how models can be debugged and to monitor errors and progress. At each step, simulate the test and training outcome values. Tabula is a free open-source application that allows you to "liberate data tables from PDF files." It is used to collect details for a visualization research.

The performance of favourable and unfavourable prediction for Testing dataset is significantly lower than that of Training datasets, as seen in the graph above. The performance rates of training and research, on the other hand, vary slightly.

Furthermore, the linear kernel SVM is built with Python's Sklearn library and the Iris Dataset, which is contained in Python's dataset library. When the data is linearly separable, i.e., when it can be divided by a single line, the Linear Kernel is used. It is amongst the most widely used kernels. It is mainly used when there are a lot of features in a single dataset. Text Classification is an example of a feature with a lot of features since each alphabet is a new feature. As a result, Linear Kernel is commonly used in Text Classification.

In this paper, it is observed from the outcomes that a linear kernel is faster than other kernels. As a result, this kernel feature achieves accuracy rates of about 70% with training data and 30% with testing data.

5 Conclusion and Future Work

Financial forecasting is a difficult task that is still being researched. Many algorithms were created with the aim of making accurate predictions and profit. The majority of these algorithms are only good on paper, but they do not work in practice. This research aims to examine the historical trend of US dollar exchange rates against various country currencies in order to forecast future values. This system is designed with a precise analysis model, and after a comparative review, it is concluded that the SVM model gives accurate results for smaller dataset. In this study, US Dollar rate was predicted against AUD and other 59 countries. The US dollar is the most commonly used currency in the world, and datasets are available for all recognized countries worldwide.

This research aims to examine the historical trend of US dollar exchange rates against various country currencies and, as a result, forecast future values. Foreign currency exchange rates can also be predicted using two more algorithms in this research: Random Forests and Decision Trees. Then, based on their findings, a comparative analysis can be performed to decide which algorithm is the best at predicting foreign currency exchange rates.

References

1. A.A. Baasher, W.F. Mohamed, FOREX trend classification using machine learning techniques. In: *Arab Academy for Science and Technology* (2010), pp. 41–47
2. C.G. Rojas, M. Herman, CS299: foreign exchange forecasting via machine learning (2018)
3. T.N.T. Thu1, V.D. Xuan, ThuongMai University, Mai Dich, Cau Giay, Hanoi, Vietnam 2 Sonla Gifted HighSchool, Sonla, Vietnam.: FoRex Trading Using Supervised Machine Learning
4. F.E.H. Tay, L. Cao, *Application of Support Vector Machines in Financial Time Series Forecasting* (2001), pp. 309–317; Omega 29
5. D. m. Nemeú, A. Butoi, Data mining on romanian stock market using neural networks for price prediction. Inf. Econ. **17**(3), 125–136 (2013)
6. H.A.R. Akkar, F.B. Ali Jasim, Intelligent training algorithm for artificial neural network EEG classifications. Int. J. Intell. Syst. Appl. **10**(5), 33–41 (2018)
7. W. Huang, Y. Nakamori, S.Y. Wang, Forecasting foreign exchange rates with artificial neural networks. Int. J. Inf. Technol. Decis. Mak. **3**(1), 145–165 (2004)
8. J. Kamruzzaman, GSCIT, Monash University School of Computer Science. Rahul A Sarker, Monash University Churchill, Victoria. Iftekhar Ahmad, Australia University of NSW.: SVM Based Models for Predicting Foreign Currency Exchange Rates. Canberra, Australia. (December 2003)
9. S.W. Sidehabi, S. Tandungan, *Statistical and Machine Learning Approach in Forex Prediction Based on Empirical Data*, vol. 978 IEEE (2016), pp. 5090–5548
10. A. Baasher, M.W. Fakhr, Arab academy for science and technology: FOREX daily trend prediction using machine learning techniques. In: Cairo/Computer Science Department. Egypt (2011)
11. S. Ramakrishnan, S. Butt, M.A. Chohan, H. Ahmad, *Forecasting Malaysian Exchange Rate Using Machine Learning Techniques Based on Commodities Prices*, pp. 1–5 (July 2017)
12. J. Zhang, J. Sun, CS229: foreign currency exchange rate forecast. (June 2019)
13. M. Fathian, A.N. Kia, Exchange rate prediction with multilayer perceptron neural network using gold price as external factor (2011)
14. S. K. Ramakrishnan, S. Butt, M.A. Chohan, H. Ahmad.: Forecasting Malaysian exchange rate using machine learning techniques based on commodities prices, in *IEEE International Conference on Research and Innovation in Information Systems (ICRIIS)* (2017)
15. G.J. Deboeck, *Trading on the Edge: Neural, Genetic and Fuzzy Systems for Chaotic Financial Markets*. Wiley, New York (1994)
16. H. He, X. Shen, Bootstrap methods for foreign currency exchange rates prediction, in *Proceedings of 2007 IEEE INNS Joint Conference on Neural Network, Orlando, FL, USA* (2007), pp. 1272–1277
17. P. Thakar, A. Mehta, A unified model of clustering and classification to improve students employability prediction. Int. J. Intell. Syst. Appl. **9**(9), 10–18 (2017)
18. W.-S. Gad, K.-H. Ng, Multivariate FOREX forecasting using artificial neural networks. IEEE Xplore (2010)
19. P. Dell Acqua, F. Bellotti, R. Berta, A. De Gloria, Time- aware multivariate nearest neighbor regression methods for traffic flow prediction. IEEE Trans. Intell. Transp. Syst. **16**(6), 3393–3402 (2015)
20. K. Kim, Financial time series forecasting using support vector machines. Neurocomputing **55**, 307–319 (2003)

Software Defect-Based Prediction Using Logistic Regression: Review and Challenges

Jayanti Goyal and Ripu Ranjan Sinha

Abstract Software analysis and prediction system development is the significant and much-needed field of software testing in software engineering. The automatic software predictors analyze, predict, and classify a variety of errors, faults, and defects using different learning-based methods. Many research contributions have evolved in this direction. In recent years, however, they have faced the challenges of software validation, non-balanced and unequal data, classifier selection, code size, code dependence, resources, accuracy, and performance. There is, therefore, a great need for an effective and automated software defect-based prediction system that uses machine learning techniques, with great efficiency. In this paper, a variety of such studies and systems are discussed and compared. Their measurement methods such as metrics, features, parameters, classifiers, accuracy, and data sets are found and discriminated. In addition to this, their challenges, threats, and limitations are also stated to demonstrate their system's effectiveness. Therefore, it was discovered that such systems accounted for 44% use of the NASA's PROMISE data samples, 68.18% metrics use of software, and also 16% use of the Logistic Regression method.

Keywords Prediction systems · Fault-proneness · Defect analysis · Machine learning · Logistic regression

1 Introduction

Automated software fault- and defect-based analysis and prediction is an important area of software engineering in the current environment. Decades ago, many such predictors and analyzers have been introduced in the industrial, corporate, and commercial spheres, but still there is a great need for an accurate and effective defect-based prediction system. During these recent years, the need for these automated models using Machine Learning (ML) techniques has grown significantly [1–8], which includes a foundation of Natural Language Processing (NLP) area

J. Goyal (✉) · R. Ranjan Sinha
RTU, Kota, Rajasthan, India

A. K. Luhach et al. (eds.), *Second International Conference on Sustainable Technologies for Computational Intelligence*, Advances in Intelligent Systems and Computing 1235,
https://doi.org/10.1007/978-981-16-4641-6_20

and Artificial Intelligence (AI) field. Software error-based prediction and analysis systems follow the steps for data acquisition, pre-processing, feature selection, metric selection, classifier training, fault prediction, and performance testing. Performance measures say the accuracy, the recall, the precision, the AUC, the clarity, and the F-measure were tested mainly in these predictors, and their studies were framed to focus on a variety of Research-based Questions (RQ).

So a review-based research contribution is designed in this paper to critically analyze various predictive systems that exist using Logistic Regression (LR). The statistical method called logistic regression is used to predict different stages, so it enters the data into an input curve. Further, LR offers the better solutions of the problems than many other methods of ML for real-time domains and applications. This review-based study next discriminates against a variety of existing systems according to six basic measurement criteria and analyzes them statistically. It also demonstrates the challenges and threats in existing approaches that bring the requirement for an automatic and effective defect-based prediction system with the base of a set of classifiers.

The authors have organized this paper in five sections. Section 2 provides the journey and systematic tour of various predictors and analyzers and illustrates these studies for the existing software systems. It compares them based on the metrics, features, parameter-sets, and classifiers. Section 3 outlines their challenges, limitations, and threats and provides a comparison of data sets with the results of the accuracy of existing systems. Section 4 sets out the performance metrics for these algorithms using % usage. Finally, Sect. 5 concludes the article with future indicators.

2 The Journey of Existing Works

The literature review presents a discussion on the various prediction models available for software that demonstrate their theoretical features and process using a variety of ML methods, particularly LR. It provides the basis for these latest and recent methods, their importance on the basis of evaluating parameters, and their applications to the practical problems.

Beginning with this journey, first software error-based analyzer and predictor method used three models, which were forward stepwise logistic regression, reverse logistic regression backward, and without the step option in logistic regression [9]. Its measure included data acquisition, the release of descriptive statistics for data use of backward logistic regression and forward logistic regression for input detection, production of error-based predictions, and lastly the training and the testing. It got very promising results of the background alarming rate with LR back and got the excellent results for sensitivity by step-based LR transmission. Next feature-oriented prediction system [10] has used the Select RUS Boost learning ensemble-based method to reduce features selection and class inequality by solving the problem of big dimensions. For this, it has incorporated feature selection and random selection methods under Sampling (RUS) with a merging (extension) process. It made a total of ten validations along with the fixed values with EMERALD method. It has obtained better results than the + RUS measurement method with Support Vector Machine (SVM), while it overrides the simple integration method with at K-Nearest Neighbor (KNN). This method got the bad side effects with no sample. Here the SRB-35 surpassed the RUS35 via SVM by a value of 0.05 in all cases in SP-1, SP-2, and SP-4, except for those in SP-1. These methods showed similar behavior in SP-3 except for the GR standard; however, the SRB-35 outperformed.

Reddy and Babu [11] introduced the reliable growth model for software failure-based prediction that measured different limits at the beginning of the testing and correction phases. It includes an additional decision-oriented concept in the test to enable it to manage its resources and deadline for the delivery of soft goods. Panichella et al. [12] used a software approach to predict the software failures with six different machine learning methods. Results were statistically unequal; therefore, the Combined DEfect Predictor (CODEP) was applied to prevent the excavation compliance. It created the data suspension by converting software metrics into new z-distribution. After that it reduced the effect of heterogeneity among different projects. Further, it had gone through ten validations and gained much better prediction accuracy using Bayesian Networks (BN) and LR than stand-alone approaches. Next it detected the cost-effectiveness by using a hybrid classifier.

Another method [13] estimated the effectiveness of binary model and multinomial model using integrated metrics in the composite object and then identified independent classes. Multinomial categorization were divided into four classes and obtained numerical calculations for different error cases, min, 25, 50, 75%, and magnitude. In addition, it has performed 18-fold validation of multi-variate software models, got excellent results in a multi-variate software model with multinomial error prediction, and found the practical results in the prioritization of different error tests. It got excellent AUC results with Eclipse-2.1. The integrated ML approach [14] included active learning, K-means integration, and algorithms of slightly monitored reading to differentiate the errors. It followed the steps of initial pre-processing including tokenization, stop word deletion and stemming; code processing; segmentation; feature extraction; and training and fault prediction. Defects and faults are labeled using active learning applications and distinguish them from data control and movement, structural defects, and code variations.

Next method [15] analyzed the performance of errors using six separate and deferred separators. It removed EPV > ten data-sets and 80% modules with feature. It repeated the procedure 1000 times with a sample-oriented bootstrap validation. It then calculated the results of AUC and included the value results for both the classifiers. With this, the random forest-based planning of the forest outperformed the discretization-based random division of the forest. Another feature-oriented prediction model used the developer's time period to convert the code changes over time to the source file. After this, it used ten folds of validation. It well-adjusted granularity at the file level works well than the page granularity measures at commit level. All this escalated when one made the changes to most of the files commits [16].

The fault-based prediction method in [17] used the feature selection after and the noise-filtering and also verified it using Chi-Square test, relief, information gain, ratio of gain, and uncertainty associated with the Maximum Likelihood LR (MLLR). Further [17] used MATLAB R2016a, WEKA 3–8, R language, and KEEL software tool using ten folds of validation. Further it got better findings in MLLR with feature-based selection method. Similarly it found better results for after-noise-filtering method than for before-noise-filtering method. Here Random Forest (RF) approach achieved better results than Multi-Layer Perceptron (MLP) and KNN strategies. In addition to this, this method obtained defect-based scores such as 6.78 in CM1; 6.16 in PDE; 3.48 in ANT; 3.84 in JDT; 32.3 in PC5; and 10.14 in Tomcat. Another proposed work tested and discriminated the classification and predictive capabilities of six machine learning methods. Here logistic regression obtained various software components. This method used ten folds of validation and also the Friedman tests to detect algorithm-based performance and error categories. It found LR and RF and LR as the worst and best alternatives, correspondingly [18].

Research based on software prediction [19] followed the data acquisition steps converting data into numbers; select a feature using the Select-K-Best method; modeling, and training, and then testing using various classifiers; and lastly performance-based analysis. Agarwal et al. [19] obtained the excellent performance on KNN having Chi-square and Select-K-Best algorithms. It detected the system efficiency using Principal component Analysis (PCA) as an extraction process. In [20], a method of selecting the cluster-based filter-wrapper features was proposed, which defined the quality coefficient of the spectral interactive factor to measure the feature redundancy and importance. It used forward options in single feature mode to find the final-level feature set. Further, it has shown much-better results than the traditional methods of calculating efficiency and accuracy.

The software prediction-oriented study and research [21] used three types of selected methods features called a single-factor rank using three subset sizes, a filter-based evaluator (relative to consistency and correlation), along with a wrapper Selection-oriented feature with learners. Next this method [21] used the WEKA for getting the system analysis with ANalysis Of VAriance (ANOVA) along with Tukey's process for significance. It saw the excellent and bad performance with the wrapper-based subset and the feature-based rank methods, correspondingly. It also obtained bad and excellent results of AUC with SVM and LR, correspondingly. Similar to [21], the approach [22] first estimated the training inequality faulty data and then classified and predicted the defective modules. Further it has used Synthetic Minority-Oversampling-TEchnique (SMOTE) for creating positive and negative classes. This method got a fault ratio from 0.7 to 32.79 and then got XT as the excellent category [22].

Another proposed system used WEKA for data inputting, the preliminary analysis, the classification and prediction, and the testing [23]. Next this system developed many risky models with the metrics of MI and mP for defining the erroneous probability-base along with impact errors, respectively. Then it operated tenfold validation, achieved good results predicting binary errors, achieved excellent results with a risk-based model, and obtained excellent formulation with MLP and RF. Then it got very promising results of the ANT-1.3 segmentation. Next method had tenfold validation for specific error prediction in the concurrent-related systems by joining twenty-four dynamic and static metrics. Further it predicted their concurrent-related errors, got a set of source code metrics for similar applications, and used dynamic metrics to predict error [24].

Next Model [25] initiated with the pre-processing steps including identification of lost and missing values and their replacement, data quality selection, dependent and independent variable declaration; statistical analysis having critical assessment and collinear tests; and analysis of possible predictions of selected parameters using classification strategies. It collaboratively analyzed the developed data of project using LR, and then it obtained the customer satisfaction in a group of 71 previously described features. The software predictor [26] included the measures for data acquisition, pre-processing; normalization, confused matrix production, and conditional use of PCA. It used five-step validation, achieved moderate PC1 accuracy levels in existing algorithms, and also found a fairly good performance in the RF cum PCA way.

In [27], the model showed the failure-factor relationships. Elements of failure in the content-oriented analysis of every project were initially identified. Further it also developed a fault prediction model and training database and finally validated it during the test. Another predictor included the voting- and stacking-based ensemble approaches to significantly improve the prediction results. Three re-sampling methods such as random sample, SMOTE, and random under-sample were used. It used ten-folds of cross-validation. Next it did the non-parametric Friedman and ANOME testing to find the significant results. So, it was noted that the modeling model surpasses other ensemble-based methods [28].

Another alternative method [29] introduced a 'Multi-Kernel Transfer Convolution Neural Network (MKTCNN)' and 'Cross-Project-based Defect Prediction (CPDP)' algorithm with AST of node-based granular strategies. It calculated the difference between the granularities of the three AST nodes and also compared the performance of the prediction to each granular scenario. Then this method has worked upon the PROMISE databases of NASA and also found that the CPDP approach goes beyond different in-depth learning approaches. With this, the process of MKTCNN has done well by reducing data on the source and targeted projects. Another such method used the steps for data inputting and preliminary analysis, use of appropriate performance metrics, determination and use of predictable bases, and the selection of ensemble classifiers. Then it used the WEKA suite along with ten-fold of validation to demonstrate the performance of ten ensemble-based predictors [30].

This tour explained and compared the processes of many of the most recent automated defect-based prediction algorithms and analyzers using LR and other ML techniques. This journey worked out that these existing procedures used multiple sets of metrics and attributes. They had a procedure to classify and predict the errors. To demonstrate the effectiveness, these approaches are discriminated based on some parameters.

Table 1 shows this comparison using five principles, called as, problem analysis, attributes, metrics, setup values, and classification. All methods have primarily used the classification techniques such as LR, Linear Regression (LinR), Defect Prediction CNN (DP-CNN), BN, Multinomial Naive Bayes (MNB), Naive Bayes (NB), SVM,

Table 1 Discriminating various existing prediction systems with five primary concerns

References number	Problem analyzed	Classifiers	Parameters, metrics and features
[9]	Prediction of defect proneness in various software-based modules	LR	Parameters p1 as 0.05 and p2 as 0.1. Used software metrics
[10]	Reduction of high-dimension and class imbalance problems using software defect-based predictor	SVM and KNN	Had significance level was 0.05 and AUC range was zero to one. Used six feature-oriented selection methods
[11]	Software reliability model with failure prediction	LR	It had overlapping chains
[12]	Cross-project defect prediction model with ML	AD Tree, LR, DT, RBF Network, MLP, and BN	PCA. Metrics were overlap, LOC, and Chidamber-Kemerer (C and K)
[13]	Efficiency comparison of software fault prediction	Binary and Multi-Nomial methods of LR	Object-Oriented (OO) metrics
[14]	Active semi-supervised defect prediction method	Multi-class type of SVM	Used fifty labeled examples
[15]	Design of defect-based predictors and classifiers using regression models	SVM, ANN, RF, LR, LinR, KNN, and CART	Feature clustering and redundancy analysis
[16]	Use of periodic developer-based metrics in software defect-based predictor	J48 in DT, IBK with KNN, RF, LR and NB	Correlation-based feature selection of WEKA API
[17]	Enhanced defect prediction using hybrid pre-processing and LR	MLP, KNN with K as 5, and RF	It used feature selection methods
[18]	To analyze the ML algorithms effectively for finding the erroneous clusters	J48 in DT, NB, RF, Bagging, Adaboost, LB and LR	OO metrics. Set Friedman test cut-off at 0.05
[19]	Review for software defect-based prediction system with statistical type learning	NB, DT, RF, KNN, and GB	Select K-Best as feature selection method

(continued)

Table 1 (continued)

References number	Problem analyzed	Classifiers	Parameters, metrics and features
[20]	Predicting the software for defects with cluster-oriented hybrid method of feature selection	RF, DT, and KNN	Feature quality coefficient, feature correlation and rank
[21]	Predicting the software for defects with metric selection	LR, MLP, KNN, NB, and SVM	Feature selection metric. 42 metrics. Cut off was 0.05
[22]	Prediction of erroneous modules of software using stack	AdaBoost, ET, Stacking, RF, DT, and XGB	Feature selection and software metrics
[23]	Prediction of software errors with the concept of probability	LR, NB, RF, J48, MLP and SVM	Object-Oriented metrics of code metrics in 'C and K'. Parameter value is 0.001
[24]	Defect prediction with concurrency for different real-world apps	RF, NB, DT, and LR	Code metrics based upon concurrency
[25]	Analyzing the prediction models in statistic mode along with user-satisfaction	LR for collinear analysis. NB, LinR, LR, ANN, RF, GBT, DT, KNN, and Nearest Neighbor	The identifier had less than 0.1 confidence at p-value
[26]	Defect prediction using ML approaches and their comparative analysis	Bagging, NB, DT, MLP, SVM, KNN, ET, RF, GB, and AdaBoost	Used PCA for reduction of twenty-one features to fifteen features
[27]	Prediction of faults and defect in software	Logistic Regression	Software metrics
[28]	Use of ensemble method for prediction of faults and defects in software	KNN, MNB, LR, DT, and NB	Software metrics and p-value less than 0.05
[29]	CNN based prediction of cross-project defects with node granularity	DP-CNN, DG, LR, TCA, DBN, TCA + , DBN + , and CNN	Twenty handcrafted features from PROMISE
[30]	Use of multi-classifiers with predictor integration and to improve the error predictions	LR, NB, and J48	It used 'C and K' metric set

Nearest Neighbor, KNN, LogitBoost (LB), Radial Basis Function (RBF) Network, Gradient Boosting (GB), Adaboost, XGBoost (XGB), Convolution Neural Network (CNN), MLP, Artificial Neural Networks (ANN), Deep Belief Network (DBN), RF, Classification And Regression Trees (CART), Decision Tree (DT), AD tree, Extra Trees (ET), bagging, Data Gravitation (DG), stacking, and Transfer Component Analysis (TCA). Therefore, all these observations are differentiated with others using a variety of methods, which assess their process concerns and performance outcomes.

It has been observed here that these existing software used software metrics such as size and LOC metrics, targeted metrics, and 'C and K' metrics. There were many approaches which used the PCA to reduce the size of their features and then set the different values for critical model parameters. With this, it was also noted that many methods applied the ensemble combinations to find the defects and errors. Further they also made the extensive use of the LR, DT, and NB classifications. In addition to the use of strong approaches, all these existing methods have illustrated the deficiencies in detecting different categories of software defects, errors, and faults. These deficiencies and failures have raised many threats, challenges, and barriers to their effective implementation, leading to the need for an effective software defect-based prediction and analysis systems.

3 Threats and Challenges Related to Software Defect-Based Analyzers and Predictors

Section 2 elaborated many of the latest software defect-based predictors and then it discriminated them against based on five key principles. Their implementations were performed on the standard databases, and they obtained their results for precision, recall, AUC, accuracy, F-measure, and other concerned measuring operations. Oppositely, they have faced many challenges and critical problems and had many limitations. Therefore, Table 2 shows some comparisons including requirements, challenges, threats, limitations, data sets, and accuracy results. It shows that they have used standard databases for the development such as Large Legacy Telecommunications Software System (LLTS); PROMISE; Eclipse; Open-NLP Library; JIRA; NASA four and five Facility Metrics Data Program (MDP); Open Science; Lucene Search Engine Library; Mahout ML Library; ISBSG Repository Release-12, and many more.

So, it is noted from Table 2 that the PROMISE databases became the first and preferred option for such prediction systems. Although these systems yielded promising results with the standard databases, but they have faced many challenges such as validity issues, limited resources, low accuracy, performance, non-balanced data, choice of classifier, big code size, dependency issues, and many more. These problems need to be addressed with good care in order to reduce them to a much higher level.

Table 2 Discriminating the challenges, threats, limitations, and data sets with accuracy findings of existing analyzers and predictors

References number	Needs, challenges, limitations and threats	Data sets and AUC (Accuracy)
[9]	Can be extended with the Halstead parameter removal, say, operands, total number of operands, operators, number of blank lines, and program length. Challenges to choose data set, fault-proneness prediction method, and evaluation metrics	CM1 data from spacecraft instrument of NASA's PROMISE sets. Average results: 87.16% accuracy, 95.36% sensitivity and 87.93% false alarm ratio
[10]	The performance of SRB-35 was better than RUS-35 with $p = 0.16$ for RF. Here the difference was not statistically significant. Two group means were found for p-value < 0.05 different from each other. There is a need to provide the solution of class imbalance problem	It used four datasets from LLTS and achieved promising results
[11]	Require to improve the performance and prediction results	It used two real-world type of data sets. The total mean square error was 92.31%. Achieved promising results
[12]	Threats: Evaluation approach, data heterogeneity, Oracle and dataset, and generalization of results. Require to enhance the cross-project prediction results using other classifiers. There is no use of normalization predictors for different hybrid approaches	Ten different open-sourced Java projects, say, LOG 4j and ANT from PROMISE data sets. Got promising prediction results
[13]	Hard to select the software metrics. Found poor results of AUC and accuracy	Used Eclipse data-sets (versions 2.0, 2.1 and 3.0). Nominal accuracy was from 58 to 65%
[14]	Used big training sample size and low AUC Risks to construct internal validity, external validity, and generic validity. Extend feature set. Need to investigate the effectiveness of different active learning, clustering, and semi-supervised learning methods	Five hundred defects from JIRA of Lucene, Mahout, and Open-NLP libraries. Found weighted precision (0.651), recall (0.669), F-measure (0.623) and AUC (0.71)
[15]	Discretized fault count was not having better performance than others methods	Average AUC: and 0.74 (KNN), 0.73 (LinR and LR), 0.78 (RF), and 0.76 (discretized defect classifiers). Seventeen Tera-PROMISE data sets

(continued)

Table 2 (continued)

References number	Needs, challenges, limitations and threats	Data sets and AUC (Accuracy)
[16]	It was implemented with the Java based projects only. It had low performance of the system due to the aggregation	File-based level: 0.705 of F-measure, 0.88 precision, and 0.007 of FPR in Lucene; and 0.826 of precision, 0.767 of F-measure, and 0.03 of FPR with RF in Jackrabbit. The Commit-based level was 0.721 of recall, 0.638 of precision, 0.677 of F-measure, and 0.402 of False-Positive Rate (FPR) with random forest for Jackrabbit. Two large-scale open source projects: Lucene and Jackrabbit
[17]	Challenge to get the optimal features. Need to consider statistical test and different performance metrics. Need of extension to evaluate the effectiveness for the un-sampled and sampled datasets	PC5, CM1, Tomcat, and Ant data sets from the project repository. It had used others from PDE and JDT. The range of results was from 74 to 99%
[18]	It was limited to Java projects only. It requires extension in this analysis with more number of methods	AUC (ML) had and 0.666–0.849 (RF), 0.619–0.868 (LB), 0.603–0.829 (Bagging), 0.669–0.854 (Adaboost), 0.413–0.758 (J48), and 0.604–0.852 (NB). Results: F-measure (0.615–0.966), AUC (0.6–0.9), recall (0.759–0.977) and precision (0.736–0.975). Seven open source data sets: Log4j, BSF, Xerxes, Click, Zuzel, Ivy, and Wspomaganiepi
[19]	Extension using different classifiers with good results	Found good results. Three open source software projects were used, say, xerses, v1.2 and xerses v1.3
[20]	Found risks to remove the irrelative and redundant features effects	Xerses 1.4 from open science. Found good results. Accuracy (%): KNN (Avg): 0.89 (recall), 0.88 (F1-score and precision) and 0.74 (AUC). Accuracy (%): 87.12 (DT Info Gain), 91.66 (GB), 89.39 (RF), 93.18 (KNN), 87.12 (DT Gini), and 79.54 (NB)

(continued)

Table 2 (continued)

References number	Needs, challenges, limitations and threats	Data sets and AUC (Accuracy)
[21]	Require to extend similarity or stability of different feature-based subset selection methods	Fifteen open datasets from Promise including Camel, Jedit, Lucene, Synapse, and Xerces. 90% confidence
[22]	Most of the datasets were written in Perl, Java, C, and C ++ only. The extension needed was to evaluate the performance of ensemble-based approaches	Most results were found in the series of 70% only. Four datasets (SP1, SP2, SP3, and SP4) from LLTS. 95% confidence
[23]	Repetition of the error reduced the accuracy to versions 1.4, 1.6, and 1.7. The training required additional information on the defective categories. The results were incorrect for all programs. Dependence on the choice of error firmness in the tester's information. Threats: Classes tend to make a lot of mistakes, number of mistakes, and appropriate levels of sharpness. Risk errors had a large code	Twelve NASA MDP datasets. Promising results
[24]	Validity externally for program represent, mutants, error coverage, and testing cases. Internal validity in implementation and tool-based faults. Noisy data say imbalanced data, random sampling, and false positives	It used from ANT1.3 to ANT1.7 versions of PROMISE datasets. Achieved very good results
[25]	Requirement to upgrade neural network parameter settings to get more promising classification and prediction results	Four compatible software projects: OpenOffice, MySQL, Apache, and Mozilla. Very promising results
[26]	Requirement to expand this analysis in many ways	One hundred and ninety one projects from ISBSG and got 82.71% accuracy with ANN
[27]	Requirement to expand this analysis in many ways	JM1, KC1, KC2, CM1, and PC1 from PROMISE. Average accuracy with and without PCA: 0.855 and 0.852; 0.82 and 0.819 (algorithmic)

(continued)

Table 2 (continued)

References number	Needs, challenges, limitations and threats	Data sets and AUC (Accuracy)
[28]	No use of ANOM test for Naïve Bayes. Dependency of results on researcher's experimental goals. Non-generalized results. Challenging classifier selection	Project failure data of Two hundred and thirty-six projects from reports, surveys, and case studies. AUC: 0.91 (training) and 0.94 (testing)
[29]	It worked upon Java projects only. Limited resources on defect-prone modules. Found validity threats such as non-appropriateness of F-measure, non-generalized experimental results, and implementation of compared models	Four datasets were used, say, KC1, CM1, PC1 and MC2 from PROMISE. Found good prediction
[30]	Threats found, say, no proper tests to prove unbalanced data, feature filtering, and generalized efficiency to see algorithmic performance. No threats in internal validity analysis	Used eleven open-source Java projects from PROMISE data sets, and achieved promising results

4 Evaluating the Performance Factors

In light of the above-mentioned comparative description, the detailed journey of existing software fault-based analyzers and predictors discriminated against many primary parameters, their risk factors, and their uses. The measurement methods were discriminated with each other in Tables 1 and 2 as illustrated in Sects. 2 and 3, respectively. Here three basic criteria are evaluated, which show the effects of parameters on comparative graphs. The % use of different metrics, data sets, and classifiers are depicted in Figs. 1, 2, and 3, respectively, in existing applications. Their observations show excellent performance of 68.18% with software metrics, best use of 44% with PROMISE data sets, and excellent use of 16% with the LR method. Such distributions are clearly illustrated in Figs. 1, 2, and 3, respectively. Additionally, all these results were seen to be the most effective and selective compared to all others.

Thus, this analysis is evident that all the research contributions of software defect-based prediction systems used many databases, metrics, and classifiers in the range of 4% to 44%, 4.54% to 68.19%, and 1% to 16%, correspondingly. Such factor evaluation depicts the scope of the strong use of these three criteria among all the systems.

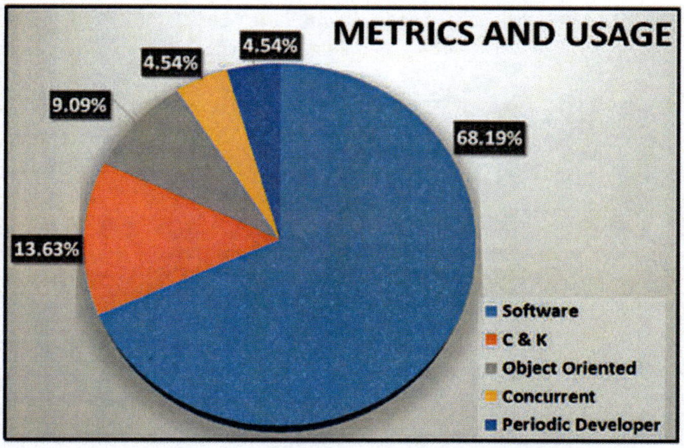

Fig. 1 Software defect-based prediction metrics and their % usage in existing ones

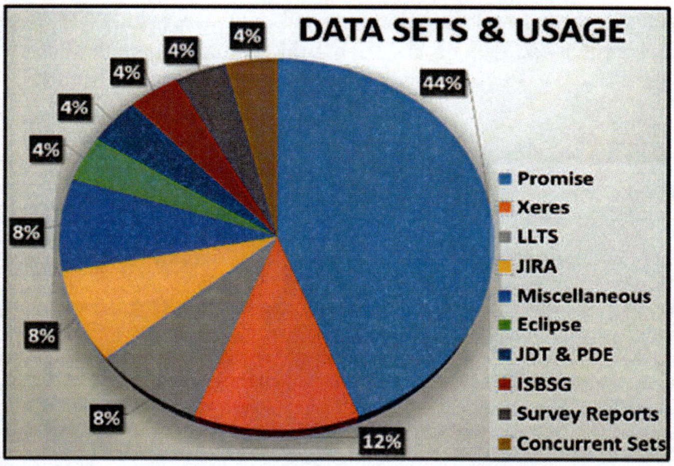

Fig. 2 Data sets and their % usage in existing predictors and applications

5 Concluding Remarks and Future Work

Focusing on the detailed insight into these existing software predictors and analyzers, this paper provided the systematic tour of many existing software defect-based prediction systems and software analysis-based systems using LR. They are compared to others according to certain basic measurement criteria, such as features, metrics, classifiers, data sets, and accuracy. Their observations concluded the use of 44%, 16%, and 68.19% of PROMISE data sets, LR classifiers, and software metrics,

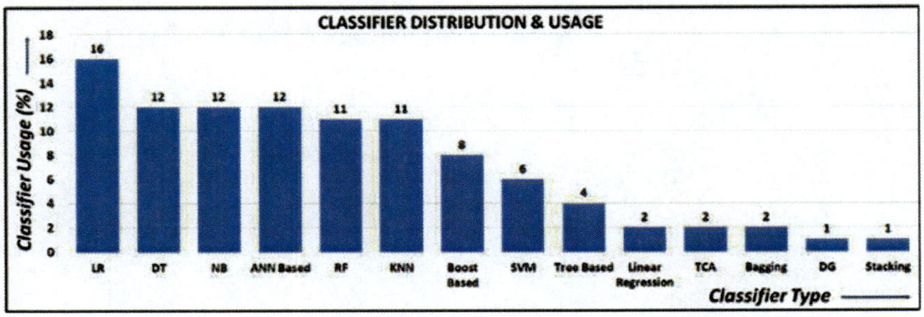

Fig. 3 Software defect-based prediction systems with % use of classifiers

respectively, among all others, where the major part of the research contributions used them as their preferred and important choices. With this, their threats, challenges, limitations, and risk factors were investigated to provide the research gaps, requirements, and improvements needed for such systems. So, it is concluded that there is a great requirement of a highly efficient, accurate, and effective predictor that can work across a wide variety of real-time domains and databases with great success. Such journey can be enhanced with other ML, hybrid ML, and ensemble-based strategies in the future.

References

1. P.K. Singh, R.K. Panda, O. Prakash, A critical analysis on software fault prediction techniques. World Appl. Sci. **33**(3), 371–379 (2015)
2. R. Malhotra, A systematic review of machine learning techniques for software fault prediction. App. Soft Comput. **27**, 504–518 (2015)
3. L. Goel, D. Damodaran, S.K. Khatri, M. Sharma, A literature review on cross-project defect prediction, in *4th International Conference on Electrical, Computer and Electronics* (IEEE, 2017), pp. 680–685
4. N. Kalaivani, R. Beena, Overview of software defect prediction using machine learning algorithms. Int. J. Pure App. Math. **118**(20), 3863–3873 (2018)
5. S. Kumar, S.S. Rathore, Types of software fault prediction, in *Software Fault Prediction, Springer Briefs in Computer Science* (Springer, 2018), pp. 23–30
6. S.S. Rathore, S. Kumar, A study on software fault prediction techniques. Art. Int. Rev. **51**, 255–327 (2019)
7. Z. Tian, J. Xiang, S. Zhenxiao, Z. Yi, Y. Yunqiang, Software defect prediction based on machine learning algorithms, in *International Conference on Computer and Communication Systems* (IEEE, 2019), pp. 520–525
8. B. Eken, Assessing personalized software defect predictors, in *40th International Conference on Software Engineering: Companion* (IEEE, 2018), pp. 488–491
9. G. Mauša, T.G. Grbac, B.D. Bašic, Multi-variate logistic regression prediction of fault-proneness in software modules, in *Proceedings of the 35th International Convention MIPRO* (IEEE, 2012), pp. 698–703
10. K. Gao, T.M. Khoshgoftaar, A. Napolitano, A hybrid approach to coping with high dimensionality and class imbalance for software defect prediction, in *11th International Conferences on Machine Learning and Apps* (IEEE, 2012), pp. 281–288

11. K.V.S. Reddy, B.R. Babu, Logistic regression approach to software reliability engineering with failure prediction. Int. J. Softw. Eng. App. **4**(1), 55–65 (2013)
12. A. Panichella, R. Oliveto, A.D. Lucia, Cross-project defect prediction models: L'Union fait la force, in *Software Evolution Week-Conference on Software Maintenance, Reengineering, and Reverse Engineering* (IEEE, 2014), pp. 164–173
13. D. Kumari, K. Rajnish, Comparing efficiency of software fault prediction models developed through binary and multinomial logistic regression techniques, in *Information Systems Design and Intelligent Applications, Advances in Intelligent Systems and Computing*, vol. 339, ed. by J. Mandal, S. Satapathy, M. Kumar Sanyal, P. Sarkar, A. Mukhopadhyay (Springer, 2015), pp. 187–197
14. F. Thung, X.D. Le, D. Lo, Active semi-supervised defect categorization, in *23rd International Conference on Program Comprehension* (IEEE Press, 2015), pp. 60–70
15. G.K. Rajbahadur, S. Wang, Y. Kamei, A.E. Hassan, The impact of using regression models to build defect classifiers, in *14th International Conference on Mining Software Repositories* (IEEE, 2017), pp. 135–145
16. S.O. Kini, A. Tosun, Periodic developer metrics in software defect prediction, in *18th International Working Conference on Source Code Analysis & Manipulation* (IEEE, 2018), pp. 72–81
17. K. Bashir, T. Ali, M. Yahaya, A.S. Hussein, A hybrid data preprocessing technique based on maximum likelihood logistic regression with filtering for enhancing software defect prediction, in 14th *International Conferences on Intelligent Systems and Knowledge Engineering* (IEEE, 2019), pp. 921–927
18. P. Singh, R. Malhotra, S. Bansal, Analyzing the effectiveness of machine learning algorithms for determining faulty classes: a comparative analysis, in *9th International Conference on Cloud Computing, Data Science and Engineering* (IEEE, 2019), pp. 325–330
19. S. Agarwal, S. Gupta, R. Aggarwal, S. Maheshwari, L. Goel, S. Gupta, Substantiation of software defect prediction using statistical learning: an empirical study, in *4th International Conference on Internet of Things: Smart Innovation and Usages* (IEEE Press, 2019), pp. 1–6
20. F. Wang, J. Ai, Z. Zou, A cluster-based hybrid feature selection method for defect prediction, in *19th International Conference on Software Quality, Reliability and Security* (IEEE, 2019), pp. 1–9
21. H. Wang, T.M. Khoshgoftaar, A study on software metric selection for software fault prediction, in *18th International Conferences on Machine Learning and Applications* (IEEE, 2019), pp. 1045–1050
22. P Singh, Stacking based approach for prediction of faulty modules, in *Conference on Information and Communication Technology* (IEEE, 2019) , pp. 1–6
23. S. Moudache, M. Badri, Software fault prediction based on fault probability and impact, in *18th International Conferences on Machine Learning and Applications* (IEEE, 2019), pp. 1178–1185
24. T. Yu, W. Wen, X. Han, J.H. Hayes, ConPredictor: concurrency defect prediction in real-world applications. IEEE Trans. Softw. Eng. **45**(6), 558–575 (2019)
25. K. Kaewbanjong, S. Intakosum, Statistical analysis with prediction models of user satisfaction in software project factors, in *17th International Conferences on Electrical Engineering/Electronics, Computer, Telecommunications and Information Technology* (IEEE, 2020), pp. 637–643
26. M. Cetiner, O.K. Sahingoz, A comparative analysis for machine learning based software defect prediction systems, in *11th International Conference on Computing Communication & Networking Technologies* (IEEE, 2020), pp. 1–7
27. M.A. Ibraigheeth, S.A. Fadzli, Software project failures prediction using logistic regression modeling, in*2nd International Conference on Information Science* (IEEE, 2020), pp. 1–5
28. E. Elahi, S. Kanwal, A.N. Asif, A new ensemble approach for software fault prediction, in *17th International Bhurban Conference on Applied Sciences and Technology* (IEEE, 2020), pp. 407–412

29. J. Deng, L. Lu, S. Qiu, Y. Ou, A suitable AST node granularity and multi-kernel transfer convolutional neural network for cross-project defect prediction. IEEE (2020), pp. 66647–66661
30. F. Yucalar, A. Ozcift, E. Borandag, D Kilinc, Multiple-classifiers in software quality engineering: combining predictors to improve software fault prediction ability. Eng. Sci. Tech. Int. J. **23**(4), 938–950 (2020)

Evaluation and Application of Clustering Algorithms in Healthcare Domain Using Cloud Services

Ritika Bateja, Sanjay Kumar Dubey, and Ashutosh Bhatt

Abstract Diseases in healthcare domain are still diagnosed at very later stage. Insights from healthcare data are still not available to clinicians or patients, which is one of the key reasons for late discovery of disease. Nature and type of healthcare data is one of the hurdles to process it for extracting insights. Clustering algorithms are proven to be effective in analyzing data and discovering patterns for effective decision making in different domains including finance, marketing, data mining, social networking, etc. Healthcare domain is one such domain where clustering algorithm can be a boon for the care providers as well as for the patients by exploring hidden pattern in data to deliver quality treatments to the patients. Clustering healthcare data can help in prediction of diseases. These predictions will not only help in saving lives of patients by early diagnosing the disease, providing timely treatments but also help in providing faster and cost effective results. This paper discusses various clustering approaches, evaluates the performance of two widely used clustering algorithms k-means and DBSCAN on healthcare datasets, and proposes a solution to implement clustering on healthcare data using cloud based services.

Keywords Clustering · Healthcare data · K-means · DBSCAN · ETL · AWS cloud

1 Introduction

We are living in the era of data explosion. A large amount of data is flooded over the internet from various domains including social networking sites, education, healthcare, etc. Management and analysis of such large data is also very tedious. Especially

R. Bateja (✉) · S. K. Dubey
Amity University Uttar Pradesh, Sec-125, Noida, India

S. K. Dubey
e-mail: skdubey1@amity.edu

A. Bhatt
Uttarakhand Open University, Dehradun, Uttarakhand, India

A. K. Luhach et al. (eds.), *Second International Conference on Sustainable Technologies for Computational Intelligence*, Advances in Intelligent Systems and Computing 1235,
https://doi.org/10.1007/978-981-16-4641-6_21

in healthcare sector, handling data and performing analysis on it was very cumbersome as data was traditionally captured in manual documents and files. Over the years, various initiatives have been taken for digitization of healthcare domain data. Digital revolution has led to new techniques being introduced to capture and store healthcare data with systems like Electronic Health Records (EHR) and Patient Health Records (PHR). Processing healthcare data to get useful insights and providing recommendations to the patients to enable patient centric self-care holds vital importance today. This healthcare data, if analyzed efficiently, can contribute to quality care by early diagnosing the disease and recommending quality treatments to patients. Data mining approaches like clustering and classification have proven to be effective in analyzing and extracting knowledge from raw data to facilitate effective decision making. Clustering is an unsupervised classification, where objects are grouped on the basis of similarity and distance functions [1]. Various applications of clustering in healthcare data include ranking of hospitals, better and smarter treatment techniques, identification of patients having high risks, controlling infections in hospitals, identifying patients having similar diseases, etc. [2]. Leveraging these modern techniques along with clustering can be of vital importance to explore pattern in healthcare data, such as:

- Grouping of patients who share similar symptoms to prescribe treatments
- Grouping of patients, undergone similar treatment but responded differently.

Various types of clustering approaches are explored in this paper, and the performances of different clustering algorithms are compared over different healthcare datasets using parameters like execution time, accuracy, no. of iterations, etc. The objective of the paper is to present a cloud-based solution to implement clustering on healthcare datasets.

2 Clustering Techniques

Clustering algorithms are classified into two categories, i.e., traditional approaches and modern approaches [3]. Traditional approaches include basic algorithms like partitioning, hierarchical, model-based, density-based, and grid-based algorithms, whereas modern approaches include quantum theory-based, ensemble-based, kernel-based, affinity propagation-based, spatial data-based, and stream-based algorithms [3]. Traditional clustering approaches are shown in Fig. 1, and modern approaches are shown in Fig. 2.

This research focuses on two widely used clustering algorithms partitioning and density-based clustering algorithms, which are further classified into various categories like k-means, k-modes, k-mediods, DBSCAN, STING, CLIQUE, etc. K-mean clustering is most widely known and extensively used partitioning based algorithm. It is the centroid-based iterative algorithm which partitions the n data points into k-clusters to achieve high intra-cluster similarity and low inter-cluster similarity [4]. It is simple and efficient clustering algorithm which works well on numerical datasets.

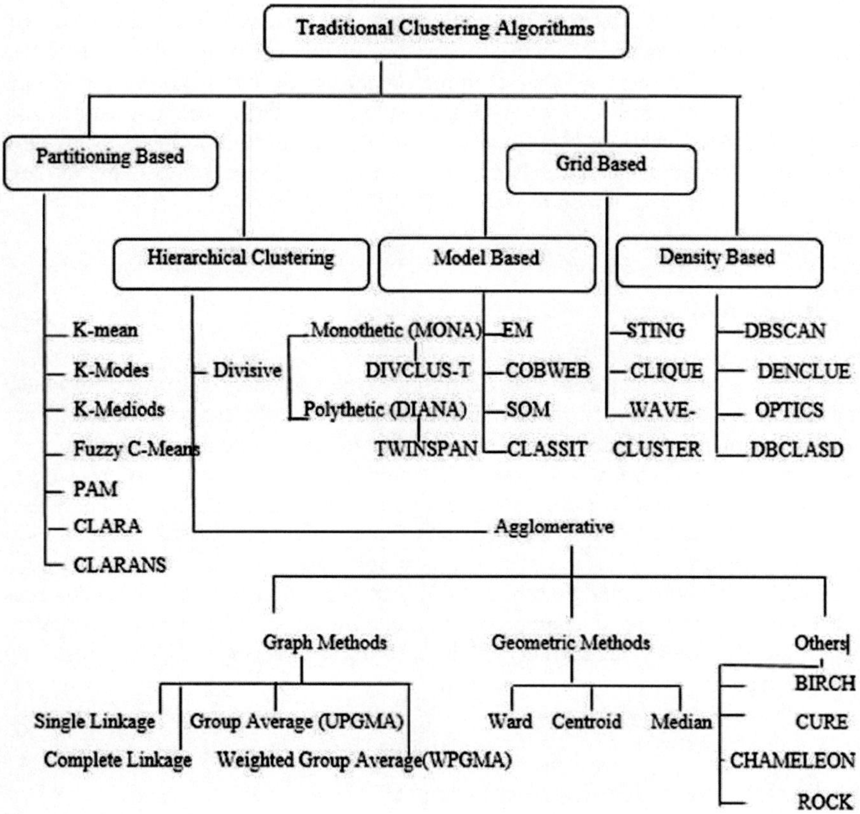

Fig. 1 Traditional clustering approaches

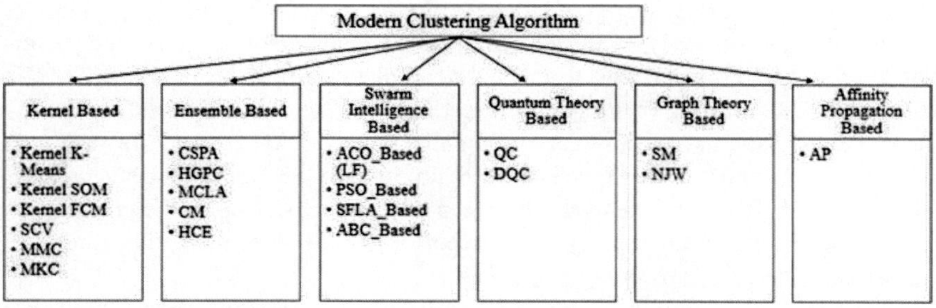

Fig. 2 Modern clustering approaches

Various variants of it also exist in the literature which is applied on other datasets like k-modes on categorical data [5], k-prototype on mixed datasets [6]. On the other hand, DBSCAN clustering is an example of density based clustering algorithm. DBSCAN is short form for Density Based Spatial Clustering of Application with Noise. DBSCAN works by exploring high-density and expanding them as clusters for similar features. Healthcare data by nature is highly dense and is likely to have arbitrary shape clusters which fit well for DBSCAN clustering.

3 Related Work

A lot of research has been done over the past few years on analyzing the performance and review the applicability of clustering algorithms on different types of healthcare data. A summary of research done by different authors is summarized in Table 1.

From Table 1 above, it was found that most of the analysis of different types of healthcare datasets like diabetes datasets, lung cancer datasets, and heart data sets were done using WEKA tool (Waikato Environment for Knowledge Analysis) due to its following advantages: freely available, ease of use, open source software under GNU GPL, and portability [20]. Although WEKA tool supports various data mining features like data preprocessing, classification, clustering, feature selection, regression, and visualization [21], it performs analysis on traditional clustering approaches.

4 Performance Evaluation Using WEKA Tool

Effectiveness of a clustering algorithm depends on various parameters like the attributes of the datasets, shapes of clusters it forms, execution time it takes, number of iterations, and accuracy of clusters. WEKA tool was used to evaluate the performance of simple K-means based on Euclidean distance measure and DBSCAN a density based Clustering on diabetes dataset [22]. As focus is on providing quality clusters to the patients and clinicians efficiently, so DBSCAN outperforms k-means in terms of time and quality of the clusters it produced. CSV file of diabetes datasets is loaded into WEKA tool, analysis is performed using clustering as shown in Fig. 3, and the results obtained after comparing the performance of both clustering algorithm on various parameters are shown in Table 2.

The above table shows that DBSCAN outperforms k-means clustering algorithm in terms of time as well as cluster quality.

Table 1 Performance evaluation of different clustering algorithm on different healthcare datasets

Year and Refs.	Authors	Datasets	Tool/programming language used	Techniques	Results
2014 [7]	Malli et al.	Rural maternity and child welfare (RMCW) Datasets	K-means Clustering Tool: WEKA tool	Clusters produces on the basis of three factors: locality, socio economic status, specifying no. of clusters	K-means clustering produces better quality clusters with large datasets
2014 [8]	Haraty et al.	Media sensor data in health monitoring system (used movies datasets)	Enhanced k-means (G-means) P.L: Java	K-means and G-means are compared on basis of entropy and F-scores and complexity	G-means outperforms K-means in terms of entropy and F-scores
2014 [9]	Anuradha et al.	Diabetic datasets of women	k-means, k-medoids, MST, k- Nearest Neighbor	Analysis of clustering on the basis of cluster quality achieved due to highest area	k-medoids outperform others on the basis of area calculated for cluster
2015 [10]	Nithya et al.	Diabetes Datasets	Hierarchical, Density and K-Means	Performance analysis on the basis of execution time and the number of clustered instances	K-means give better results than the other two
2015 [11]	Vijayarani and Sudha	hemogram blood datasets	k-means, Fuzzy C-means, Weighted k-me ans, ProposedWeighted k-means algorithm	Comparison on the basis of time, cluster accuracy and error rate	Weighted k-means performs well, achieve high accuracy

(continued)

Table 1 (continued)

Year and Refs.	Authors	Datasets	Tool/programming language used	Techniques	Results
2015 [12]	Dharmarajan and Velmurugan	Lung Cancer Datasets	k-means, Farthest First Tool: WEKA tool	Performance analysis on the basis of time taken by algorithm on different instances	k-means is efficient in terms of computational complexity
2016 [13]	Merlin	Heart Datasets	CLARA, k-means, PAM Silhoutte width measure	Compared the performances of three clustering algorithm using Silhoutte width measure	CLARA shows better performance than K-means and PAM
2017 [14]	Mirmozaffari et al.	Patient datasets from hospital in Iran	EM, Farthest First, Filtered Clusterer, Make Density Based Clusterer, Simple k-means Tool Used: WEKA tool	Analysis of datasets on basis of clustering accuracy, time taken to build model, SSE and no. of iteration using various clustering algorithm	Filtered Clusterer, Make Density Based Clusterer, and Simple K-Means performs better in terms of accuracy, time, SSE and no. of iteration
2017 [15]	Silitonga	Patient's disease data at Haji Adam Malik Hospital in Medan	K-means Clustering Tool Used: WEKA tool	Analysis of datasets on the basis of k-means clustering using WEKA tools	Septicemia Disease was found to have high tendency pattern
2018 [16]	Ogbuabor and Ugwoke	Movement Activity dataset from "MyHealthAvatar" domain	DBSCAN, K-means Silhoutte analysis	Performance of algorithms are compared using Silhouette score values	k-means performs better than DBSCAN in terms of accuracy and execution time

(continued)

Table 1 (continued)

Year and Refs.	Authors	Datasets	Tool/programming language used	Techniques	Results
2018 [17]	Shanthipriya and Prabavathi	Diabetic datasets of girl students	K-Means, Hierarchical and Density-Based algorithms	Performance of algorithms are compared on basis of cluster accuracy and execution time	k-means clustering Algorithm gives better prediction than Hierarchical and Density Based clustering algorithms
2019 [18]	Kodati et al.	Heart Disease Datasets	k-means farthest first, filtered cluster hierarchical cluster, OPTICS Tool: WEKA	Performance of algorithms are compared on basis of time taken by them	
2020 [19]	Singh et al.	Diabetes Datasets	DBSCAN, k-means, Filtered Cluster Tool Used: WEKA tool	Analysis of diabetes datasets on the basis of Error rate, Computation time and accessing time	DBSCAN per forms well as compared to other two algorithms

5 Use of Cloud Services in Healthcare Domain

Due to the exponential growth of data, population, ever growing demands of the patients for providing effective patient centered healthcare to them and for improving quality of life, a lot of pressure has been put on improving healthcare by using cloud based services in healthcare for providing scalability in storage and processing of data, for effective sharing and integration, and for providing more security, reliability, and serviceability [23].

Healthcare cloud is of utmost importance for providing cost effective centralized services, and its components include servers, virtual desktops, networks, hardware, application, and software platform. Based on the access to the users of the network, cloud infrastructure is classified into public cloud, private cloud, and hybrid cloud, out of which hybrid cloud is the most promising cloud deployment model used in healthcare as it is a blend of both public and private cloud [24]. It can be accessed by both: the general public (patients) as well as within the organizations (hospitals, government authorities, doctors, etc.). Cloud-based Electronic Health Records (EHR), Personal Health Records (PHR), and Electronic Medical Records (EMR) are

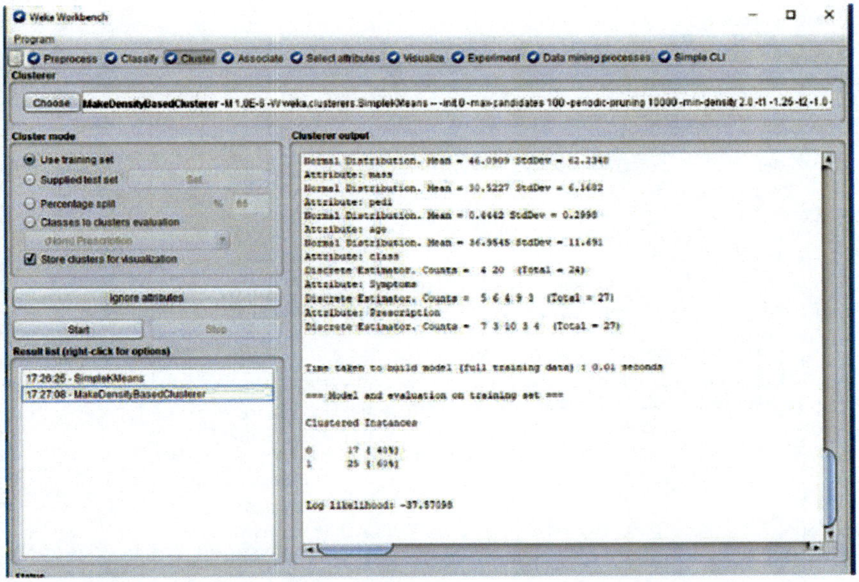

Fig. 3 Cluster based analysis of DBSCAN

Table 2 Results on diabetes datasets

Parameters	k-means	DBSCAN
No. of clusters	2	2
Computation time	0.02 s	0.01 s
Clustered instances	Cluster 0–48% Cluster 1–52%	Cluster 0–40% Cluster 1–60%
No. of iterations	5	5

integrated and stored in centralized location giving a unified view of data and are made accessible to the patients and the providers.

6 Proposed Work

In this section, a solution based on cloud based services is proposed to implement DBSCAN clustering.

- **Step 1—Get raw healthcare data**: In this step, data will be sourced from various healthcare data sources. As data standards for healthcare data is still evolving, so data is likely to be in various formats/shapes/size.

- **Step 2—Bringing the data in consistent form**: In this step, data is further cleaned up to remove inconsistency in terms of formats, attributes, and shape. In this stage, python based scripts can be applied to read the data from files/data sources and convert them into consistent format and structure.
- **Step 3—Load the data in healthcare repository**: In this step, structured data is loaded into data repository so that it can be accessed by clustering algorithm. In order to efficiently store the data, cloud based data storage solution is used.
- **Step 4—Apply appropriate clustering algorithm to discover patterns in the data**: This is the core step of the overall process. In this step, clustering algorithm is applied to discover the patterns/groups from the data. A program can be developed to connect to data repository, extracting the data and applying the clustering algorithm.
- **Step 5—Clusters of patient data**: In this step, patients are clustered into groups based on attributes like nature of disease, treatment, demographic attributes, etc.
- **Step 6—Recommendations**: The clustered data can be used for providing recommendations to the patients.

6.1 Logical View of the Process

In this section, a logical view of the process is demonstrated, which is agnostic of any technology/programming language. Logical view of the solution is also shown in Fig. 4.

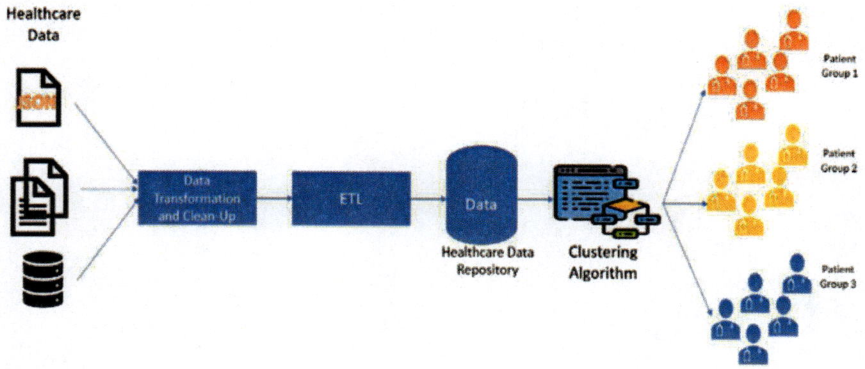

Fig. 4 Logical view of the solution

6.2 Physical View of the Process

In this section, technologies are discussed that are used in proposed solution. In order to get 24 * 7 access to data, AWS Cloud is used to store the data. However existing EHR, EMR, and PHR are likely to be hosted on premise, and moving large amount of data from on premise to cloud is challenging. In order to overcome this challenge, "Talend integration cloud" is used, which is cloud based ETL Service to move on premise data to the cloud. Basically, Talend is an open source integration platform used for ETL, data warehousing, and Business Intelligence and supports many features like real time debugging, GUI based features, and robust execution [25]. Talend cloud integration is a scalable and secure cloud integration platform-as-a-service (iPaaS) is used for making data driven decisions. It provides platform and tools for hosting and managing virtual infrastructure, security and compliance, data warehousing, and integration [26]. Following is the list of different technologies that can be used:

- Talend Integration Cloud as ETL
- AWS EC2 to host the job schedule pulling data from Talend
- AWS Redshift as data warehouse solution
- AWS Personalize for extracting recommendations out of the data warehouse
- AWS Elastic Search for faster retrieval of data.

There are two data flows demonstrated in Fig. 5.

- Flow 1: Step 1 to 7, to bulk ETL data from EHR, EMR, and PHR to AWS Redshift.
- Flow 2: Step 8 to 11, to push incremental data from patient devices (PHR) to S3/AWS Redshift.

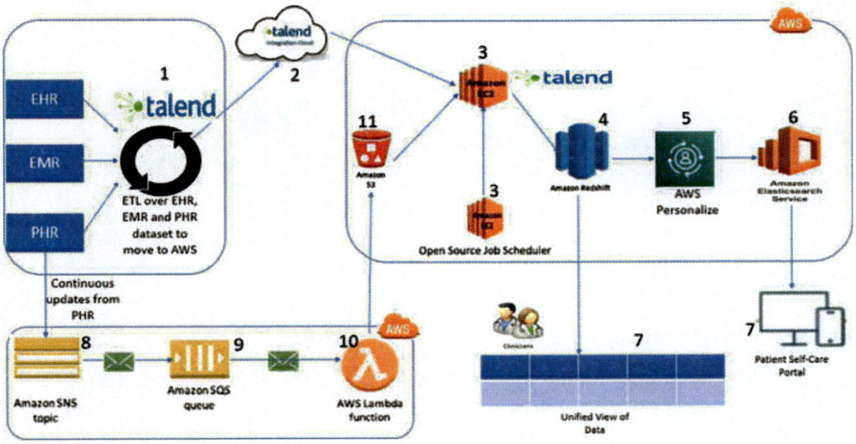

Fig. 5 Proposed Technologies used for integration, ETL, storage, and recommendations

AWS Redshift used here is cloud-based data warehouse service. It supports various features like faster loading of data, query optimization procedures, efficient data compression, horizontally scalable, and more secure [27]. Continuous updates from PHR are fed to Amazon SNS which then store it in queue using SQS (Amazon Simple Queue Service). SQS is used to send, receive, and store messages between software components [28]. AWS Sage Maker tool is used for implementing clustering on data consolidated from ETL process [29]. AWS personalize is used to deliver high quality recommendations by selecting the right algorithm and train the model after examining the data. It is basically using state-of-the-art recommendation which also solves cold start problem [30]. AWS Elastic Search is an open source, scalable enterprise-based search engine which perform effective search by searching through index rather than searching the text using Apache Lucene. Apart from the advances, it includes in terms of speed, security, scalability, and hardware efficiency; it also supports clustering and leader selection [31].

7 Exposing Clustered Data to Clinicians and Patients

All the insights drawn by proposed solution is only relevant if it is easily accessible to patients and clinicians. For this, the data is exposed via APIs and access it using different user interfaces as shown in Fig. 6. The summary of different steps in this solution is as follows:

- Clustered data in elastic search is exposed to patients and clinicians
- Java based Drop wizard framework is used to expose the data
- Different APIs endpoints are developed to provide access to the data
- API endpoints can then be accessed via native mobile apps, desktop apps, and by other means so that patient and clinicians can access the data
- Open API specification is recommended to be used to expose the data.

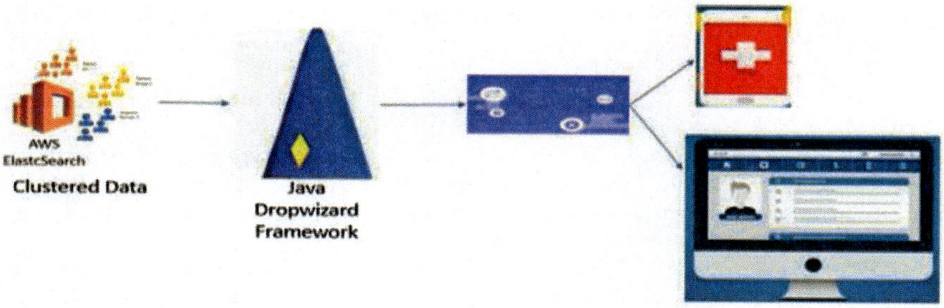

Fig. 6 Exposing grouped data to clinicians and patients

8 Conclusion and Future Scope

In this paper, the performances of two widely used clustering approaches, DBSCAN, and k-means are evaluated on diabetes datasets using WEKA tool. It was found that DBSCAN outperforms k-means in terms of quality of clusters produced and time. Further, Talend Integration Cloud is used for ETL, AWS Personalize for recommendations, AWS Redshift for storage, and AWS Sage maker for clustering. Data is made accessible to patients and clinicians via APIs. Clustering healthcare data gives various benefits to patients and clinicians. Clinicians can utilize the data to analyze treatments which have worked for similar patients in the past and accordingly prescribed relevant treatment to other patient as results of the diagnosis. Patients can also use such clustered data as self-care to find out relevant treatment for them when they search by providing different attributes as inputs. In future, this solution can be further enhanced and optimized. Data processing proposed in this solution can be optimized and made efficient by using parallel processing techniques such as map-reduce. These techniques can be applied by using mappers at the ETL layer and clustering steps, followed by use of reducers whilst grouping the patients, which can reduce overall time taken to process the data.

References

1. A. Saxena et al., A review of clustering techniques and developments. Neurocomputing **267**, 664–681 (2017)
2. M. Durairaj, Data mining applications in healthcare sector: a study. Int. J. Sci. Technol. Res. **2**(10), 29–35 (2013)
3. D. Xu, A comprehensive survey of clustering algorithms. Ann. Data Sci. **2**(2), 165–193 (2015)
4. A. Choudhary, Survey on K-mean and its variants. Int. J. Innov. Res. Comput. Commun. Eng. **4**(1), 949–953 (2016)
5. K. Lakshmi, Clustering categorical data using k-modes based on cuckoo search optimization algorithm. ICTACT J. Soft Comput. **8**(1), 1561–1566 (2017)
6. S. Harous, M.A. Harmoodi, H. Biri, A comparative study of clustering algorithms for mixed datasets, in *2019 Amity International Conference on Artificial Intelligence* (AICAI) (IEEE, 2019), pp. 484–488
7. S. Malli, H.R. Nagesh, H.G. Joshi, A study on rural healthcare data sets using clustering algorithms. Int. J. Eng. Res. **3**(8), 546–548 (2014)
8. R.A. Haraty, M. Dimishkieh, M. Masud, An enhanced k-means clustering algorithm for pattern discovery in healthcare data. Int. J. Distrib. Sens. Netw. **11**(6), 1–11 (2014)
9. S. Anuradha et al., Comparative study of clustering algorithms on diabetes data. Int. J. Eng. Res. Technol. (IJERT) **3**(6), 922–926 (2014)
10. R. Nithya, P. Manikandan, D.D. Ramyachitra, Analysis of clustering technique for the diabetes dataset using the training set parameter. Int. J. Adv. Res. Comput. Commun. Eng. **4**(9), 166–169 (2015)
11. S. Vijayarani, S. Sudha, An efficient clustering algorithm for predicting diseases from hemogram blood test samples. Indian J. Sci. Technol. **8**(17), 1–8 (2015)
12. A. Dharmarajan, T. Velmurugan, Lung cancer data analysis by k-means and farthest first clustering algorithms. Indian J. Sci. Technol. **8**(15), 1–8 (2015)

13. J.K. Merlin, Srividhya: performance analysis of clustering algorithms on heart dataset. Int. J. Modern Comput. Sci. **5**(4), 113–117 (2016)
14. M. Mirmozaffari, A. Alinezhad, A. Gilanpour, Heart disease prediction with data mining clustering algorithms. Int. J. Comput. Commun. Instrum. Eng. (IJCCIE) **4**(1), 16–19 (2017)
15. D.P. Silitonga, Clustering of patient disease data by using K-means clustering. Int. J. Comput. Sci. Inf. Secur. (IJCSIS) **15**(7), 219–221 (2017)
16. G. Ogbuabor, F.N. Ugwoke, Clustering algorithm for a healthcare dataset using silhouette score value. Int. J. Comput. Sci. Inf. Technol. **10**(2), 27–37 (2018)
17. M. Shanthipriya, G.T. Prabavathi, Healthcare predictive analytics. Int. Res. J. Eng. Technol. (IRJET) **5**(2), 1459–1462 (2018)
18. S. Kodati, R. Vivekanandam, G. Ravi, Comparative analysis of clustering algorithms with heart disease datasets using data mining weka tool, in *Soft Computing and Signal Processing. Advances in Intelligent Systems and Computing*, vol. 900, ed. by J. Wang, G. Reddy, V. Prasad, V. Reddy (Springer, Singapore, 2019)
19. L. Singh, P. Singh, N. Dhillon, Diagnosis and prediction of diabetes patient data by using data mining techniques. Int. Res. J. Eng. Technol. **7**(10), 1564–1568 (2020)
20. Ekta, S. Dhawan, Classification of data mining and analysis for predicting diabetes subtypes using WEKA. Int. J. Sci. Eng. Res.**7**(12), 100–103 (2016)
21. S.B. Jagtap, B.G. Kodge, Census data mining and data analysis using WEKA, in *International Conference in Emerging Trends in Science Technology and Management, ICETSTM – 201, Singapore* (2013), pp.35–40
22. https://www.kaggle.com/uciml/pima-indians-diabetes-database
23. G. Aceto, V. Persico, A. Pescape, Industry 4.0 and health: internet of things, big data, and cloud computing for healthcare 4.0. J. Ind. Inf. Integr. **18**(1), 1–14 (2020)
24. M. Parekh, Designing a cloud based framework for healthcare system and applying clustering techniques for region wise diagnosis. Proc. Comput. Sci. **50**, 537–542 (2015)
25. J. Sreemathy, S. Nisha, R.M. Gokula Priya, Data integration in ETL using TALEND, in *2020 6th International Conference on Advanced Computing and Communication Systems (ICACCS), Coimbatore, India* (2020), pp. 1444–1448
26. https://www.talend.com/resources/what-is-cloud-integration/
27. A. Gupta, et al., Amazon redshift and the case for simpler data warehouses, in Proceedings of the 2015 ACM SIGMOD International Conference on Management of Data, pp. 1917–1923, Melbourne, Victoria, Australia (2015)
28. J.P. Buddha, R. Beesetty, The definitive guide to AWS application integration: with amazon SQS, SNS, SWF and Step Functions (2019)
29. E. Liberty, Z. Karnin, B. Xiang, L. Rouesnel, B. Coskun, R. Nallapati, J. Delgado, A. Sadoughi, Y. Astashonok, P. Das, C. Balioglu, S. Chakravarty, M. Jha, P. Gautier, D. Arpin, T. Januschowski, V. Flunkert, Y. Wang, J. Gasthaus, L. Stella, S. Rangapram, D. Salinas, S. Schelter, A. Smola, Elastic machine learning algorithms in amazon sagemaker, in Proceedings of the 2020 ACM SIGMOD International Conference on Management of Data, SIGMOD '20, New York, USA (2020), pp. 731–737
30. https://dius.com.au/2020/20/08/amazon-personalize-helping-to-unlock-the-power-of-advanced-recommendation-solutions-for-the-lean-digital-business/
31. P. Gupta, S. Nair, Survey paper on elastic search. Int. J. Sci. Res. **5**(1), 333–336 (2016)

Prediction of Stock Movement Using Learning Vector Quantization

Anand Upadhyay, Santosh Singh, Ranjit Patra, and Shreyas Patwardhan

Abstract Predicting stock market movement is one of the most sought after and valuable area in finance. Machine learning has evolved in the past many years and is widely used in financial modeling and prediction. There has been an increased curiosity among researchers in using these models to solve for many stock related prediction. In this paper, stock market pattern recognition is studied using Learning vector quantization (LVQ). This paper describes the model used for the research and provides experiment results and conclusion based on the finding. The authors have used many readily available attributes of a stock such as open and close price, high and low price of a stock along with few derived attributes such as strength index and moving averages. In this research, various moving averages are used and accuracy is studied with these averages. Experiment result shows that LVQ can give a promising accuracy with moving averages where the time period used for calculation of moving average is less.

Keywords Stock prediction · Learning vector quantization · Machine learning · Moving average

1 Introduction

In India, only two percent of the population own stocks, whereas in developed countries like United States, over 50% of the population own stocks. There are various factors such as lack of trust, capital and knowledge, responsible for this tepid response among the Indian population when it comes to investing in stock market. Majority of Indian population do not have appetite to take risk. Despite their higher returns, equity investment has not seen the foot fall as much as it should have. There are various macroeconomic factors which plays an important role in determining the trend in equity market. It focuses on the aggregate changes in the economy such as growth rate, inflation, unemployment, foreign markets, government policies, exchange rates,

A. Upadhyay (✉) · S. Singh · R. Patra · S. Patwardhan
Department of Information Technology, Thakur College of Science and Commerce, Thakur Village, Kandivali (East), Mumbai, Maharashtra 400101, India

A. K. Luhach et al. (eds.), *Second International Conference on Sustainable Technologies for Computational Intelligence*, Advances in Intelligent Systems and Computing 1235,
https://doi.org/10.1007/978-981-16-4641-6_22

and gross domestic product and inflation. It focuses on behavior and decision making of an economy as a whole. For an individual stock, micro economic factor plays an important role. It deals with individuals, business, or a sector. While making a decision on which stock to invest in, one should study the micro economic data, of the business, available in public domain. Before investing in any stock a thorough research should be done about the company, the market share of the company, products, and services provided by the company, suppliers, competitor, and general public sentiment about the company and customer base.

Along with the macro economic and micro economic factors there are key fundamental factors of a company, which comes into play when determining if a stock is worthy of the price tag. These factors are Industry performance, investor sentiments or confidence, media coverage on profits and earnings, future growth of the company, distribution of dividends, securing large contracts, launch of new products, anticipated mergers and acquisitions, change of management, scandals or employee layoffs. The authors of this research paper have tried to decompose the complexity and identify the key technical attributes of a stock which can aid in predicting a given stock based on past performances over last five years. Using these technical attributes, the authors have applied soft computing techniques to predict the movement of a given stock based on the historical data. The technical indicators taken into considerations are close price, open price, maximum price of a stock on a given day, minimum price of the stock, moving averages, and strength index.

The four attributes, opening price of a stock, closing pricing of the stock, maximum price of the stock in a given day, and minimum price of the stock in a given day and average price, are readily available in the public domain. The other two attributes, strength index and moving averages, are derived based on the price of a stock at the closure of the market. Strength index is similar to psychological line. It determines public's sentiment of the stock and is calculated by dividing the sum of price ups by sum of price ups and downs over certain past period. Price ups and downs can be determined by doing a relative compare of a given day's closing price with the previous day closing price. When calculating a moving average, a mathematical analysis of the stock's average value over a predetermined time period is made. As the stock price changes over a period of time, its moving price moves up and down. Important information about the price point of a stock can be derived using the direction of a moving average, a moving average going north shows the price are generally increasing, while a moving average going south indicates that the price, on an average, is falling.

While a price above the long term moving average reflects an overall uptrend, the price below a long term moving average reflects an overall downtrend. A 'buy' signal is generated when the stock's price rises above its moving average, while a 'sell' signal is generated when the equity's price drop below the moving average.

In this paper the authors will do a comparison of various prediction accuracy based on 5, 10, 15, and 30 days moving averages.

2 Motivation

For majority of Indian house hold, investment has been always related to safer bets like gold, fixed deposits, public provident fund, post office fixed deposits, or government bonds. Though these instruments are safe, they often fall short of inflation. The motivation behind this research paper is to understand attributes of stock in more details and identify attributes which can be used along with various machine learning algorithm to find patterns. This can help in identifying stocks which can be purchased or sold based on historical data. The authors in the paper want to demystify relationship between various input attributes, thereby identifying patterns and prototypes. The lack of knowledge and interests in the equity market has motivated this team of researchers to understand the subject better using machine learning algorithm.

3 Literature Review

Tremendous amount of work has gone into the area of stock market prediction using machine learning and soft computing techniques. Ishita Parmar et al. in their research paper [2] used a regression-based model using long short-term memory to predict accuracy of a given stock. The trend line between predicted and actual result shows similarity. However, there is a significant error for majority part of the predicted output. V. Kranthi Sai Reddy in his research paper has used radial basis function and support vector machine algorithm by having stock price volatility, stock momentum, index volatility, and index momentum as features. This model gives a better result, and error is reduced significantly [3]. Similar studies have been done using Artificial Neural Network using back propagation by the author Bikramaditya Ghosh [4]. Aparna Nayak M. et al. in their paper have predicted the stock market movement based on sentiments from tweets and news, and it also uses stock related information from Yahoo finance site. This is a holistic approach for prediction of the stocks [7]. Kumari A et al. in their research paper had used a model to extract information from social network site for analysis. This can be used to further progress this research paper by using market sentiment for better accuracy [10]. The authors of this paper also studied paper which has a model to study impact of demonetization on the India stock market [9]. Majority of the research has been done using traditional machine learning models like feed forward, back propagation, SVM, and LSTM methodology [1, 1]. The authors will use Learning Vector Quantization which has not be used widely in stock market prediction.

4 Methodology

There are many machine learning algorithms and models used in various research papers. For this paper, the authors have selected Learning Vector Quantization algorithms to predict the stock movement by training the algorithm with historical data and making necessary assumptions about the relationships of various attributes with one another.

Learning Vector Quantization is a proto-type based supervised learning method useful in classifying patterns. LVQ takes in number of input vectors and assigns one or more prototype representing one or more class. Since this network uses supervised learning, the network will be provided with training data with known pattern of input vectors and output classes.

Learning Vector Quantization uses the following algorithm:

- Initialize reference (weight) vectors from set of training vectors, take first "m", numbers of clusters and utilize them as weight vectors, remaining vectors can be used for training.
- Assign initial weights and classification.
- Assign learning rate alpha.
- Calculate square of Euclidean distance for j = 1 to m and i = 1 to n.

$$D(j) = \sum_{i=1}^{n} \sum_{j=1}^{m} (x_i - w_{ij})^2$$

- Find winner neuron where D(j) is minimum.
- Update the weight on winning unit using the following conditions:
  If T = C_j
  then $w_j(new) = w_j(old) + \alpha[x - w_j(old)]$
  If T $\neq C_j$
  then $w_j(new) = w_j(old) - \alpha[x - w_j(old)]$
- Stop when maximum number of epoch is reached or learning rate is reduced to a negligible value.

4.1 Model

For the duration of research, the following model was created to aid the process and help in evaluating the findings. This model consists of multiple phases. The process starts with the phase 'Getting raw historical data for a given stock'. In this phase we evaluated various internet sites to get historical stock related information. In this digital transformation age, historical stock data is readily available. The next phase is 'Attribute selection and Feature extraction' where the data attributes are selected based on their ability to help in the prediction, in this phase the data is transformed and features like 'Strength Index' and 'Moving Averages' are derived. The next step is to identify and classify the prototypes for training vectors. In the next phase the input dataset will be treated to Machine Learning Algorithm—Learning Vector Quantization where the data will be trained. In the next phase a part of input vectors will be simulated. Once the simulation is complete the output is evaluated for correct classifications for the vectors.

5 Experimental Results

The proposed model is trained and tested with four different datasets. Each dataset has variation to the 'Monthly Average' and 'Strength Index' attributes with all other attributes such as opening price, closing price, maximum price, minimum price, and average price remaining the same. Monthly average is calculated as sum of average price for 'n' no of days divided by 'n'. For this research moving averages is calculated with 5, 10, 15, and 30 day's average. Strength index is calculated as number of stock rises divided by number of stock rises and lows for a given 'n' days.

Table 1 displays 'Buy' and 'Sell' confusion matrix for given moving averages with a test size of 33% selected randomly from the input datasets. For 5 days moving averages the model predicted buy and sell signal with over 80% accuracy. Accuracy for 10 days moving average went down to 72% followed by 15 days accuracy to 69% and 30 days accuracy to 68%. The trend of accuracy remains the same in Table 2 where for 5 days moving average the accuracy is over 77% followed by 73% in 10 days moving average and 70% in 15 days moving average 66% in 30 days moving average.

Table 1 Confusion matrix and accuracy percentage across moving averages for 33% test data

	Moving Averages—5 Days		Moving Averages—10 Days		Moving Averages—15 Days		Moving Averages—30 Days	
Accuracy%	80.22		72.20		69.05		68.19	
	Buy	Sell	Buy	Sell	Buy	Sell	Buy	Sell
Buy	146	31	120	48	122	35	118	35
Sell	38	134	49	132	73	119	76	120

Table 2 Confusion matrix and accuracy percentage across moving averages for 20% test data

	Moving averages—5 days		Moving averages—10 days		Moving averages—15 days		Moving averages—30 days	
Accuracy%	79.71		73.11		70.28		66.98	
	Buy	Sell	Buy	Sell	Buy	Sell	Buy	Sell
Buy	78	22	64	24	61	23	61	25
Sell	21	91	33	91	40	88	45	81

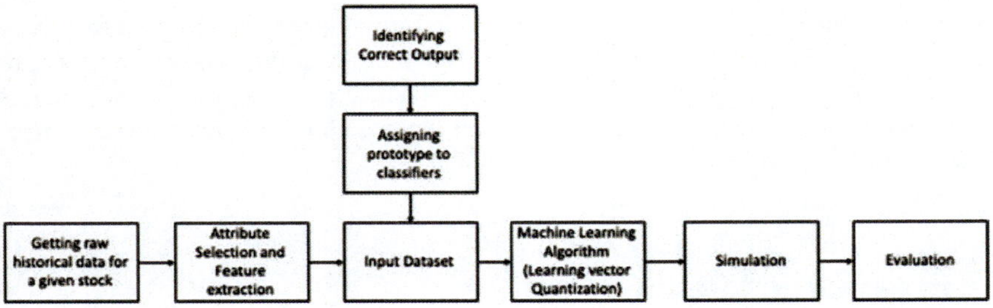

Fig. 1 Research model used for training and testing data

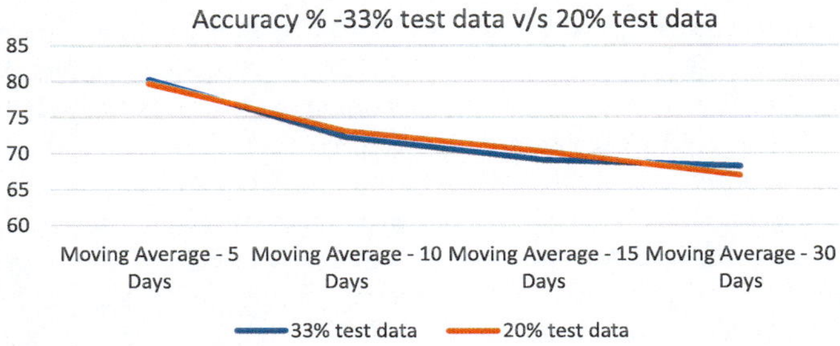

Fig. 2 Accuracy percentage versus moving averages

Figure 2 above shows trend line of accuracy across the four datasets for 33% randomly selected test data versus 20% randomly selected test data. As can be seen in the above graph, for both the sample size for test data the accuracy percentage is in line with each other and linear. The accuracy drops with the increase in days used to calculate moving average. When the moving average of a stock price is calculated using 5 days as time period it helps in predicting the stock movement better. It can be deduced that the current market sentiment over past 5 days prevails over the market sentiment over 10 days. The more historic data is used, and the market sentiment gets diluted and thus decreases the accuracy.

In below Fig. 3, the results of the experiment are represented by ROC curve for 33% of test data. In this graph TPR and FPR are plotted for the said moving averages. TPR and FPR is determined using the following formulae:

$$TPR(TruePositiveRate) = \frac{TP(TruePositive)}{TP(TruePositive) + FN(FalseNegative)}$$

$$FPR(FalsePositiveRate) = \frac{FP(FalsePositive)}{FP(FalsePositive) + TN(TrueNegative)}$$

Performance of a model can be measured using AUC (Area under the ROC Curve). AUC is the entire two-dimensional area underneath the ROC curve. AUC helps in

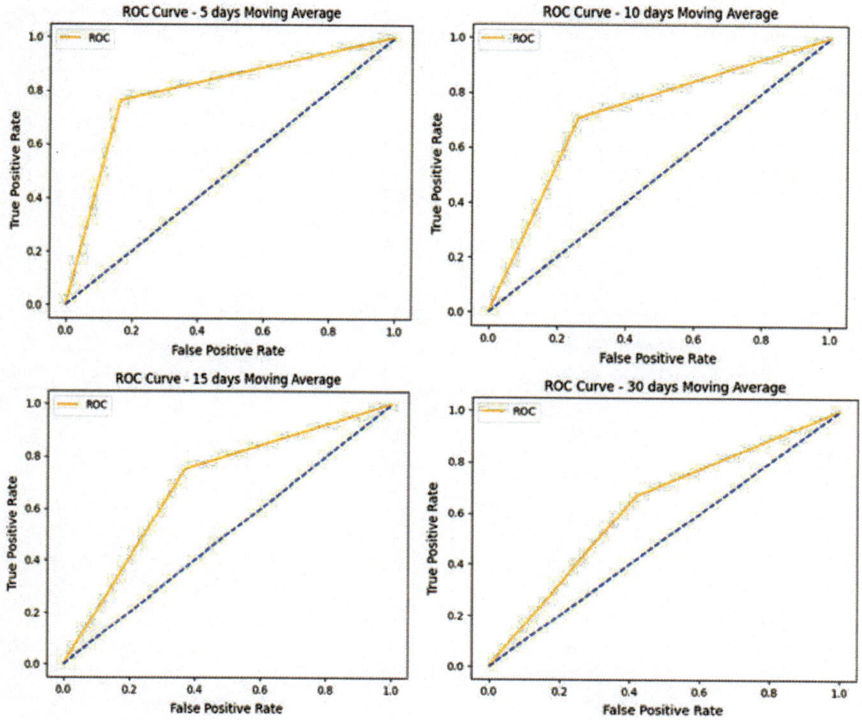

Fig. 3 ROC Curve with 33% test data

analyzing the strength and predictive power of a classifier. A good classifier should have a ROC closer to the top left of the curve because a good classifier has a better true positive rate at lower false positive rate. This also helps in comparing models and selecting the most suitable model.

In the below Fig. 3 AUC of 5 days moving average is better than the other samples selected. With AUC of 0.80 in 5 days, 0.72 in 10 days, 0.69 in 15 days, and 0.66 in 30 days it can be clearly inferred that the lower the number of days used to determine the moving average the better is the predictive model.

Similar trend can be seen when the test data is reduced to 20% of entire population in Fig. 4. Area under curve increases with the decrease in number of days taken for calculating moving averages. AUC is 0.79 for 5 days moving average, followed by 0.72, 0.69, and 0.66 for 10, 15, and 30 days respectively.

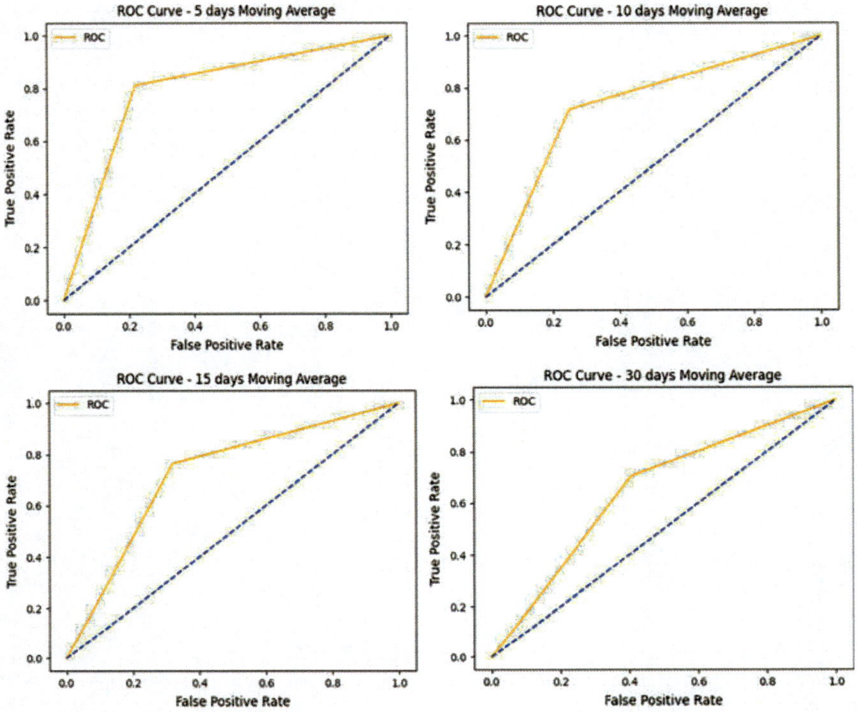

Fig. 4 ROC curve with 20% test data

6 Conclusion

The focus of the paper is to understand the accuracy of Learning vector quantization machine learning model in classifying the 'Buy' and 'Sell' signal across various datasets in which only moving averages varies. A comparison is made by varying the size of the test data for 5, 10, 15, and 30 days moving averages. The experimental result shows that the accuracy decreases when the moving averages are calculated for a larger number of days. The accuracy is inversely proportional to the 'n' in the number of days used to calculate moving average. Also, as the test data decreased, the accuracy also fell slightly for 5 days moving averages. It can be concluded that for a better accuracy there should be adequate test data and the moving averages should be shorter. Organizations can use this model in tandem with other micro and macroeconomic factor to make a sound decision on the value of a given stock.

7 Future Work

This model focuses mainly on moving averages and its correlation with the accuracy of classification of 'Buy' and 'Sell' signal. Also, the size of test data was varied to do a two-way comparison. The historical data was taken for last four years. This work can be further used to identify any improvement of accuracy when the historical data is taken for more recent history period. Apart from the two signals used, further research can use an additional signal for 'Hold'. An additional classifier can be used to give a better prospective of the stock. Future research can be done using additional attributes like price to earnings ratio, dividend yield percentage, and market cap ratio in the same segment.

References

1. M. Hiransha, E.A. Gopalakrishnan, V.K. Menonab, K.P. Soman, *ICCIDS 2018: NSE Stock Market Prediction Using Deep-Learning Models*
2. I. Parmar, N. Agarwal, S. Saxena, R. Arora, S. Gupta, H. Dhiman, L. Chouhan, *04–2018: Stock Market Prediction Using Machine Learning*
3. V.K.S. Reddy, *Stock Market Prediction Using Machine Learning*
4. B. Ghosh, *Comparative Predictive Modeling on CNX Nifty with Artificial Neural Network*
5. K. Zhanga, G. Zhonga, J. Donga, S. Wanga, Y. Wanga, *Stock Market Prediction Based on Generative Adversarial Network*
6. A. Bhardwaj, Y. Narayan, V. Pawan, M. Dutta, *Sentiment Analysis for Indian Stock Market Prediction Using Sensex and Nifty*
7. A. Nayak M.M.M. Pai, R.M. Pai, *Prediction Models for Indian Stock Market*
8. P. Mohapatra, A. Raj, T.K. Patra, *Indian Stock Market Prediction Using Differential Evolutionary Neural Network Model*
9. S. Chopra, D. Yadav, A.N. Chopra, *Artificial Neural Networks Based Indian Stock Market Price Prediction: Before and After Demonetization*
10. A. Kumari, R.K. Behera, K.S. Sahoo, A. Nayyar, A. Kumar Luhach, S. Prakash Sahoo, Supervised link prediction using structured-based feature extraction in social network, in *Concurrency and Computation: Practice and Experience* (2020), p e5839

Tree Hollow Detection Using Artificial Neural Network

Anand Upadhyay, Jyotsna Anthal, Rahul Manchanda, and Nidhi Mishra

Abstract Tree hollow is a semi-enclosed cavity in any kind of tree. The detection of a tree hollow is important not only for a tree but also for the species who use a tree hollow for their survival and settlement. A tree hollow plays a vital role for bird ecology for their survival, growth, and population; therefore, there is a need for detection of a tree hollow. This research paper worked on the same principle of detection of a tree hollow to make people aware of a tree hollow. Here, a feed forward neural network with back propagation of error neural network-based machine learning algorithm is used to automatically detect a tree hollow. The proposed algorithm is implemented using sklearn python-based packages. The implementation shows an accuracy of 82% for the detection of a tree hollow which is good results for detection of a tree hollow.

Keywords Tree hollow · Tree · Multilayer perceptron · Artificial neural network · Python · Machine learning

1 Introduction

Tree hollow is a semi-enclosed cavity that is formed naturally or artificially on the trunk of a living tree. It is formed due to the destruction of the internal tissues in a tree trunk. This resultant opening through the trunk of the tree leaves the interior part of the wood exposed to the external environment. The formation of tree hollows is usually the result of the vital activity of several species of saprophytic fungi and bacteria, often with the aid of ants and birds (woodpeckers). The tree that bears the hollow form habitat for varieties of species of birds, mammals, reptiles, and amphibians for breeding or shelter. Sometimes, the hollow may be the result of fungal

A. Upadhyay (✉) · J. Anthal · R. Manchanda · N. Mishra
Department of Information Technology, Thakur College of Science and Commerce Thakur Village, Kandivali(East), Mumbai, Maharashtra 400101, India

R. Manchanda
e-mail: rahulsmanchanda@live.com

A. K. Luhach et al. (eds.), *Second International Conference on Sustainable Technologies for Computational Intelligence*, Advances in Intelligent Systems and Computing 1235,
https://doi.org/10.1007/978-981-16-4641-6_23

or bacterial infections and dynamic events such as storms and wildfire. Studies show that hollows on trees can be formed in any trees, but there are more chances of hollow being formed in matured trees. The reason behind the formation of hollow on the branch of the tree can also be due to the self-pruning of the lower branches of the trees as they mature and exposes an area that results in a hollow or cavity. The size of the hollow of the tree depends on the age of the tree. The more the tree becomes older the more the hollow gets deeper. Trees which bear the hollow grow poorly and since they grow poorly they lose their solidity. If the hollow on trees is treated at an early stage, the life of the tree can be prolonged. For the purpose of detection of the tree hollow, machine learning techniques have proven to be effective [1]. Artificial neural networks (ANNs) are amongst the widely used machine learning techniques which have the capability to learn and it is capable to model non-linear and complex relationships. The capabilities of ANN have motivated us to develop an ANN-based classification model for the detection of the hollow on trees.

2 Literature Review

Machine learning enables computers to learn without explicit software development. Using machine learning the data can be made understood to the computers and new algorithms can be created for devising new techniques. The machine learning algorithm that is used for our data set is a multi-layer perceptron. MLP belonging to supervised learning is suitable for the classification and prediction problems where the inputs are assigned a class. Multilayer perceptron has a provision to solve the problems which are stochastic in nature for deriving the output for complex problems. There was a study wherein a neural network along with IoT was used for indexing the health of a tree, wherein multiple sensors are used for monitoring and assessing the health of the tree and the indexing was then used to assist with the preventive measures to be applied [2]. There were other studies that used additional devices like a ground-penetrating radar for the purpose to monitor the health of a tree, which could be applied using handheld antennas in real-time for monitoring purposes [3], and a Raman spectrometer along with chemometric analysis for identifying and detecting cankers and blight on the trees [4]. A study also exhibited the capabilities of remote sensing technologies to characterize the health of individual trees [5]. Various other studies also consider the usage of neural networks for detecting decomposition (rot) on the tree's stem [6] and for detecting a disease in plants [7]. Neural networks were also used in studies for predicting tree survival and mortality [1] and for mapping endangered species of trees [8]. However, there was not a study that was carried out in detecting the hollow(s) present on the trunk of the tree in an efficient manner without spending any cost for the inspection of the tree by an arborist. This study works on the principle of making the people aware of the hollow and for taking remedial actions for saving a tree from dying.

3 Methodology

Multilayer Perceptron (MLP)

Multilayer Perceptron is a class of feed forward neural network which is useful for classifying data that are linearly separable. Multilayer perception contains more than one hidden layer in the network and is useful for training data that are labelled. MLP has proven to be reliable for the analysis of hyperspectral data. In this study, MLP has been trained with the back propagation network. An MLP consists of an input layer, a hidden layer, and the output layer as shown in Fig. 1.

Input Layer: Initial data which requires further processing is brought to this layer. This layer works as a means for the data to be entered into the network.

Hidden layer: This being the second layer is responsible for the required computations to be performed on the data. This layer is the intermediary between the input layer and the output layer.

Output layer: This layer is responsible for the production of outputs based on the given inputs.

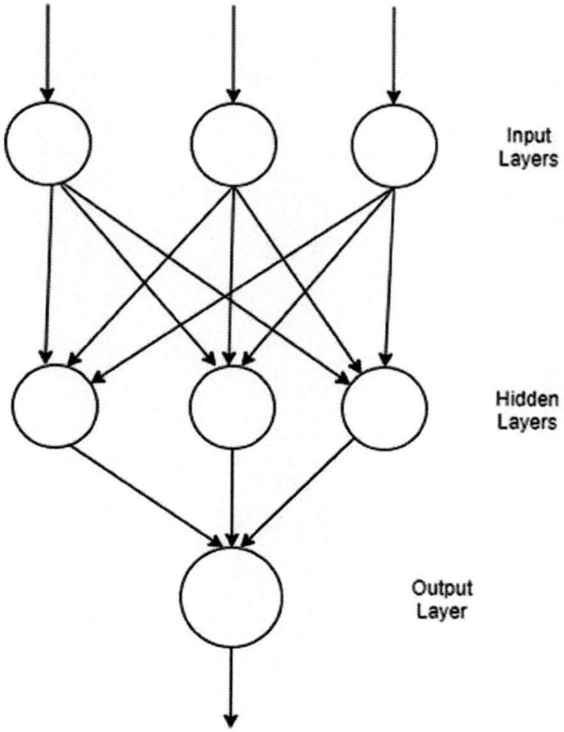

Fig. 1 Architecture of multilayer perceptron

The datasets that have been used for this research consist of images of trees with hollow and images of trees without hollow. These images were manually captured and cleaned before using it as an input was training the machine for the required classification. Following is the design of the process used after the image is cleaned (Fig. 2).

1. Image selection of trees.

This sub-process is the very initial step consists of primarily selecting a particular image from the system. Here for selecting a particular image, the user is given a prompt via a dialog box through which the user can navigate to the desired directory or a file in which the user has stored the desired image for further computations as shown in Fig. 3.

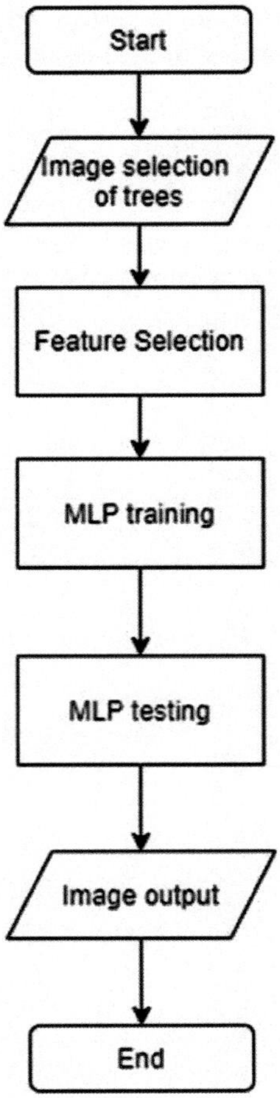

Fig. 2 Multilayer perceptron design

Fig. 3 Input image selection

2. Feature Selection.

This step is responsible for selecting the area of hollow from the trunk of the tree. This step plays a vital role since it selects the region of interest from the image and uses certain inputs obtained from it to train the model.

3. MLP Training.

To train the multilayer perceptron, we have used two input nodes. One node contains the input values derived from the images which contained the hollow, and the other node consisted of the values of the images which did not have a hollow. The outlier data from the input nodes due to their tendency to affect the overall accuracy of the model were then scaled using min–max transformation and then the MLP model was designed with a maximum iteration set to 300.

4. MLP Testing.

Based on the model which was trained, testing is then carried out for making the required classification. The more the model is trained the more accurate the result shall be from the testing.

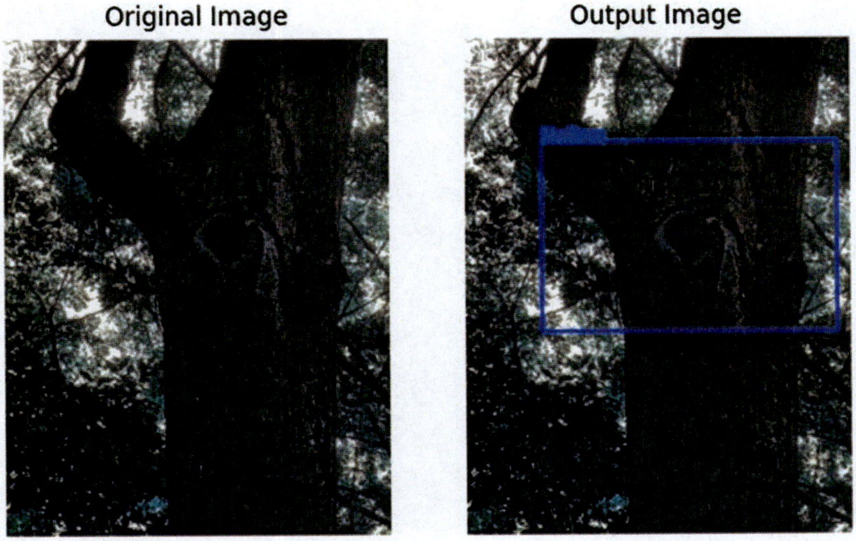

Fig. 4 The output

5. Output.

The output is then displayed to the user. The output window consists of an original image which was selected as an input and the second image (output image) with a frame that says whether a hollow is present (output: hollow) or whether it is absent (output: no hollow) as shown in Fig. 4.

4 Expected Result

As explained in the methodology an output was displayed via an image to the user. An accuracy of 82% was achieved. A confusion matrix was also created in order to summarize the performance of this classification algorithm.

Confusion matrix.

A confusion matrix is a tabular representation of the number of predictions being in the correct and in the incorrect category made by a classifier. Since it has the capability of presenting the correct and the incorrect classified instance(s) of a class one can thus get a better insight into the performance of the classifier (Table 1).

Above is the table of the confusion matrix. The respective accuracy can be calculated based on the values obtained from their respective cells.

Table 1 Table of the confusion matrix

	True positive values	True negative values
Predicted positive values	41	6
Predicted negative values	7	21

Receiver Operating Characteristic Curve.

Receiver Operating Characteristic Curve is a graph that shows the performance of the classification model at different threshold settings. The followings are its parameters.

True Positive Rate.

True Positive Rate is the proportion of the units with a known positive condition for which the predicted condition is positive. This rate constitutes the Y-axis on the ROC curve. The images of the tree which will contain a hollow shall be accurately reported. The formula of the true positive rate, which is synonymous with recall and sensitivity is

$$TPR = \frac{TP}{TP + FN}$$

Equation 1. True Positive rate.

False Positive Rate.

This rate is the proportion of units with a known negative condition for which the predicted condition is positive. The images of the tree which does not contain a hollow shall not be accurately reported. This rate constitutes the X-axis of the ROC curve (Fig. 5).

$$FPR = \frac{FP}{FP + TN}$$

Equation 2. False Positive rate.

Following is the ROC curve that was obtained using the above parameters.

5 Conclusion

Our planet Earth has lost 11.9 million hectares of tree cover in the year 2019. An impact has also been caused to the vertebrate and the invertebrate species which depend on the hollows of the trees causing a significant downfall in their population.

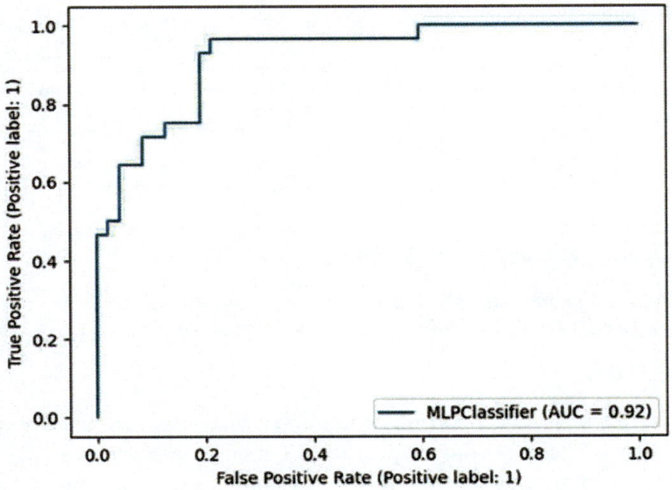

Fig. 5 Roc curve

Maintaining the tree and making people aware of the tree hollow hence becomes critical. This proposed paper utilizes an idea via the usage of machine learning to achieve the required classification thesis based on the hollow feature on the tree. Usage of perceptron learning gave a good result in terms of accuracy.

6 Future Scope

One can further look into a location-based classification based on a specific geographical area for a better understanding of the trees in that particular perimeter. One can also look into the implementation of different algorithms if it helps to increase the overall accuracy. Automation concerning this classification can also be programmed in machines for further computations regarding the health of a tree.

References

1. M. Bayat et al., Application of artificial neural networks for predicting tree survival and mortality in the Hyrcanian forest of Iran. Comput. Electron. Agric. **164**, 104929 (2019)
2. C.K. Wu et al., An IoT tree health indexing method using heterogeneous neural network. IEEE Access **7**, 6617666184 (2019). https://doi.org/10.1109/ACCESS.2019.2918060
3. I. Giannakis et al., Health monitoring of tree trunks using ground penetrating radar. IEEE Trans. Geosci. Remote Sensing **57**(10), 8317–8326 (2019)
4. L. Sanchez et al., Detection and identification of canker and blight on orange trees using a hand-held Raman spectrometer. J. Raman Spectrosc. **50**(12), 1875–1880 (2019)

5. I. Shendryk et al., Mapping individual tree health using full-waveform airborne laser scans and imaging spectroscopy: a case study for a floodplain eucalypt forest. Remote Sens. Environ. **187**, 202–217 (2016)
6. P. Ahmadi et al.,Early detection of Ganoderma basal stem rot of oil palms using artificial neural network spectral analysis. Plant Dis. **101**(6), 1009–1016 (2017)
7. K. Golhani et al., A review of neural networks in plant disease detection using hyperspectral data. Information. Process. Agric. **5**(3), 354–371 (2018)
8. G. Omer et al., Performance of support vector machines and artificial neural network for mapping endangered tree species using WorldView-2 data in Dukuduku forest, South Africa. IEEE J. Sel. Top. Appl. Earth Obs. Remote. Sens. **8**(10), 4825–4840 (2015)

Accident Identification and Alerting System Using ARM7 LPC2148

Palanichamy Naveen, A. Umesh Chandra Reddy, K. Muralidhar Reddy, and B. Sandeep Kumar

Abstract Nowadays vehicle accidents are happening often, and it's terrible to assist them in maximum conditions. So, this research paper would like to build an automatic system to observe and inform, whenever accidents occur while traveling. This system will be designed by using ARM, GPS, MAX232, and GSM. Whenever an accident takes place, the automatic and manual alarms are realized. The vehicle position and user information are going to be transmitted to the pre-set of treatment centers through a global system for mobile communication in the form of messages along with a global positioning system location. Once the treatment center will open the message it will provide direction to the victim's location, so with this system, we can able to save the accident victim's life in maximum cases.

Keywords Accident identification · Accident alert · GSM · GPS · ARM 7 LPC2148 · MAX232 · MEMS sensor

1 Introduction

Nowadays, it is exceptionally troublesome to discover a mishap. It is much more troublesome for the victim's lives, indeed anything the individual knows the data and reports it to Crisis vehicles such as ambulances or healing centers and in case they happen in inaccessible ranges that there will be no trust of survival. To dodge this, there are diverse innovations such as GSM/CDMA, and Worldwide situating frameworks are utilized. Depending on GPS occurrence distinguishing proof unit contains Miniaturized scale Electro-mechanical framework (Micro-electro-mechanical system), collision sensor, infrared sensor, fire sensor, and global positioning system unit associated with arithmetic device. At the minute of the mischance, collision Sensor, Micro-electro-mechanical system identifies that a mishap giving the data is sent to the ARM 7 LPC 2148, who will show the data on the liquid crystal display screen, turn on the alarm and send the data to rescue vehicle to the police

P. Naveen (✉) · A. Umesh Chandra Reddy · K. Muralidhar Reddy · B. Sandeep Kumar
Kalasalingam Academy of Research and Education, Krishnankoil, India

A. K. Luhach et al. (eds.), *Second International Conference on Sustainable Technologies for Computational Intelligence*, Advances in Intelligent Systems and Computing 1235,
https://doi.org/10.1007/978-981-16-4641-6_24

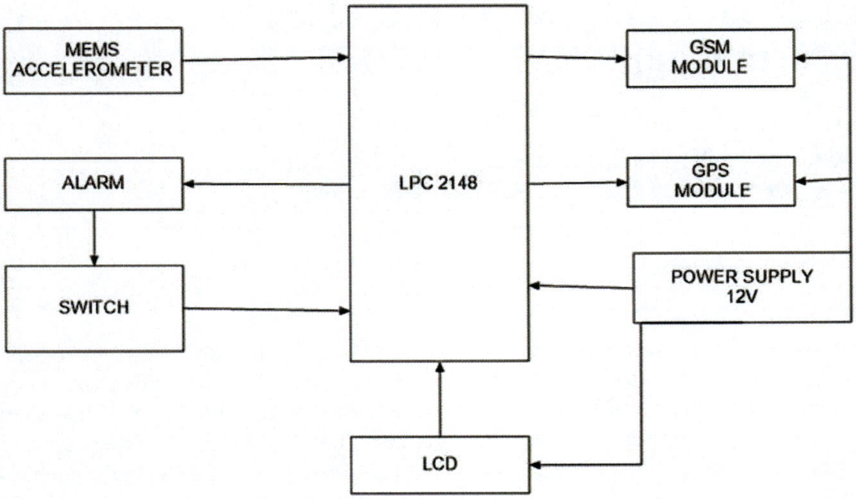

Fig. 1 Proposed block diagram

and the proprietor/guardians through the GSM network. Here, the framework moreover gives the client with the capacity to track a Vehicle location. This gives quick reaction and rescue the accident victims. The entire paper depends upon ARM-based microcontroller. This controller is used for coordinating all activities in the system. Ingredients details are ARM 7 (LPC 2148), the accelerometer (Mems), GPS, and GSM (Figs. 1 and 2).

2 Problem Statement

At present situation, we are unable to find where the mishap has happened and consequently no data is correlated to it driving to the decease of a person. Still some research is going on for locating the co-ordinates of the automobile indeed in dim awkward areas, in which we are unable to establish the communication. In our venture, global positioning system is utilized for following the co-ordinates of the automobile. global system for mobile communication is utilized for conveying the information, and the Automatic voltage regulator controller is utilized for preserving our contacts within the memory and sending the information to global system for mobile communication when a mischance has been identified. Hence, with this venture usage, we can readily locate the co-ordinates of the automobile where the mischance has happened accordingly, and we can furnish the primary help as soon as conceivable (Figs. 3 and 4).

Fig. 2 Flow chart of the proposed system

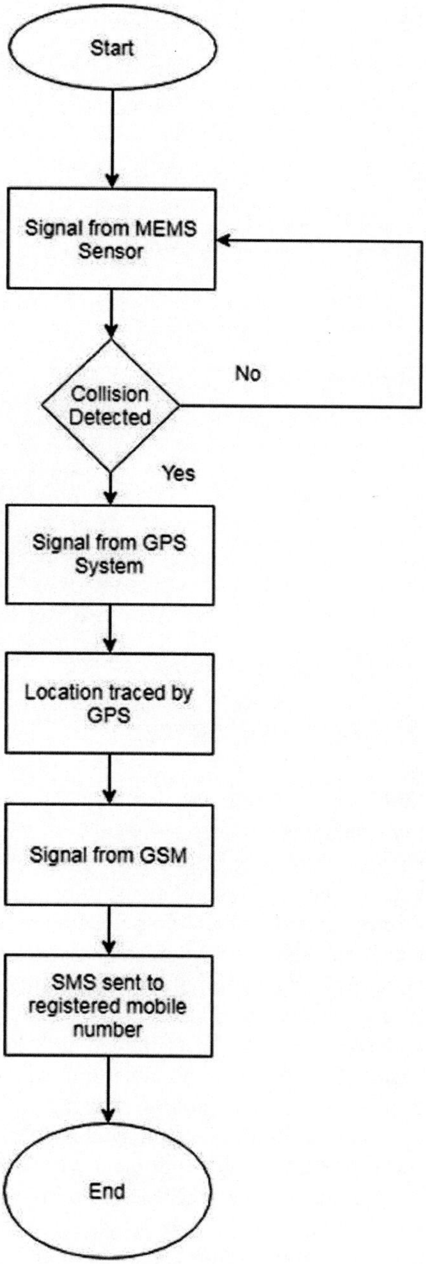

Fig. 3 LPC2148

Fig. 4 GPS

3 Literature Survey

In this paper, Bankar Sanket Anil [1] provides an idea how first aid can be provided to mishap victims. Here, mishap can identify using global system of mobile communication, global positioning system, collision sensor and co-ordinates of a mishap can be shared to our relatives, police services, emergency services like ambulance through global system of mobile communication in the form of messages. This paper can check status of the mishap victims by keeping camera inside the automobile. Here, this paper can able to detect the co-ordinates of an automobile using global positioning system [2, 3].

Ramadan et al. [4] explained anti-theft using global positioning system, global system for mobile communication, embedded hardware systems. Here also the automobile theft can be monitored using global system for mobile communication, and automobile communication can be identified by using global positioning system. In this project, the used equipment will be very low price [5, 6].

Wankhade [7] established theft control system for an automobile. In this work, they had used microcontroller global positioning system and global system for mobile communication. When the automobiles are theft by another person then the global system for mobile communication gives the information in the form of message. SIM in global system for mobile communication and global positioning system gives the exact co-ordinates to the user [8, 9].

Dhole [10] proposed that this paper is totally based on smart mishap's identification and providing the user with an automobile observing system. The current position of the automobile can reache the user through the server [11].

Lakshmi [12] explained how information will reach in a fraction of seconds after automobile mishap. Here additionally provided a buzzer system, position of the mishap victim can reach the user using API software and co-ordinates of mishap can be known using global positioning system [13].

Noopur Patne [14] implemented as smart element in order to decrease the road mishaps. Because, nowadays after mishaps people may die due to lack of shield to their head. We can able to stop the vehicle if and only if we wear the smart element. Additionally, they had included alcohol sensor also [15, 16].

Nowadays road mishaps are happening day-to-day life due to the rise of population. The death rate can be reduced if we can able to reach the mishaps victim's position in a less period. This paper shows how we have to overcome this issue from road accidents [17].

In this paper, they are going to use a technology called "SONAR" which will work based on wireless technology in order to prevent the mishaps. Here also mishaps victim co-ordinates can be received using global positioning system through global system for mobile communication [18].

In this paper, they have additionally implemented Android app through which the user can able to observe the state of the automobile and mishap's victim. This methodology can also be used in finding the vehicle from the theft [19].

In this proposed paper, they have completed the project with low cost. They also included alcohol sensor in order to decrease the drunk and drive mishaps. Here, the user can able to drive the automobile if and only if he has completed the alcohol test. If the user takes alcohol, he may not able to start the vehicle. Here, co-ordinates of the automobile can be known using global positioning system [20].

Road traffics are directly related to the road accidents, and increase in the road traffics may cause mishap often. Software Defined Networking (SDN) can able to monitor the road traffics which may decrease mishap deaths [21] system (Figs. 5 and 6).

Fig. 5 MAX232

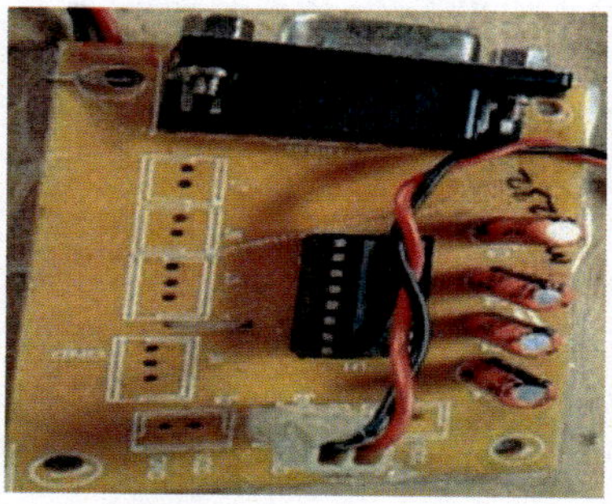

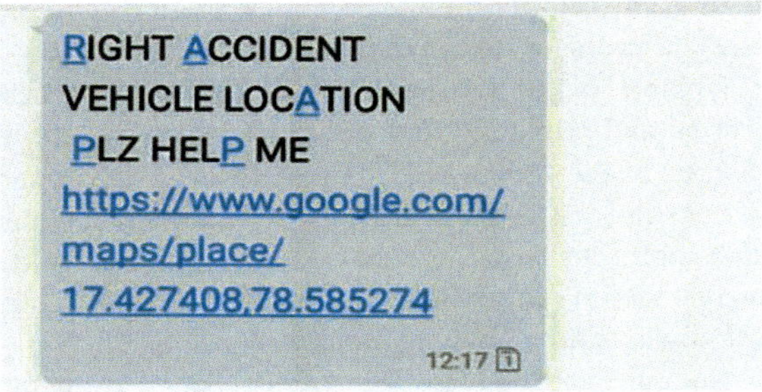

Fig. 6 Sample SMS received

4 Methodology

1. Our research paper demonstrates a programmed vehicle mishap location and informing utilizing GSM and GPS electronic equipment utilizing ARM7 operating are created among the taking below steps:
2. An electricity detector can observe the event of a mischance and supply its yield to the microcontroller.
3. Global positional system acknowledges the co-ordinates of an automobile.
4. The co-ordinates of the vehicle are distributed as message through the global system for mobile communication.
5. The signal is pre-recorded among the electronically erasable programmable read-only memory.
6. At no matter if a mischance is going on, the location is known, and a data have been sent to already registered mobile numbers.

4.1 Block Diagram

See Fig. 1.

4.2 Flowchart

See Fig. 2.

4.3 Algorithm

Step 1 Begin the program.
Step 2 Wait for signal from MEMES sensor.

Step 3 If collision condition is detected, go to next step, otherwise go to previous step and start the iteration once again.
Step 4 Get signal from GPS system.
Step 5 Get the location co-ordinates of the victim using GPS.
Step 6 Get signal from GSM and send messages to pre-registered mobile numbers.
Step 7 END the program.

4.4 Working Procedure

Our entire project work can be explained in two ways.

Step 1 Signal detection from MEMS sensor.
Whenever an accident takes place, MEMS sensor gets activated and sends data to the ARM7 LPC2148 about vehicle collision. ARM7LPC2148 sends the data (signal) to the GSM besides it gets location co-ordinates from GPS. The LCD display gives the value of tilted position angle of our vehicle.
Step 2 Location detection and sending co-ordinates through GSM via messaging system.

5 Proposed Work

The given document proposes the plan and execution of mishap alert framework built with remote arrange communications based on global positioning system, advanced RISC machines, and global system for mobile communication. The setup of medication will be provided to the mischance area. The fundamental portion is the setup of medication unit, which acts as a Data handling unit. Person vehicle is prepared with a framework called as mischance sense framework, which comprises global system for mobile communication and global positioning system. When the mishap happened, vehicles state and areas will be transmitted to the pre-set of treatment middle although remote communication innovations of global system for mobile communication in message form system.

6 Components and Figures

The given document proposes the plan and execution of mishap alert framework built with remote arrange communications based on Global positioning system, Advanced RISC machines and Global system for mobile communication, setup of medication will be provided to the mischance area. The fundamental portion is the setup of medication unit, which acts as a Data handling unit. Person vehicle is prepared with a framework called as mischance sense framework, which com-prises of Global system for mobile communication and Global positioning sys-tem. When the mishap happened, Vehicles state and areas will be transmitted to the Pre-set of treatment middle although remote communication innovations of Global system for mobile communication in message form system [22].

6.1 LPC2148

LPC 2148 has an inbuilt chip memory of 512 kb which works on the operating voltage of 3.3 V, and it has a crystal of 12 MHz. It is a 16- or 32-bit microcontroller which has a flash memory of 512 kb and 40 kb of on chip static.

6.2 Global Positioning System (GPS)

1. Global positioning system is an electronic device which provides location to the user.
2. It has user, space, and control segments.
3. Nowadays we are going to use global positioning system in automobiles, mobiles, etc.
4. By using this feature, we can able to detect our theft automobiles and mobiles.

Applications

1. Navigation.
2. Remote sensing.
3. Mapping and surveying.

6.3 GSM

1. Nowadays GSM service is easily available to everyone.
2. You need not to carry a different remote.
3. It is used to send SMS.
4. It provides serial interface.
5. It can be controlled through serial AT commands.

6.4 LCD

Liquid crystal display is an electronic device which indicates the status of our venture output.

6.5 MEMS Sensor

MEMS sensor is known as vibration sensor or collision sensor, which is used to detect the vehicle collision. It can detect the collision over three axes, i.e.,X-axis, Y-axis, and Z-axis. In real world it can detect back, front, right, left, and accident of vehicle.

6.6 MAX232

It is a double transmitter or dual receiver that regularly is utilized to change over the signals. It is an co-ordinates circuit which changes over the signals from the serial port to the right flag which are utilized within the TTL consistent computerized advanced circuits (Tables 1 and 2).

Table 1 Comparison with our proposed work

Parameters	Existing work (%)	Proposed work (%)
Speed	92.01	96.84
Logic accuracy	98.45	99.05
Cost	80.15	72.4
Modem	80	100

Table 2 Integrated test result

Case	Expected output result	Observed output result	Test output result
When accident is detected load cell can be able to detect the vibration	Can detect	Can detect	Pass
GPS module can be able to detect location co-ordinates of vehicle	Exact co-ordinates of the vehicle should be detected	Exact co-ordinates of the vehicle should be detected	Pass
GSM module should send SMS	SMS will be sent	SMS will be sent	Pass
Microcontroller can be able to send data to server via Wi-Fi module and retrieve the information through server	It can send and retrieve information	It can send and retrieve information	Pass

Table 3 System test result

Case	Expected output result	Observed output result	Test output result
User can be able to get SMS	SMS should be received	SMS should be received	Pass
User can be able to identify victim accident co-ordinates	Accident co-ordinates should be identified	Accident co-ordinates should be identified	Pass
Ambulance can be able to locate the accident spot	Ambulance must identify location with proper notification	Gets the required proper notification	Pass
Admin should be able to update server periodically	Can update the server data	Can update the server data	Pass

6.7 EEPROM

Our relative's phone numbers can be stored in EEPROM by the admin which can be changeable at any point of time. The data can be restored or retained even though power supply is in off state for longer duration (Table 3).

7 Comparison with Our Proposed Work

8 Result and Conclusion

The modern examine proposes and implements a clever coincidence detection and rescue gadget for busy metropolitan cities. This gadget guarantees clever coincidence detection and rescue mechanism which could offer the precise area of the coincidence. It additionally aids any ambulance to get direction. Ambulance motive force can take the shortest direction to the mishap spot and thereby rescue the sufferers with inside the least possible timeframe. Besides, our gadget places no less than financial stress at the consumer to put into effect than many different systems, but with a greater dependable output.

References

1. B. S. Anil, Kale Aniket Vilas, Prof. S. R. Jagtap, Intelligent System for Vehicular Accident Detection and Notification, 978–1–4799–3358–7114/$31.00 ©2014 IEEE International Conference on Communication and Signal Processing, April 3–5,2014, India
2. Abdullah Mubarak, Nirmala Murali, Harish Anantharaman, "MCAS: A collision sustenance mechanism for motorcyclists", Smart Systems and Inventive Technology (ICSSIT) 2018

International Conference on, pp. 66–71, 2018.
3. Sonjoy Rana, Shounak Sengupta, Sourav Jana, Rahul Dan, Mahamuda Sultana, Diganta Sengupta, "Prototype Proposal for Quick Accident Detection and Response System", Research in Computational Intelligence and Communication Networks (ICRCICN) 2020 Fifth International Conference on, pp. 191–195, 2020.
4. Montaser N. Ramadan, Mohammad A. Al-Khedher, Senior Member, IACSIT, and Sharaf A. Al-Kheder "Intelligent Anti-Theft and Tracking System for Automobiles" Vol. 2, No. 1, Feb 2012.
5. Yang, G-X., Li, F-J., (2010) Investigation of Security and Defense System for Home Based on Internet of Things, IEEE International Conference on Web Information Systems and Mining (WISM), 2010. https://doi.org/10.1109/WISM.2010.32
6. Souvik Roy, Akanksha Kumari, Pulakesh Roy, Rajib Banerjee, "An Arduino Based Automatic Accident Detection and Location Communication System", Convergence in Engineering (ICCE) 2020 IEEE 1st International Conference for, pp. 38–43, 2020.
7. Pravada P. Wankhade and Prof. S.O. Dahad Government College Of Engineering/Department of Electronics and Telecommunication, Amravati (Maharashtra), India "Real Time Vehicle Locking and Tracking System using GSM and GPS Technology-An Anti-theft System ", Jan –March 2011- Vol.2.No.3
8. Ayten Ozge Akmandor, Hongxu YIN, Niraj K. Jha, "Smart Secure Yet Energy-Efficient Internet-of-Things Sensors", Multi-Scale Computing Systems IEEE Transactions on, vol. 4, no. 4, pp. 914–930, 2018.
9. Neel Desai, Pratha Kulkarni, Lijin Joy, Prachi Raut, "IoT based Post Crash Assistance System", Trends in Electronics and Informatics (ICOEI)(48184) 2020 4th International Conference on, pp. 734–739, 2020.
10. Pranav Dhole1, Saba Shaikh2, Nishad Gite3, Vijay Sonawane4, Sandip Institute of Technology and Research Centre, Nashik, India. International Journal of Emerging Trends in Science and Technology,IJETST- Vol.||02||Issue||04||Pages 2285–2288||April||ISSN 2348–9480.
11. Mohd Aliff Mushthalib, Hasmah Mansor, Zulkifli Zainal Abidin, "Development of eCall for Malaysia's Automotive Industries", Mechatronics Engineering (ICOM) 2019 7th International Conference on, pp. 1–5, 2019.
12. Vidya Lakshmi, J.R.Balakrishnan, Anand Institute of Higher Technology, Affiliated to Anna University. International Journal of Scientific and Research Publications, Volume 2, Issue 4, April 2012.
13. Pranto Karmokar, Saikot Bairagi, Anuprova Mondal, Fernaz Narin Nur, Nazmun Nessa Moon, Asif Karim, Kheng Cher Yeo, "A Novel IoT based Accident Detection and Rescue System", Smart Systems and Inventive Technology (ICSSIT) 2020 Third International Conference on, pp. 322–327, 2020.
14. Noopur Patne, Mangala Madankar, G.H. Raisoni College of Engineering,Nagpur, India, IRF International Conference, 13th April-2014, Pune, India.
15. Shubhankar Lipare, Prasenjit Bhavathankar, "Railway Emergency Detection and Response System using IoT", Computing Communication and Networking Technologies (ICCCNT) 2020 11th International Conference on, pp. 1–7, 2020.
16. P. Naveen, P. Sivakumar, Adaptive morphological and bilateral filtering with ensemble convolutional neural network for pose-invariant face recognition. J Ambient Intell Human Comput (2021). https://doi.org/10.1007/s12652-020-02753-x
17. P.Kaladevi, T.Kokila, S.Narmatha, V.Janani "Accident Detection Using Android Smart Phone" March 2014
18. Rashida Nazir, Ayesha Tariq, Sadia Murawwat*, Sajjad Rabbani "Accident Prevention and Reporting System Using GSM (SIM 900D) and GPS " 2014
19. Pratiksha R. Shetgaonkar, VijayKumar NaikPawar , Rajesh Gauns "Proposed Model for the Smart Accident Detection System for Smart Vehicles using Arduino board, Smart Sensors, GPS and GSM" July- August 2015
20. Mr.Dinesh Kumar HSDK, Shreya Gupta, Sumeet Kumar, Sonali Srivastava "Accident Detection and Reporting System Using GPS and GSM Module." May 2015

21. Kumar, A., Krishnamurthi, R., Nayyar, A., Luhach, A. K., Khan, M. S., & Singh, A. (2021). A novel Software-Defined Drone Network (SDDN)-based collision avoidance strategy for on-road traffic monitoring and management. Vehicular Communications, 28, 100313
22. Kumar, A., Krishnamurthi, R., Nayyar, A., Luhach, A. K., Khan, M. S., & Singh, A. (2021). A novel Software-Defined Drone Network (SDDN)-based collision avoidance strategy for on-road traffic monitoring and management. Vehicular Communications, 28, 100313.

Skin Cancer Detection and Severity Prediction Using Computer Vision and Deep Learning

Sangeeta Parshionikar, Renjit Koshy, Aman Sheikh, and Gauravi Phansalkar

Abstract Amongst all the diseases prevalent in today's world, dermatological diseases take its place at the top of the ladder. Although it is ubiquitous, its diagnosis isn't quite as easy as its occurrence and experience plays a crucial role. The most common malignancy affecting humans is skin cancer. Traditional diagnosis of this disease comprises clinical screening and biopsy. Categorization of skin ruptures using images is a difficult task because of minute granular variations in the appearance of skin ruptures. This paper dwells on creating a system that efficiently combines Computer vision and Deep Learning on dermoscopic images and identifies the type of skin cancer. We used Computer Vision which recognizes and exercises images in the same manner that human vision does. We further use Convolution Neural Networks (CNNs) to recognize skin cancer based on specific peculiar pathological attributes observed in analyzing the skin. Based on its severity, we further divide it into five classes. The attributes to distinguish the images into different classes are the colour and size of the affected region. The system also contains a one-class classifier that eliminates any false image and processes only dermoscopic images to obtain precise image analysis results. The proposed method, which uses Transfer Learning to predict the presence and severity of dermatological diseases, proves to be efficient and promising. We achieved an accuracy of 79.96% compared to 77.5% in the ensemble learning method and 94.37% compared to the SVM-based system. Severity being a new domain in this field, we have achieved an accuracy of 79.54%.

Keywords Skin cancer · Melanoma · Transfer learning · Mobilenet

S. Parshionikar · R. Koshy (✉) · A. Sheikh · G. Phansalkar
Department of Electronics, Fr. CRCE, Mumbai University, Mumbai, India

S. Parshionikar
e-mail: sangeeta@frcrce.ac.in

A. K. Luhach et al. (eds.), *Second International Conference on Sustainable Technologies for Computational Intelligence*, Advances in Intelligent Systems and Computing 1235,
https://doi.org/10.1007/978-981-16-4641-6_25

1 Introduction

The skin has three basic layers [1]: an outer layer (epidermis), middle layer (dermis) and the innermost layer (subcutaneous layer). The epidermis is constantly exposed to damage from factors like exposure to harsh environmental conditions, wear and tear. The skin cells intact have the job of multiplying rapidly to replace the damaged skin cells. Sometimes these cells begin to reproduce or multiply extensively at a very rapid rate leading to a skin tumour that may either be benign or skin cancer.

There are many variants of skin conditions [2]. Out of these variants, the ones we are focusing on are as follows.

1.1 Actinic Keratosis

It is a pre-cancer formation on the skin which occurs due to long term exposure to ultraviolet radiation. (Appearance: Small, dry, scaly patches of skin; Its colour may be red, light or darn tan, flesh toned, white, pink or sometimes a combination of all these colours and may sometimes be raised).

1.2 Basal Cell Carcinoma

It arises from irregular, unmanageable growth of basal cells [3]. (Appearance: It looks like open sores, red patches, reddish pink bumpy growths. It may be elevated, have rolled edges and a central indentation).

1.3 Benign Keratosis

It is a skin growth which is not cancer (benign), and it is causes due to constant exposure to sunlight. Its colour can range from white to brown or black raised areas.

1.4 Dermatofibroma

It is a small benign painless growth whose reason for occurrence is not known. They can be pink, grey, red or brown in colour and their colour may change over the years. Their size is lesser than an inch in diameter.

1.5 Melanocytic Nevi

Melanocytic nevi are benign neoplasms (new and abnormal growth) that constitutes melanocytes (the pigment-producing cells that constitutively settle in the epidermis).

1.6 Melanoma

Melanoma is a grave variant of skin cancer that begins in body cells known as melanocytes. It is a very hazardous because it spreads to other organs uncontrollably if not treated early. Melanoma occurs when damage in the DNA from burning exposure to UV radiation triggers mutation in the melanocytes causing alarming cell growth. Melanoma visually appears in various sizes, colours and shapes [4, 5].

1.7 Vascular Skin Lesions

They are of various types and consist a broad category. This includes pyogenic granuloma which is an acquired lesion, some which are present at birth just following birth called vascular birthmarks and vascular irregularities. Vascular irregularities are like permanent scars, having localized defects.

The system predicts the skin cancer type based on the features of the skin lesion. Firstly, the input image is processed to determine its class, i.e. if it belongs to skin lesions or an unknown class. If the input image belongs to known class of skin lesions, it is further processed to detect the type of skin cancer. As the prime focus of the project is on detection of Melanoma, if the system detects the image as Melanoma, it is further processed to detect the severity of Melanoma.

The arrangement of the remaining part of this paper as follows: Section II provides a six-step proposed methodology for the proposed system where we discuss the preprocessing techniques for image enhancement and system compatibility, the model development, and model optimization phases in detail. It also discusses the one-class classification method in detail. Section III showcases the results to support the approach, and finally, Section IV concludes the work and discusses its future scope.

2 Proposed Methodology

2.1 Dataset Preparation

The HAM10000 dataset was used to train and verify the model. It is an openly available dataset and can be used to develop classification models for different skin lesions. The dataset was divided in the proportion of 80:20, the training set consists of 8912 images, and the validation set consists of 1103 images [6].

2.2 Image Preprocessing

We preprocess the training images using the following steps before we utilize them in our deep learning algorithm:

Image Resizing—Image resizing is essential to enhance or diminish the cumulative number of pixels, resize or distort an image from a one-pixel grid to another. This instance is used for uniformity to increase the computational speed when used in Convolution Neural Network. The images are resized to pixel sizes of 224*224 px as supported by the MobileNet architecture used in this system.

Denoising—The resized images are further denoised using two filters [7].

Sharpening Filter—A sharpening filter is employed to the resized and denoised images to sharpen the details of the infected region.

Median Filter—After the sharpening filter, a median filter reduces impulse noise in the digital images; furthermore, this filter reduces brightness and reflection.

Contrast Enhancement using Histogram Equalization—This approach increases the global contrast of images when close contrast values represent its essential data which allows for areas of lesser local contrast to gain a more significant contrast [8].

2.3 Model Development

This phase involves remodelling the MobileNet architecture by adding a dropout and a compact output layer. We use dropout to avoid overfitting the data. After adding this layer, we add a final, dense layer with an activation function attached to it. This function uses all of the feature maps that it has accumulated and then renders us the prediction. On making these changes to the network structure, we halt every layer other than the last 23 layers to make the training events a lot faster. Hence our image data is trained on only the last 23 layers of architecture [9].

2.4 *Training Model*

We train the model for 20 epochs using an Adam optimizer beginning with a learning rate of 0.01 and Categorical Cross Entropy as loss function. The Call backs we employ in the training process are Model Check point and Reduce LROn Plateau. Model Check point supports the best version of the model to be saved automatically. Reduce LROn Plateau grants us a decrease in the learning rate by a factor of 0.5 if Validation Top 3 Accuracy does not improve after tolerance of 2 epochs.

2.5 *Severity Approach*

A similar approach as mentioned above implemented on a dataset consisting of 4199 images in Training Set and 44 images in Validation Set is used to classify the identified melanoma image into which stage of melanoma it lies under. The model is trained for 50 epochs with Validation Accuracy as the performance metric. Optimizer, Loss Function and Callbacks are same as defined for Skin Lesion Classification model.

2.6 *One Class Classification*

Deep learning has made significant advances in many machine learning obstacles. However, despite all its credit, if an unknown class object (image) is presented or given for prediction, the neural network foretells this class object as any on the n classes which it knows, i.e. it assigns it to belong to the class which it has a maximum correlation to which means it misclassifies it thus logically causing an incredible reduction in the accuracy of classification [10].

In order to deal with this problem, we incorporate a one-class classifier that predicts whether the input image belongs to the train class of skin lesions or of an unknown class by allowing images that have predicted probability below defined threshold. If the input belongs to the train class, the image can be further processed and if not, it will detect that the image does not fit into the domain being classified (Fig. 1).

Figure 2 depicts the implementation course of the proposed model.

3 Result

The performance of the model is displayed in Figs. 3, 4, and 5, which exhibits the plots for different performance metrics.

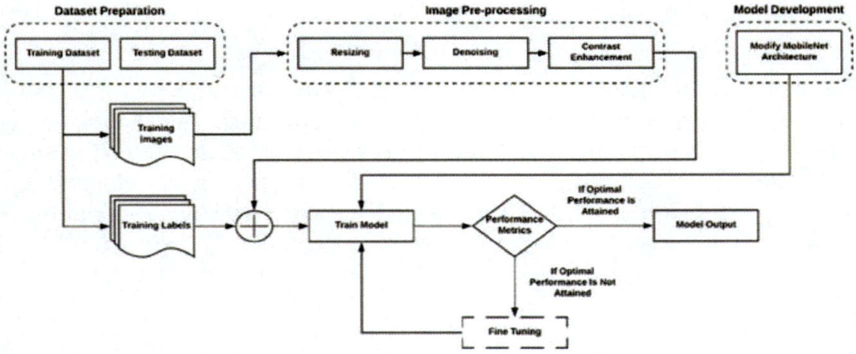

Fig. 1 Overview of the proposed methodology

After running the experimental analysis of our model on the validation set, the epoch saved having the highest Top 3 Accuracy contained the following performance metrics (Fig. 6).

In Fig. 7, we see the confusion matrix of the model from which we can determine the various performance metrics of the model.

As we deal with multiclass classification where foretold values are interdependent, an accuracy score does not determine much regarding the model. Thus, we determine other metrics, namely macro-averaging score and weighted averaging score, to determine the performance of the model as follow (Tables 1, 2 and 3).

Similarly, on running the test analysis of Severity Prediction model on the validation set, the epoch saved having the highest Validation Accuracy contained the following results.

Comparison of our work with related works in this domain has revealed stark differences in the implementation and performance. We achieved an accuracy of 79.96% as compared to 77.5% in ensemble learning method [8] and 94.37% in comparison with SVM-based system [10].

4 Conclusion and Future Scope

Skin cancer prediction using computer vision and deep learning is highly beneficial in saving human lives. It enables a commoner to avoid overlooking a skin disease considering it to be normal just because it looks like a rash. This method also avoids a biopsy, a process rather costly, painful, invasive and time-consuming. Through the proposed work, we intended to help dermatologists and aid them in making more accurate diagnoses, ultimately saving lives. The proposed method, which uses Transfer Learning to predict dermatological diseases, proves to be efficient and accurate by achieving a categorical accuracy of 79.96%. We obtained results using 80%

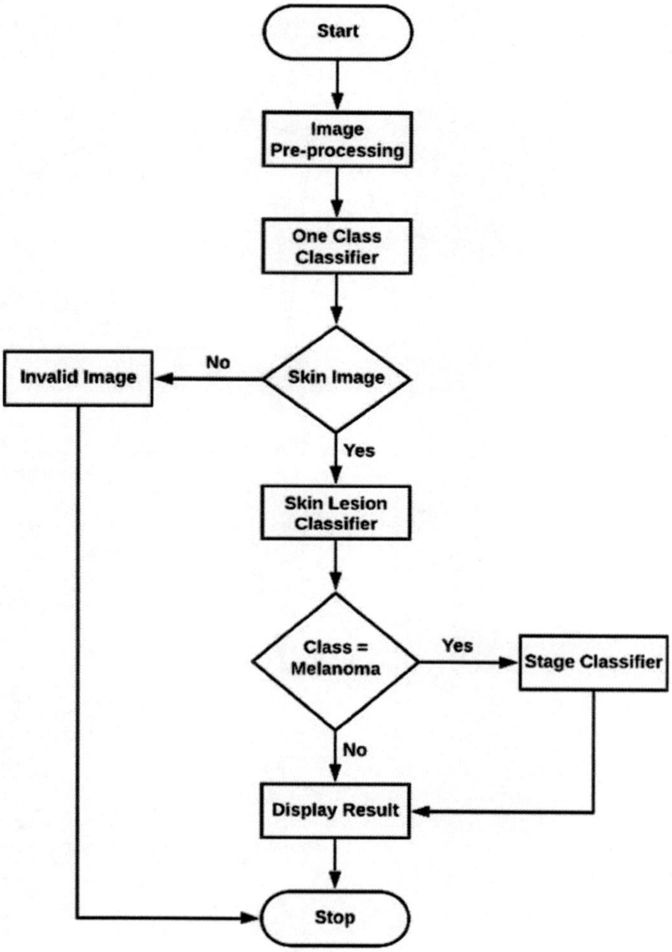

Fig. 2 Implementation flow of the system

of the data for training and 20% data for testing. Severity being a new domain in this field, we have achieved an accuracy of 79.54%.

The scope of the project is very specific and directed towards the identification and classification of a fixed set of skin diseases and the severity is predicted particularly for melanoma. Future scope sees a model which can predict all skin diseases along with their severity. As for every project, the accuracy could be improved upon more removing the most minute possibility of error. With the addition of the features mentioned above this study/application has the potential to facilitate clinical judgement making for dermatologists.

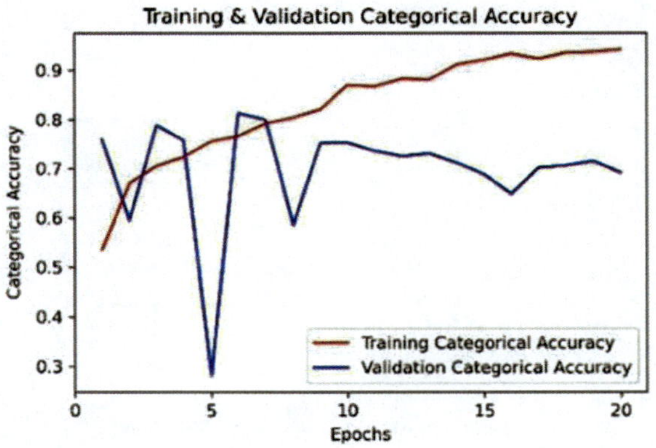

Fig. 3 Training and validation categorical accuracy

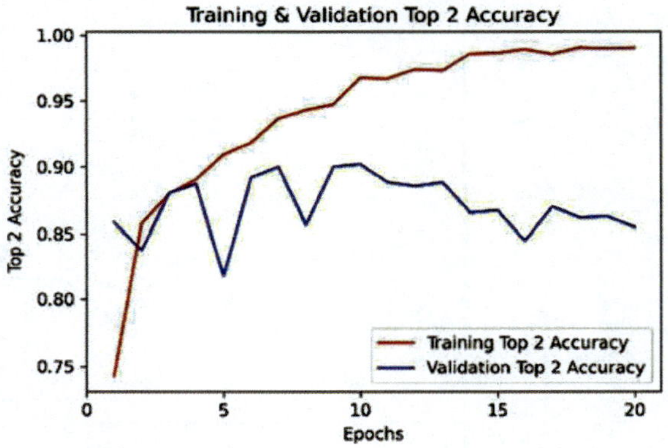

Fig. 4 Training and validation top 2 accuracy

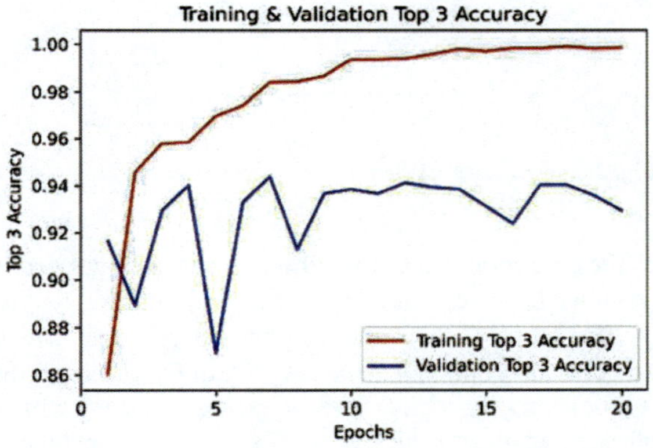

Fig. 5 Training and validation top 3 accuracy

Fig. 6 Training and validation loss

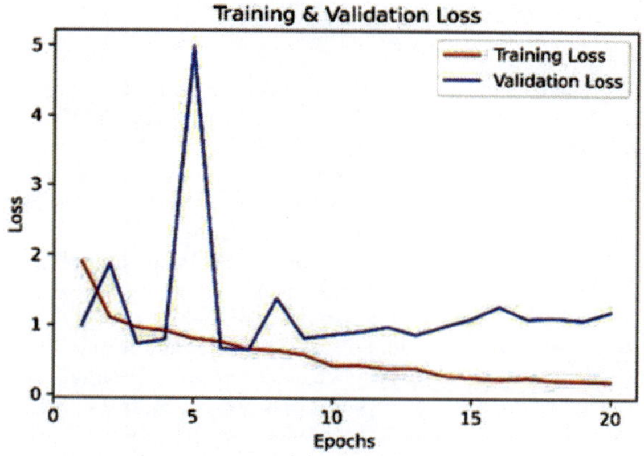

Fig. 7 Confusion matrix of the proposed model

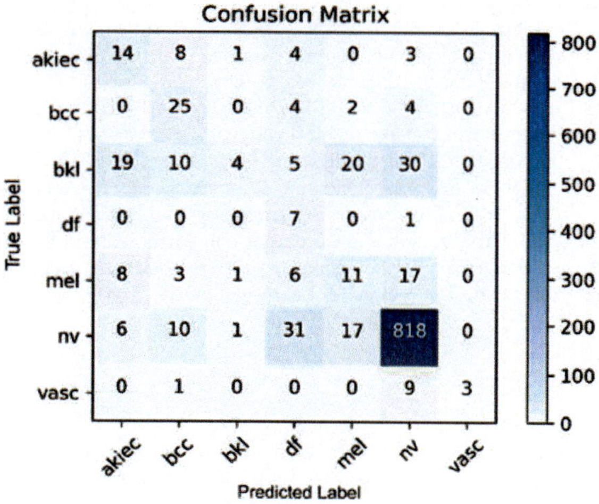

Table 1 Performance of the proposed model

Validation categorical accuracy	79.96%
Validation top 2 accuracy	90.02%
Validation top 3 accuracy	94.37%

Table 2 Overall metrics of the model

	Precision (%)	Recall (%)	F1 score (%)
Macro Avg	51.11	49.96	39.11
Weighted Avg	83.19	79.96	79.14

Table 3 Performance of the proposed method

Validation accuracy	79.54%
Validation loss	3.28%

References

1. Stacey Feintuch, "What Is Skin Cancer?" Healthline, 2018
2. Rowan Prichard Jones, "Skin Anatomy and Types Of Skin Cancer," BAPRAS, 2015
3. Azadeh Noori Hoshyar, Adel Al-Jumaily and Afsaneh Noori Hoshyar, "The Beneficial Techniques in Preprocessing Step of Skin Cancer Detection System Comparing," International Conference on Robot PRIDE 2013–2014 - Medical and Rehabilitation Robotics and Instrumentation, ConfPRIDE 2013–2014
4. Soniya Mane and Dr. Swati Shinde, "A Method for Melanoma Skin Cancer Detection Us- ing Dermoscopy Images," Fourth International Conference on Computing Communication Control and Automation (ICCUBEA), 2018
5. Rahul Sarkar, Chandra Churh Chatterjee and Animesh Hazra, "Diagnosis of melanoma from dermoscopic images using a deep depthwise separable residual convolutional net- work," IET Image Process., 2019, Vol. 13 Iss. 12, pp. 2130–2142
6. Rahul Sarkar, Chandra Churh Chatterjee, Animesh Hazra. "Diagnosis of melanoma from dermoscopic images using a deep depthwise separable residual convolutional network", IET Image Processing, 2019
7. Vinayshekhar Bannihatti Kumar, Sujay S Kumar and Varun Saboo, "Dermatological Dis- ease Detection Using Image Processing and Machine Learning," Third International Con- ference on Artificial Intelligence and Pattern Recognition (AIPR) 2016
8. Anabik Pal, Sounak Ray and Utpal Garain, "Skin disease identification from dermoscopy images using deep convolutional neural network," Challenge Participation in ISIC 2018: Skin Lesion Analysis Towards Melanoma Detection, 2018.
9. Chigozie Enyinna Nwankpa, Winifred Ijomah, Anthony Gachagan and Stephen Marshall, "Activation Functions: Comparison of Trends in Practice and Research for Deep Learn- ing," The 27th International Conference on Artificial Neural Networks (ICANN 2018).
10. Chuanqi Tan, Fuchun Sun, Tao Kong, Wenchang Zhang, Chao Yang and Chunfang Liu, "A Survey on Deep Transfer Learning," arXiv, 2018.

Investigating the Value of Energy Storage Systems on a Utility Distribution Network

Xolisa Koni, M. T. E. Kahn, Vipin Balyan, and S. Pasupathi

Abstract South African's own monopoly utility company is experiencing huge energy crises, and this directly affects both consumers the indigent and the privileged. Therefore, radical action must be taken to consider diverse alternative energy sources. This paper presents a method used in modelling and simulating an Energy Storage System (ESS) for an industrial customer connected on a Utility Distribution Network. This paper used Digsilent Power Factory simulation software. The functionality used of the software was the Quasi-Dynamic Simulation Language (QDSL). Furthermore, the programmable logic enables the author to define how the energy storage system operates, including setting limits and measurements that are calculated autonomously. A simple example is the coding of how the State of Charge (SOC) of the battery is determined, including the charging/discharging operation status of the battery and how it should behave. This paper analysed the existing network and identified network violations and constraints and proposed the Energy Storage System (BESS and FC) as the preferred solution. Moreover, to exploit the benefits offered by the energy storage sources when working in parallel. These benefits include peak shaving, reducing electrical line losses, maintaining balanced voltage levels and the desired fault levels. The purpose of this study was to evaluate the value of energy storage for an industrial customer by seeking to achieve the following: alleviating thermal loading on transformers, achieve peak shaving, reducing electrical losses and operating the network within the voltage limits.

Keywords Energy storage · Battery energy storage · Fuel cell · Quasi-dynamic simulation · Distribution utility network

X. Koni · M. T. E. Kahn · V. Balyan (✉)
Cape Peninsula University of Technology, Symphony way, Bellville, Cape Town, South Africa

S. Pasupathi
HySA Systems at University of the Western Cape, Private Bag X17, 7535 Cape Town, South Africa

A. K. Luhach et al. (eds.), *Second International Conference on Sustainable Technologies for Computational Intelligence*, Advances in Intelligent Systems and Computing 1235,
https://doi.org/10.1007/978-981-16-4641-6_26

1 Introduction

South Africa is facing ever-escalating energy crises, and this is include but not limited to load shedding, infrastructure vandalism, rising energy cost, high Greenhouse Gas (GHG) emissions, etc. Therefore, this has necessitated the need to investigate and explore alternative energy source that can be connected near to customers, decrease GHG emissions, reduce network losses and subsequently minimise associated energy costs. In this way, energy management strategy to reduce power losses and to maintain the balance between supply and demand depending on the availability of power can be achieved [1]. There are various types of alternative renewable source such as solar, wind and hydropower. But due their intermittency character, energy storage systems interconnected to the utility electrical network is the most preferred solution [2]. Generally, Battery energy storage and Fuel cells do rise to the occasion when adequately planned and designed. Fuel cells are expected to play an important role in future power generation. The benefits of utilising a fuel cell to provide the chemical-to-electrical energy conversion are its high fuel-to-electrical energy efficiency of about 50% depending on the type of fuel cell technology employed including system losses [3]. Nevertheless, despite that, fuel cell technology in terms of cost effectiveness is after all deemed too expensive compared to incumbent technologies such as Internal Combustion Engines (ICEs) and batteries. This is of course partly due to the early stage of development [4]. And some interprets this as an indictment of the technology as it is often written off as another one of those technologies that is perpetually five years from commercialisation [5]. Moreover, fuel cell has received less attention in the literature, but could potentially generate low-carbon electricity while avoiding some of the practical consume acceptance issues faced by other low-carbon technologies [6]. There are several types of Fuel cells depending on the type of electrolyte such as, Solid Oxide Fuel Cell (SOFC), Alkaline Fuel Cell (AFC), Molten Carbonate Fuel Cell (MCFC), Phosphoric Acid Fuel Cell (PAFC), Proton Exchange Membrane Fuel Cell (PEMFC) and Direct Methanol Fuel Cell (DMFC). In order obtain higher voltage needed out of fuel cells, several cells must be connected in series to form a fuel cell stack [2]. Proton Exchange Membrane Fuel cell (PEMFC) due to its benefits such as the ability to operate at low temperatures, being cost effective, highly efficient and long life span is the most promising alternative energy source for electric power systems [7]. Efficiency and effectiveness operation of Fuel cells is best obtained when combined with other energy sources with fast dynamics, such as energy storage source, of which in the purpose of this research is a Battery Energy Storage System (BESS). Over the years, Batteries have gained public acceptance, and they show almost no restrictions. Furthermore, the escalated induced voltage drop as a result of the high current mutation increases the design difficulty of the back-stage converter and subsequently shortening the life cycle of the fuel cell. This necessitates the need for an auxiliary power sources which its function is to absorb or release momentary power [8]. In this regard, battery energy storage is the preferred auxiliary source. Moreover, for high peak power period that remain for a long duration, the battery energy storage is suitable for such application. This is due to its extra ordinary properties of high power and capacity. In this paper, battery is the chosen auxiliary power supply, for corresponding energy management strategy and control strategy to achieve better bus steady state and dynamic characteristics. Successful introduction and integration of Fuel Cell (FC) with Battery Energy Storage System (BESS) interconnected to the distribution utility network will alleviate the burden experienced by the industrial customer. This provides an alternative and sustainable renewable energy source of supply that inherits reliability and security energy characteristics. A structural review of thermoelectricity for fuel cells CCHP applications is given in [9].

2 Modelling of the System

2.1 System Description

The system consists of Battery Energy Storage System (BESS), an electrolyser-fuel cell system, hydrogen tank, power conversion system, an inverter, interconnected to the utility distribution network and load as illustrated in Fig. 1. The utility distribution network is the primary source that fulfils the load requirements. The BESS and electrolyser-fuel cell system are storage system that aims to cater for power cuts, and peak demands. As mentioned in the previous chapter, the utility distribution network, whilst supplying the load, it also uses its surplus power to charge the BESS and produces enough hydrogen through the electrolyser for the fuel cell stack (Fig. 2).

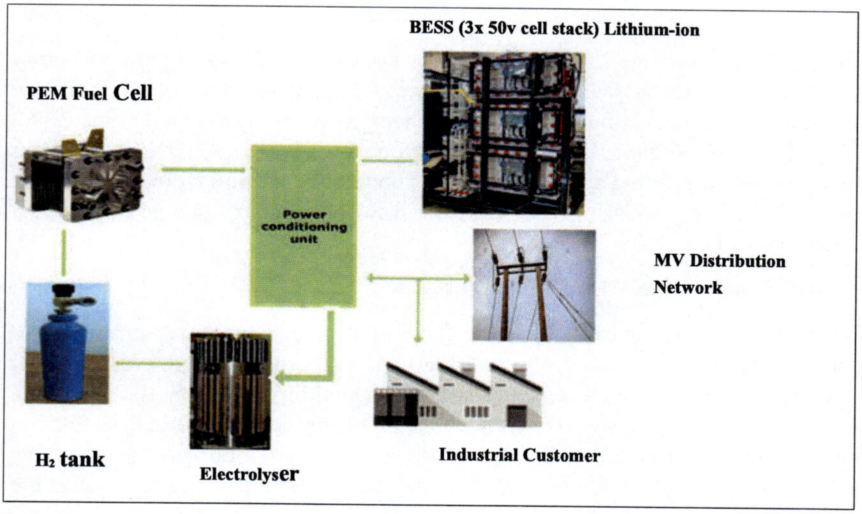

Fig. 1 Full description of the BESS (cell stack) & FC (electrolysis) system

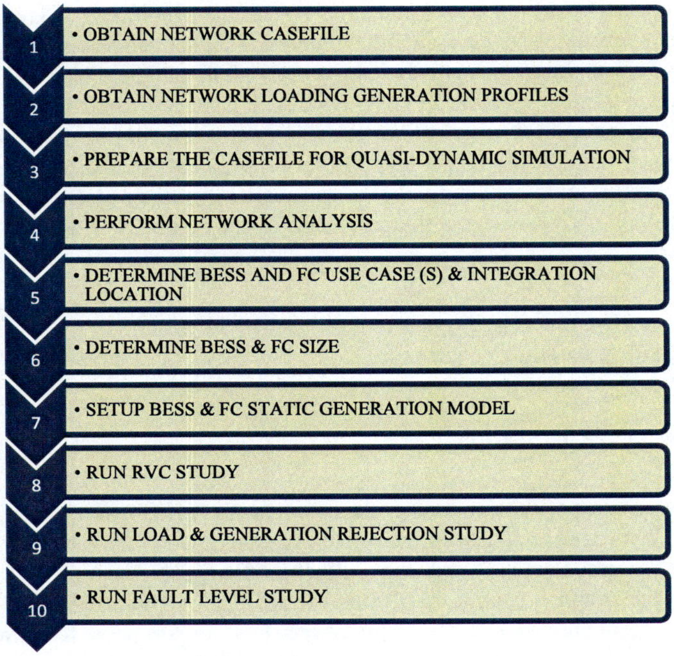

Fig. 2 ESS modelling and simulation procedure flow chart

2.2 *Digsilent Method*

The simulation software that is used in this work is the Digsilent Power Factory due to the fact that the author is mostly being accustomed to the tool, and that the software has a functionality to model Energy Storage System (BESS and FC) using a functionality named Quasi-Dynamic Simulation language (QDSL). Digsilent Power Factory is also the approved network modelling and simulation tool used by popular utility companies such as (Eskom in SA), Municipalities and by private sector companies.

Quasi-Dynamic Simulation Language (QDSL)

The energy storage quasi-dynamic simulation language model is effectively a static generator model with programmable logic (an assigned model definition or QDSL type) and can therefore be utilised for the majority of steady state and time series Energy Storage System (ESS) integration studies. Quasi-Dynamic Simulation language is the programming language used in the model definitions to give existing network models logic, which is dependent on the manner in which one requires the model to operate.

The programmable logic enables the user to define how the energy storage system operates, including setting limits and measurements that are calculated autonomously. A simple example can be the coding of how the state of charge of the battery is determined, including the charging and discharging operation status of the battery and how it should behave. Input signals and output signals can also be coded into the model.

Summarised Procedure

The following is the summarised snapshot of the procedure followed in the simulation of energy storage system (BESS and FC) using the static generator models on Digsilent. Once the network model without any ESS integration is set up for quasi-dynamic simulation and analysis, the next step is to run a quasi-dynamic simulation for the base year and forecasted years of the study. This is to analyse the network for any existing constraints and future network constraints for the forecasted years. The procedure also facilitates to run the Rapid Change Voltage (RVC) against the Draft Grid Connection Code for Battery Energy Storage Facilities (BESF), Utility standards, and NRS048-2 and NRF048-4 guidelines as well as simulation of fault current levels.

As mentioned above, below is the flow chart, which summarises and depicts the advocated process to analyse BESS and FC on Medium Voltage (MV) networks.

3 Results and Discussion

3.1 *Load and Generation Profiles*

This includes checking whether there are any embedded generation that are connected to the network. In cases where there are no embedded generation connected to the network, no metering data is requested with respect to embedded generation. Likewise, in this paper, there are no existing embedded generators.

Load Forecasting

When planning the energy storage system integration to a network, forecasting is an important factor. Below is the formulae used to determine the percentage growth of the network for both short and long-term scenarios.

$$\left(\frac{kVA_f}{kVA_p}\right)^{\left(\frac{1}{n}\right)} - 1 \tag{1}$$

where

kVA_f = Future kilo Volt-Amps.

kVA_p = Present Kilo Volt-Amps.

n = Number of years.

Figure 3 shows that the total load estimated to grow by at least 2.45% from 2020 up until 2025, which translates to an increase of about 282kVA per year. As shown in Figs. 4 and 5, the main feeding transformer (T1) is expected to exceed its maximum thermal limit of 80% in the year 2025 from 75.5% to 86.39%, and (T2) is currently exceeding its thermal loading by 2.4% in year 2020, respectively.

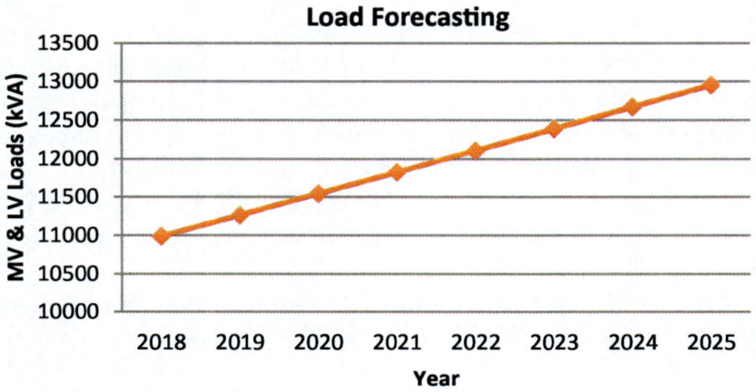

kVA/yr = 282 kVA

Fig. 3 Short and long-term load forecast

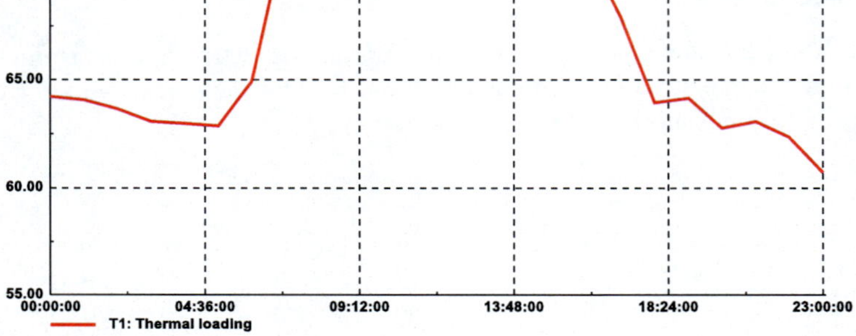

Fig. 4 Thermal loading in year 2020

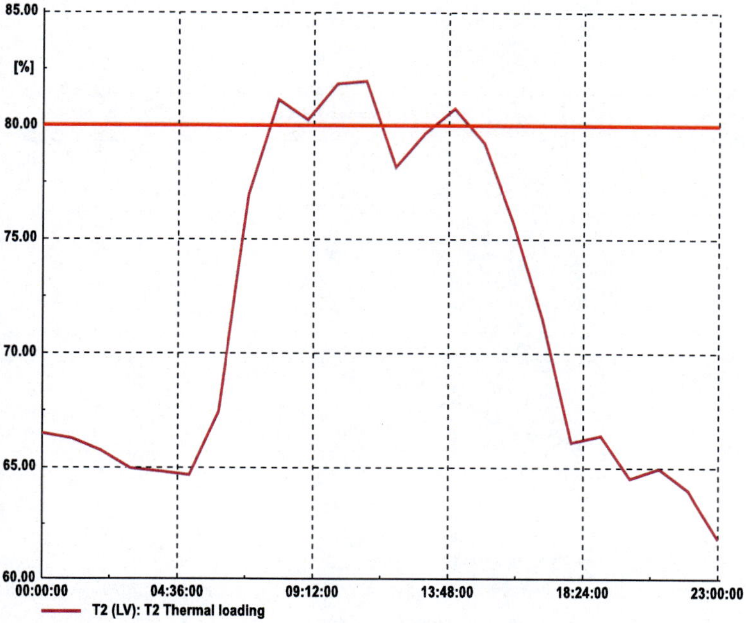

Fig. 5 T2 Thermal Loading in year 2020

3.2 *Results Showing Energy Storage System Integrated to the Network*

The introduction of energy storage offers many benefits when integrated to an electrical system. It has the ability to reduce peak load demand by means of peak shaving or load levelling. This subsequently minimises energy costs, which leads to cost energy saving for the customer. In the diagrams (Figs. 6 and 7) exhibited, are the results illustrating benefits of interconnecting energy storage system (BESS and FC) with the distribution network for a medium voltage customer.

When the energy storage system is integrated to the network, the peak demand drops from 75.5% to 62.616%, showing a tremendous drop of 15%, therefore provides less strain on the network, and can cater for any load growth in the future.

Similarly, T2 feeds Low Voltage (LV) loads of the network, the peak demand sinks from 82.4% to 72.3%, totalling to a drop of about 10.1% and peak shaving achieved.

Electrical Losses (kW) After Energy Storage System Integration

Electrical losses on the lines and cables also form part of the benefits exploited by the customer due to the inclusion of the battery energy storage and fuel cells in the

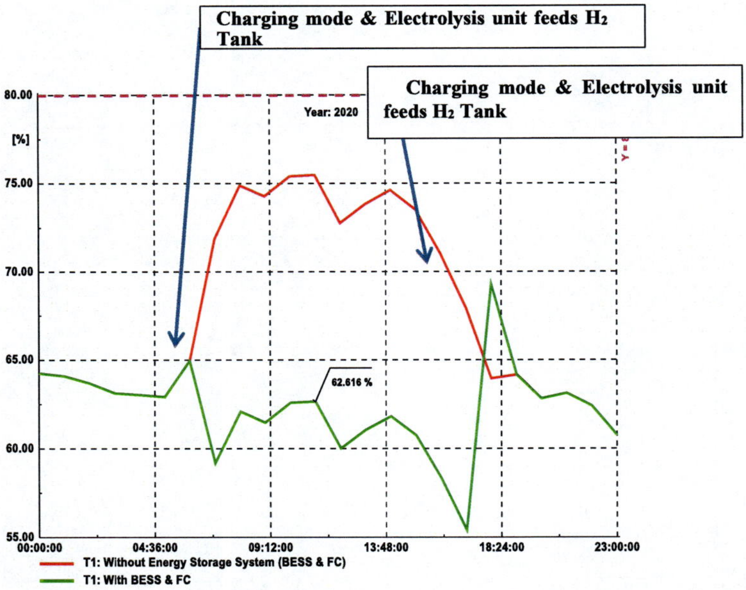

Fig. 6 T1 with Energy Storage System integrated to the network

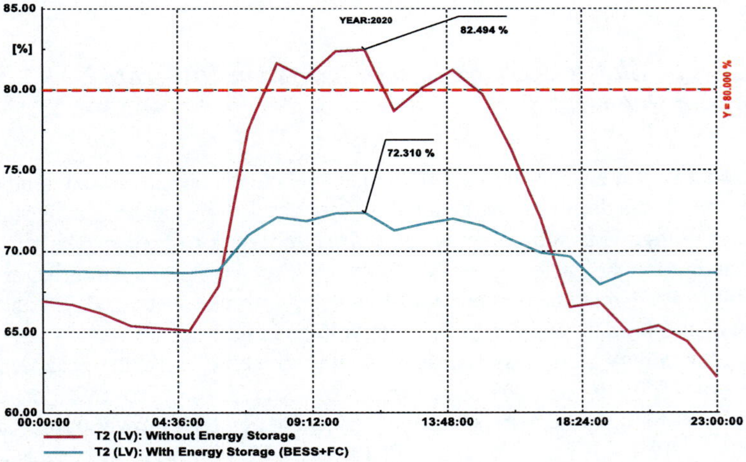

Fig. 7 T2 Peak Shaving through Energy Storage System

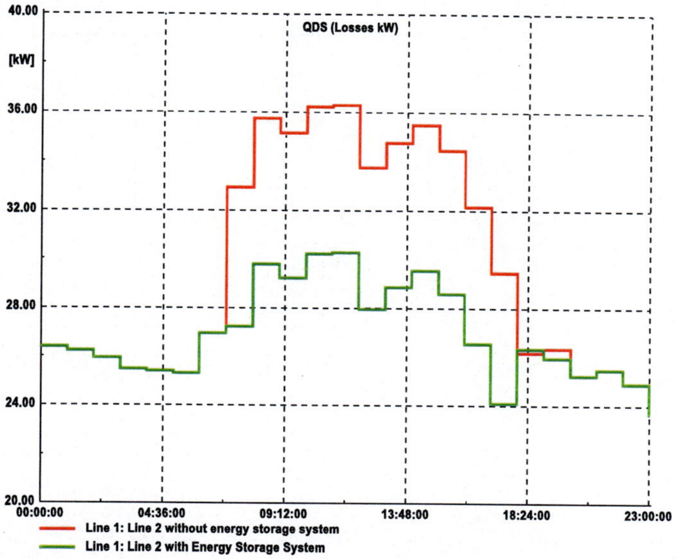

Fig. 8 Electrical line losses (kW)

network. Figure 8 depicts that losses dropped almost by 16.6% from 36.3 kW to 30.2 kW.

Network Voltage Improvement

The energy storage also plays an important role in improving the magnitude of the voltage levels in the network. The influence of the Distributed Generator voltage magnitude is such that when the voltage angle is zero (i.e. $\theta = 0$) and $V_{DG} > V_{Grid}$, the reactive power Q is positive. This is reactive power transferred from the Battery and Fuel cell to the utility distribution grid.

Figure 9 shows an improved busbar voltage magnitude due to the BESS and FC integration to the utility distribution grid, recording 0.99p.u in the steady state operation. The red line indicates the minimum allowable voltage value on any busbar of the network.

Fault levels Analysis

Fundamentally, the fault level philosophy ensures that the predefined manufactures ratings of the equipment is not subjected to higher fault levels. In this case, no equipment is pushed to work above its limit.

Moreover, power quality is achieved by maintaining adequate network fault levels. Furthermore, in order to safeguard a network against electrical faults, sufficient fault levels are required. Fault levels are responsible for the correct operation of protective devices. Inadequate fault levels will create grading difficulties for the protection

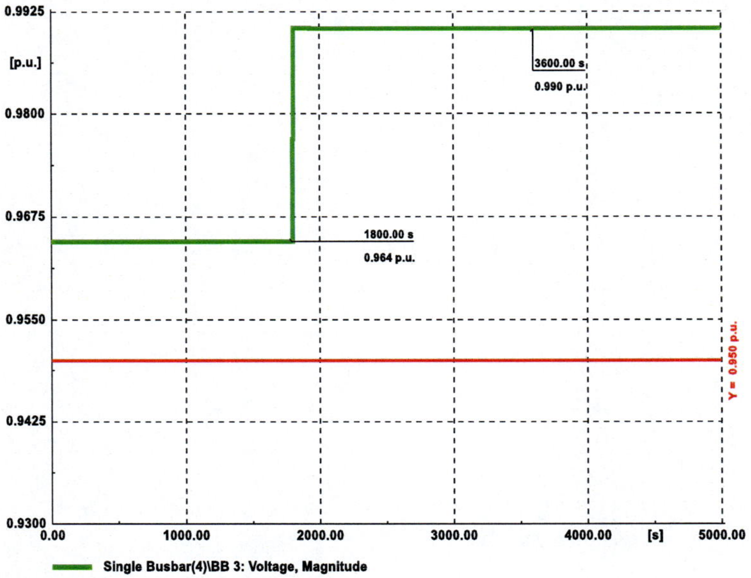

Fig. 9 Improved voltage levels

settings. Too low fault levels complicate the differentiation between fault current and load current since their values may be close to each other.

As seen in Fig. 10 the more the Busbars are located far from the main source, the fault levels decreases. Busbar 4 is the furthest from the main supply; hence, short-circuit levels have dropped much lower than other Busbars to approximately 33%. Busbar 3 has dropped almost by 16.9% from Busbar 2. The fault levels indicate that all equipment assessed will not be stressed beyond their carrying capacity.

Fig. 10 Short-circuit output results

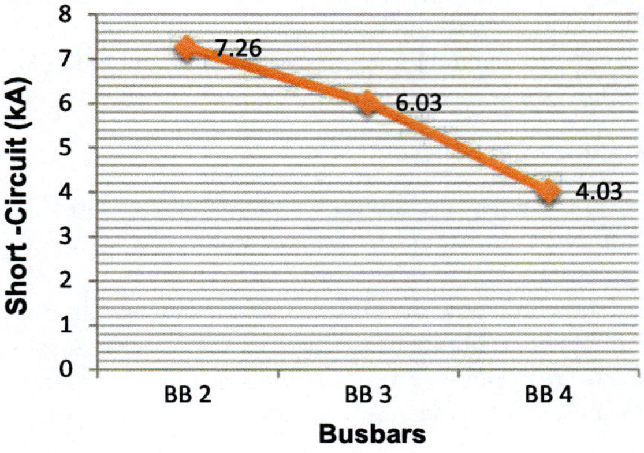

4 Conclusion

This paper investigated the value of energy storage system for an industrial customer connected to the utility distribution network. This paper focused mainly on how the industrial consumer can benefit and improve the power quality from integrating energy storage systems to its network. The energy storage system considered for this study was the Battery Energy Storage System (BESS) working in parallel with the Fuel Cell Storage System (FC).

The study used the Digsilent functionality named Quasi-Dynamic Simulation Language (QDSL) to model the network. The programmable logic enables the user to define how the energy storage system operates, including setting limits and measurements that get calculated autonomously. Quasi-Dynamic Simulation was coded to determine when the BESS and FC should charge and discharge. This paper successfully achieved the investigation on the exploitation of the benefits offered by Battery Energy Storage System connected in parallel with fuel Cell interconnected to the electrical distribution network. This paper further showed the Quasi-Dynamic Simulation results by means of graphs and tables, specifically showing successful Peak Shaving of the load, improvement of the Busbar voltage magnitudes and reduced electrical line losses as well as fault levels of the network.

References

1. D. N. Luta and A. K. Raji, “Energy management system for a remote renewable fuel cell system,” Proceedings. 27th International. Conference. Domestic Use of Energy, pp. 20–24, AIUE, Cape Town 2019
2. P. Alain, “Energy Management of Battery-PEM Fuel Cells Hybrid Energy Storage System for Electric Vehicle,”Proceedings. 2016 International Renewable and Sustainable Energy Conference (IRSEC), pp. 1–70, IEEE, Marrakech, Morocco, 2016
3. A. K. Raji, "Modelling and development of fuel cell off grid power converter system".Thesis, Cape Peninsula University of Technology, bellville, Cape Town, 2008
4. S. Hardman, A. Chandan, R. Steinberger-Wilckens, Fuel cell added value for early market applications. J. Power Sources **287**, 297–306 (2015)
5. J. M. Schneider, “Fuel cells: an electric utility perspective,”Proceedings. IEEE Power Engineering Society General Meeting, pp. 1634–1636, IEEE, Denver, CO, USA, 2005
6. P. Ekins, I. Staffell, P. Grünewald, P. E. Dodds, F. Li, A. D. Hawkes, and W. McDowall, “Hydrogen and fuel cell technologies for heating: A review,” Int. J. Hydrogen Energy, vol. 40, no. 5, pp. 2065–2083, ELSEVIER Ltd, UK, 2015
7. M. Derbeli, A. Charaabi, O. Barambones, and L. Sbita, “Optimal Energy Control of a PEM Fuel Cell/Battery Storage System,” 2019 10th International. Renewable. Energy Congress. (IREC), pp. 1–5, IEEE, Sousse, Tunisia, 2019
8. C. Qian, Z. Jing, Q. Peng, L. Yi, H. Xiaoming, Di. Chao, and N. Xiaojun, “Smooth Control Strategy for Fuel Cell-Battery Power System,” 2019 3rd IEEE Conference. Energy Internet Energy System Integration. Ubiquitous Energy Network Connect. Everything, EI2 2019, pp. 1139–1143, IEEE, Changsha, China, 2019
9. N.P. Bayendang, MTE Kahn, V. Balyan,“A Structural Review of Thermoelectricity for Fuel Cells CCHP Applications,” J. Energy, vol. Volume 2020, no.Article, pp.23 pages, Hindawi 2020, ID 2760140, 23 pages, https://doi.org/10.1155/2020/2760140

Predicting the Impact of Android Malicious Samples Via Machine Learning

Archana Lopes, Sakshi Dave, and Yash Kane

Abstract The Android operating system, as well as smartphone apps for just about everything, is widely available in the official Google Play store and several third-party markets. Furthermore, the critical function of smartphones contributes to the storage of a large amount of data on computers, not only personal but also corporate and technical data, attracting malware creators and developers to create resources that can permeate user's devices to take out data. The antivirus programme today counts on a database that is signature-based and only fruitful in detection of known malware. To combat the threats of Android malware, the paper proposes an Android malware detection system that employs machine learning techniques.

Keywords Android malware · Malicious samples · Hybrid machine learning · Algorithms · Datasets and classifiers

1 Introduction

There are multiple functions that can be performed using a smartphone like checking email, messaging companions, making appointments, monitoring the weather and news bank transactions, etc. The Android took over around 85% of the overall smartphone volume [1]. Ericsson has forecasted that there could be an increment of 6.4 billion smartphone using individuals by the end of 2020. Smartphones are an inviting focus for criminals present online. By far the majority are malicious, with malware accounting for 77% of all submitted cases [2]. G DATA protection experts identified a large amount of advanced malware for Android every day at the start of 2018. This demonstrates that a new piece of malware appears every 10 s. G DATA analysts reported that 3.4 million new Android malware was created in 2018. The number of

A. Lopes
Fr. Conceicao Rodrigues College of Engineering, Bandra, Mumbai, India
e-mail: archana_lopes@fragnel.edu.in

S. Dave · Y. Kane (✉)
Fr. Agnel Ashram, BandstandMumbai, Bandra (W)Maharashtra 400050, India

A. K. Luhach et al. (eds.), *Second International Conference on Sustainable Technologies for Computational Intelligence*, Advances in Intelligent Systems and Computing 1235,
https://doi.org/10.1007/978-981-16-4641-6_27

usable applications on the Google Play Store was most recently set at 3.3 million Apps in March 2018 [3].

Although signature-based antivirus is very good at detecting known malware, it falls short when it comes to detecting new malware. They can also be defeated by altering the procedure for launching an attack. Heuristic scanning, which uses rules to check for commands that can indicate an intent that is malicious, is proposed to resolve such signature-based limitations. Machine learning methods are proposed to solve the problem of heuristic search. Machine learning techniques can classify a series of trends that can lead to the discovery of previously undetected malware. The growing complexity of Android malware demands a new strategy for detection that is robust enough to be outwitted.

Problem Statement and Justification

The growing popularity of Android malware has exposed end users to significant protection and privacy risks. Malicious Android samples pose a protection and privacy risk to billions of smartphone users. As a result, being able to suggest a solution for automatically detecting malicious samples with greater security and privacy implications is critical. The paper manually identifies two metrics for malicious Android samples. As a result of our research, a new dataset of Android malware is created with ground reality (high/low impact). This latest dataset is being used to analyse the characteristics of low and high impact malicious samples empirically. The innovation in this project is reflected by the fact that Hybrid Techniques in Machine Learning Algorithms have been used in the course of the project.

2 Implementation

2.1 Block Diagram

The proposed structure is divided into three stages. To begin, low/high impact truth is collected with the training Android malware samples. Next, based on the present malware resources, reverse engineering is done to take out features from the disassembled binary groups AndroidManifest.xml and dex codes. These acceptable features which are extracted are employed into representing the characteristic impact of every sample of Android malware. The string features would then be converted into vector numerical representations [2]. Lastly, these vector representations are used to train the prediction model for impact, which is used to classify particular Android malware samples as benign or malicious (Fig. 1).

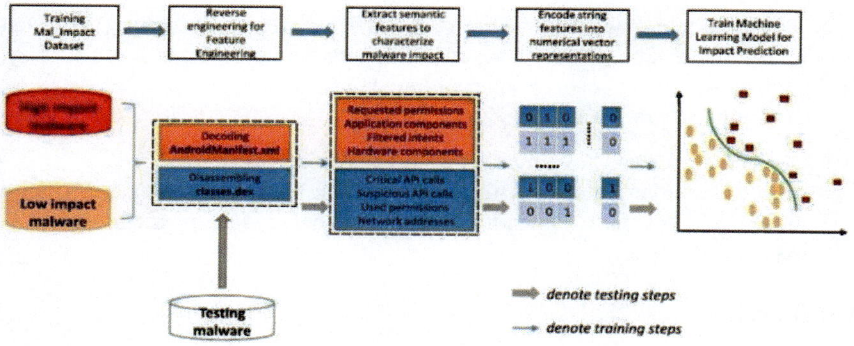

Fig. 1 Block diagram

2.2 *Working*

Two data sets have been used which consist of Benign and Malign samples respectively (Fig. 2).

The work begins by applying a Feature selection model on these datasets which will help us fetch all important parameters which determine whether a particular sample is benign or malign.

Feature selection—Select from Model

The generation process, evaluation function, stopping criteria and validation procedure are the four aspects of feature selection. To begin, create a function subset from the entire set. Second, use the evaluation function to determine the feature subset. Third, compare the outcome to the criteria for stopping. If the result meets the criteria, the process will end; otherwise, another function subset will be created and used to pick the feature. Finally, the process of validating the selected subset is completed. The search algorithm is divided into three categories: complete, heuristic and random in the Android app's generation procedure. For its backtracking, the best first quest, which is a greedy complete algorithm, is chosen. Select N features from the feature set as a subset, place it in an unlimited-length priority list, then remove the subset with the highest score from the queue, enumerate all the feature sets provided by adding a feature to the subset and place them in the queue. Filter and wrapper make up the evaluation feature. The first examines the interior characteristics of the function subset regardless of classification choice [4].

Fig. 2 Two datasets

After all the important features are fetched, five different algorithms are tested wherein each algorithm generates a result with a particular percentage of accuracy, the figures of which are generated during implementation.

Each algorithm produces:

- Predicted Label: that shows number of true and false predictions.
- Receiver Operating Characteristic (ROC) Curve: True Positive Rate Vs False Positive Rate. The more perpendicular curve the better is the prediction.
- Classification Report

Algorithms

AdaBoost Classifier Algorithm

See Fig. 3.

It operates in the following manner:

- At first, AdaBoost chooses a subset at random for training.
- It trains the AdaBoost model iteratively by selection of training set depending upon the correct prediction of the previous session.
- It associates the higher weight to observations that are wrongly classified.
- In each iteration, it assigns a weight to the classifier that is trained, based on the classifier's accuracy. The classifier with the highest accuracy will be given the most weight.

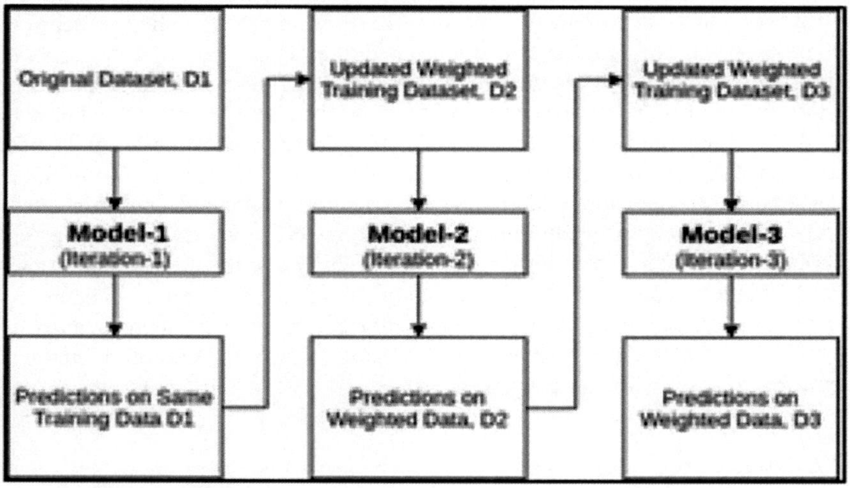

Fig. 3 AdaBoost flowchart

- This method repeats until all of the training data suits perfectly or until the given maximum number of estimators is reached.
- To classify the learning buildings, cast a vote across all of them.

Random Forest Classifier

Random forest algorithm, a supervised classification algorithm, makes a forest with a no. of trees. In general, the higher the number of trees in a forest, the more full-bodied it appears. Similarly, in the random forest classifier, the more the number of trees in the forest, the more accurate the results are. This algorithm combines many algorithms which are of similar kind [5].

The following are the steps of random forest algorithm:

- N records (random) are picked from dataset.
- Based on these particular N records, a decision tree is built.
- Setting number of trees in the algorithm and repetition of step 1 & 2.
- If there is a problem of regression, given a new record, every tree predicts a value for output of Y. By combining all of the values expected by all of the trees in the forest, the final value can be determined. Alternatively, if there is a classification issue, each tree in the forest predicts which group the new record belongs to. After that, the new record is assigned to the group that receives the most votes.

The following are some of the benefits of using Random Forest Classifier:

- The overall bias of the algorithm is less.
- This algorithmic software has a high level of stability. Even if a new piece of information is added to the knowledge set, the algorithm's overall programme is not affected.
- The random forest algorithm performs fine when there are categorical and numerical features in the data.
- The random forest algorithm often performs well when there are missing values or when the data is not scaled properly.

Decision Tree

One of the most efficient and well-known methods for classification and prediction is the decision tree. A decision tree is a flowchart that resembles a tree, with each internal node representing a test on an attribute. Here, each branch represents the test's result, and each leaf node, or terminal node, holding a class name (Fig. 4).

In the example above, a tree can be "realized" by dividing the source set into subsets. This process is called recursive partitioning because it is repeated on each subset (derived) in a recursive manner. The decision tree classifier is generally accurate. A popular inductive approach to learning classification information is decision tree induction [3].

Decision trees carry out simple classification of instances by sorting them down the tree from the root node to a certain number of leaf nodes, which effectively classifies the instance. The decision tree determines if it is appropriate for playing

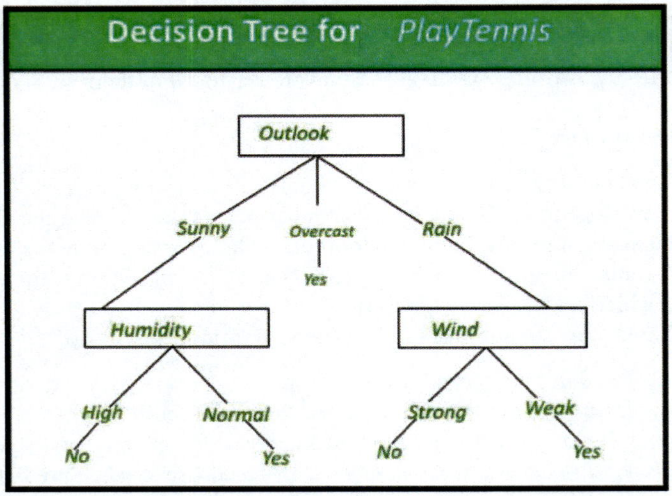

Fig. 4 Decision tree example

tennis (as seen in the diagram) and returns the classification associated with that leaf. Yes or No, depending on the circumstances.

Gradient Boosting

Gradient boosting is a machine learning technique that produces a prediction model in the form of an ensemble of weak prediction models for both regression and classification problems and tasks. Gradient Boosting is a technique that builds a model stage by stage and then generalises it by allowing the optimization of any differentiable loss function.

Gradient boosting is an iterative process that combines weak learners into a single strong learner. A new model is fitted as each poor learner is incorporated (added), resulting in a more reliable estimate of the response variable. The negative gradient of the loss function associated with the entire ensemble is maximally correlated with the new weak learners. The concept of gradient boosting is that one is able to perform integration of a group of weak prediction models (relatively) to create a comparatively stronger prediction model.

It's a very effective technique for creating predictive models. Gradient boosting may be used to optimise the prediction accuracy of a variety of risk functions, which is a benefit over traditional fitting approaches. This allows us to be more creative with our models. It also addresses questions about multi collinearity, which is an issue in which two or more predictor variables have strong correlations.

Naïve Bayes Classifier

A Naive Bayes classifier is defined as a machine learning model that is probabilistic and used for classification task. The concrete basis of this particular classifier is

Fig. 5 Application window

predominantly based on the Bayes theorem.

$$P(A|B) = P(B|A)\,P(A)/P(B)$$

By using Bayes theorem, one is able to find probability of (in this case) A happening, given that B has already occurred.

Classification

After all five algorithms are tested, the system uses the best results to predict whether the given sample is benign or malign. An application window is then seen which has the option to browse for test APKs (i.e. samples which can be either benign or malign). Once one of these test APKs is browsed and chosen, the application window takes some process time and shows if the sample induced consists of some malware or is legitimate.

The application window looks like the following (Fig. 5).

2.3 Code Explanation

While Pandas is used to work with Datasets, Sklearn is used for implementing Machine Learning Algorithms.

The project starts with a benign and a malign data set. The paper has assigned benign as 1 and malign as 0. Firstly, the data frames of these two sets are concatenated. Then, a feature selection model is applied on this fetch the most important parameters which will classify between benign and malign android samples.

Now, the five algorithms are tested one after the other wherein there is a confusion matrix (predicted label) and accuracy of the test set which gives us a thorough insight as to which algorithm works the best. Along with a unique confusion matrix, we also plot Receiver Operating Characteristic Curve and Classification Report. The best working algorithm is saved in classifier directory along with the list of most important features [2].

For the application window, a button is created to browse and two text boxes—one which shows the path of the test APK files and the other which shows whether the added sample is benign or malign after APK analysis occurs. If result variable is 0, then output is malign and if it is 1, then benign (as assigned previously).

To make the result evident, 2 png files are used—one showing a green tick indicating benign while one showing a red cross indicating malign in the frame window.

2.4 Software Used

Python 3.6.8—8th and last maintenance release of Python 3.6 (Fig. 6).

While the two most important supplementary tools to work with Python were as follows (Fig. 7).

Fig. 6 Python 3.6.8

Fig. 7 Sklearn for Python

3 Results

After feature selection is performed, each one of the five algorithms is tested as follows:

```
Actual Dimenions of feature space is  (1000, 4064)
==========================Feature Selection==============================
385 features identified as important:
1. feature perm_73 (0.104086)
2. feature perm_17 (0.069278)
3. feature perm_4 (0.033277)
4. feature perm_48 (0.027329)
5. feature recv_14 (0.025929)
6. feature perm_31 (0.025296)
7. feature perm_72 (0.025253)
8. feature perm_21 (0.025244)
9. feature perm_261 (0.015322)
10. feature perm_18 (0.014504)
11. feature perm_190 (0.014365)
12. feature perm_28 (0.012711)
13. feature serv_1436 (0.011158)
14. feature perm_9 (0.010976)
15. feature perm_19 (0.010821)
16. feature serv_10 (0.010468)
17. feature serv_9 (0.010247)
18. feature perm_30 (0.010226)
19. feature perm_12 (0.009561)
20. feature perm_5 (0.009308)
21. feature perm_13 (0.009254)
22. feature perm_33 (0.008340)
23. feature perm_102 (0.007490)
24. feature perm_34 (0.007450)
25. feature recv_1263 (0.007342)
26. feature perm_8 (0.007030)
27. feature perm_35 (0.006881)
28. feature perm_7 (0.006589)
29. feature perm_16 (0.006312)
30. feature perm_262 (0.006205)
31. feature perm_42 (0.006200)
32. feature serv_1425 (0.006123)
33. feature perm_47 (0.005797)
34. feature perm_52 (0.005773)
35. feature perm_134 (0.005639)
36. feature perm_77 (0.005532)
37. feature recv_12 (0.005439)
38. feature perm_66 (0.005406)
```

```
367. feature recv_1402 (0.000261)
368. feature recv_954 (0.000257)
369. feature recv_1338 (0.000257)
370. feature serv_1726 (0.000255)
371. feature recv_401 (0.000253)
372. feature recv_1426 (0.000252)
373. feature serv_402 (0.000252)
374. feature recv_1 (0.000252)
375. feature recv_247 (0.000251)
376. feature serv_1649 (0.000251)
377. feature serv_1725 (0.000251)
378. feature serv_1664 (0.000250)
379. feature serv_1452 (0.000249)
380. feature serv_465 (0.000247)
381. feature recv_696 (0.000247)
382. feature recv_1524 (0.000247)
383. feature serv_1601 (0.000247)
384. feature recv_1566 (0.000246)
385. feature recv_1327 (0.000246)
```

(1) Decision Tree Algorithm (Figs. 8, 9, 10 and 11).
(2) Random Forest Classifier Algorithm (Figs. 12, 13, 14 and 15).

```
DecisionTree:Confusion Matrix:  [[348   3]
 [ 18 331]]
DecisionTree:Accuracy :  97.0
DecisionTree:Test set
DecisionTree:Confusion Matrix:  [[135  14]
 [ 18 133]]
DecisionTree:Accuracy :  89.33333333333333
AUC: 0.91
              precision    recall  f1-score   support

         Yes       0.88      0.91      0.89       149
          No       0.90      0.88      0.89       151

    accuracy                           0.89       300
   macro avg       0.89      0.89      0.89       300
weighted avg       0.89      0.89      0.89       300
```

Fig. 8 Overall result of decision tree algorithm testing

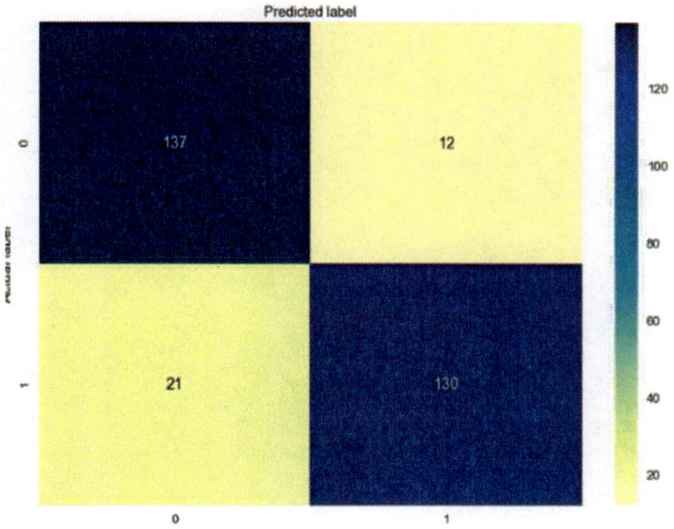

Fig. 9 Decision tree predicted label

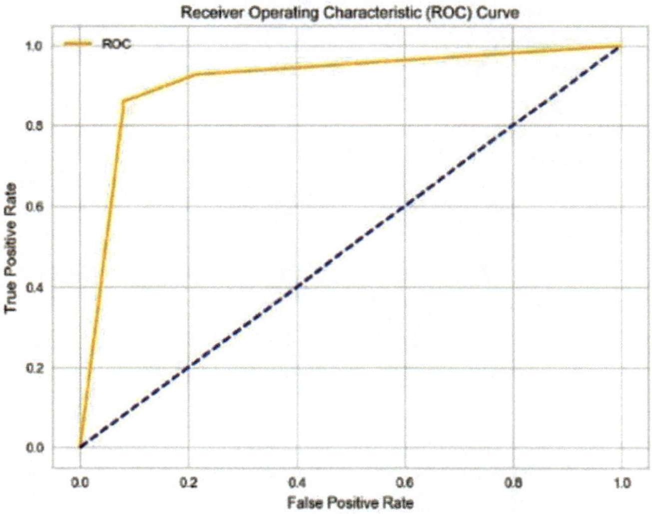

Fig. 10 Decision tree ROC curve

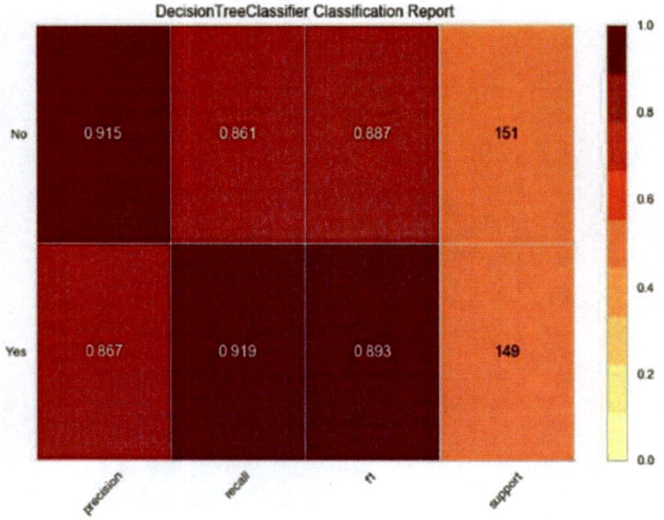

Fig. 11 Decision tree classification report

```
RandomForest:Confusion Matrix:  [[350   1]
 [  0 349]]
RandomForest:Accuracy :  99.85714285714286
RandomForest:Test set
RandomForest:Confusion Matrix:  [[137  12]
 [  5 146]]
RandomForest:Accuracy :  94.33333333333334
AUC: 0.97
              precision    recall  f1-score   support

         Yes       0.96      0.92      0.94       149
          No       0.92      0.97      0.94       151

    accuracy                           0.94       300
   macro avg       0.94      0.94      0.94       300
weighted avg       0.94      0.94      0.94       300
```

Fig. 12 Overall result of random forest classifier algorithm testing

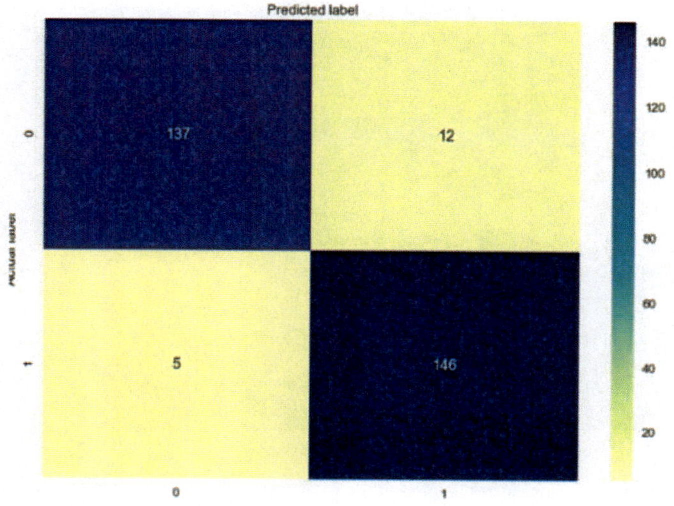

Fig. 13 Random forest classifier predicted label

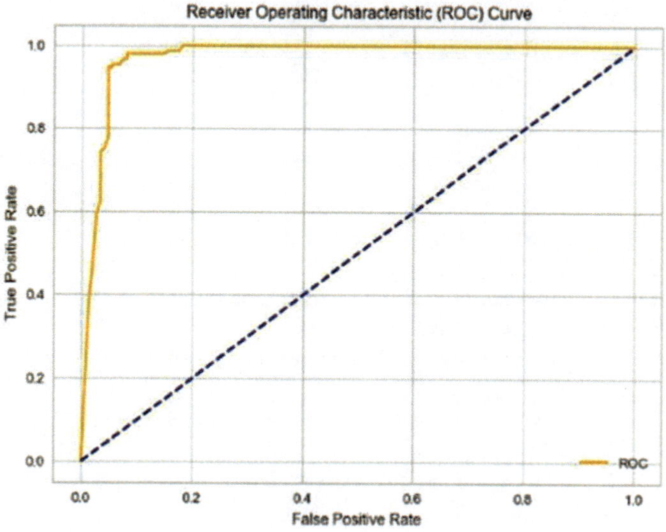

Fig. 14 Random forest classifier ROC curve

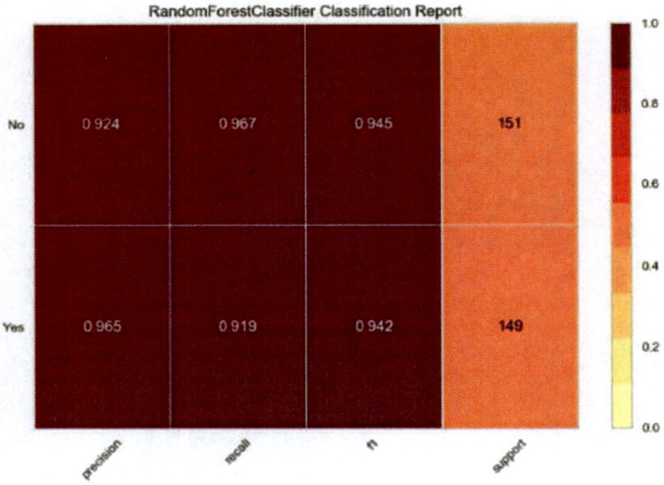

Fig. 15 Random forest classifier classification report

(3) Gradient Boosting Algorithm (Figs. 16, 17, 18, 19 and 20).
(4) AdaBoost Algorithm (Figs. 21, 22, 23, 24, 25, 26, 27).
(5) Naïve Bayes Classifier Algorithm (GNB).

When the Test set Accuracy of each of these five algorithms is observed, it is evident that Random Forest Classifier gives us the best accuracy equal to 94.3333.

```
GradientBoosting:Confusion Matrix:  [[340  11]
 [ 14 335]]
GradientBoosting:Accuracy :  96.42857142857143
GradientBoosting:Test set
GradientBoosting:Confusion Matrix:  [[137  12]
 [  7 144]]
GradientBoosting:Accuracy :  93.66666666666667
AUC: 0.97
              precision    recall  f1-score   support

         Yes       0.95      0.92      0.94       149
          No       0.92      0.95      0.94       151

    accuracy                           0.94       300
   macro avg       0.94      0.94      0.94       300
weighted avg       0.94      0.94      0.94       300
```

Fig. 16 Overall result of gradient boosting algorithm testing

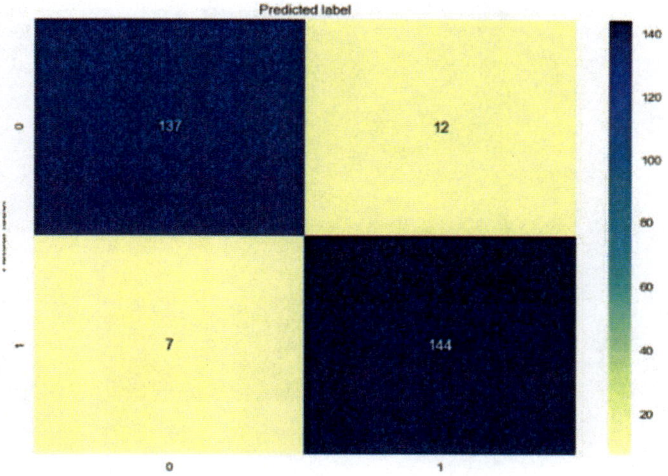

Fig. 17 Gradient boosting predicted label

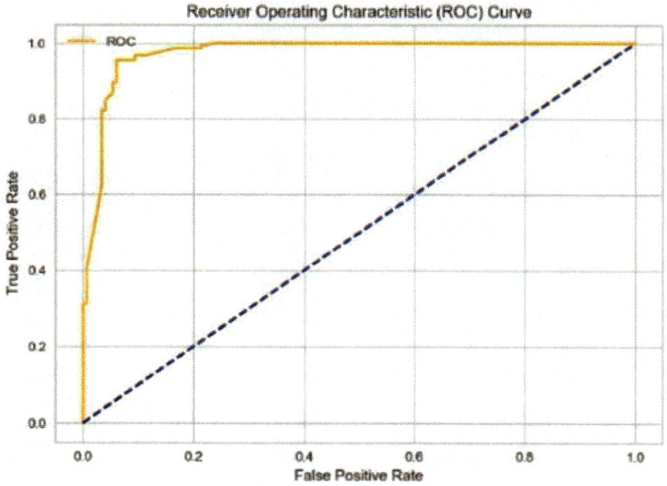

Fig. 18 Gradient boosting ROC curve

Hence, the system uses this particular algorithm to classify android samples. Now an Application window is seen wherein test APK files are browsed and added (Benign or Malign).

For Benign sample—Here, a Facebook File has been added (Figs. 28 and 29).

For a Malign Sample—Here, a Malware File has been added.

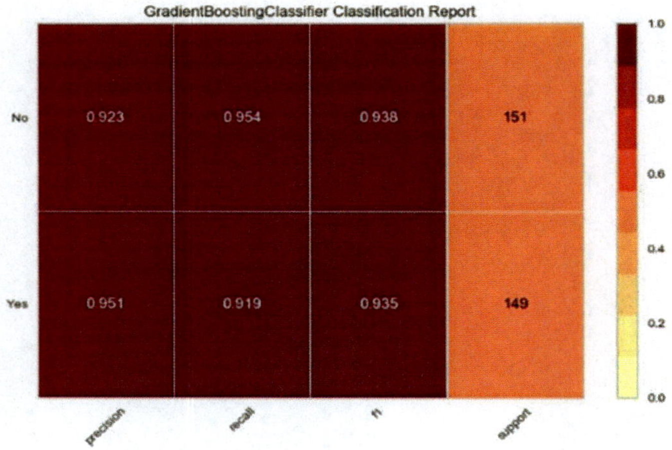

Fig. 19 Overall result of adaboost algorithm testing

```
AdaBoost:Confusion Matrix:  [[347   4]
 [  4 345]]
AdaBoost:Accuracy :  98.85714285714286
AdaBoost:Test set
AdaBoost:Confusion Matrix:  [[132  17]
 [  3 148]]
AdaBoost:Accuracy :  93.33333333333333
AUC: 0.99
              precision    recall  f1-score   support

         Yes       0.98      0.89      0.93       149
          No       0.90      0.98      0.94       151

    accuracy                           0.93       300
   macro avg       0.94      0.93      0.93       300
weighted avg       0.94      0.93      0.93       300
```

Fig. 20 Gradient boosting classification report

4 Conclusion and Future Scope

Aside from the flaws in the current security system, the explosive growth of Android malware in recent years necessitates an effective solution to keep Android malware at bay. The paper proposes a deep neural network-based method for detecting Android malware that allows use of permissions combinations, API calls as feature to construct

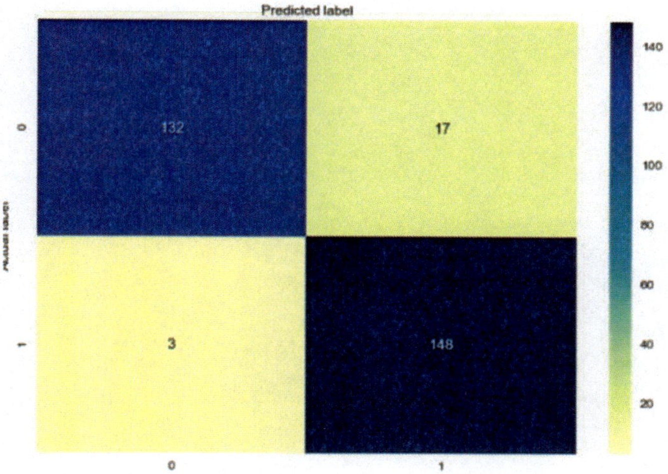

Fig. 21 Overall result of AdaBoost algorithm testing

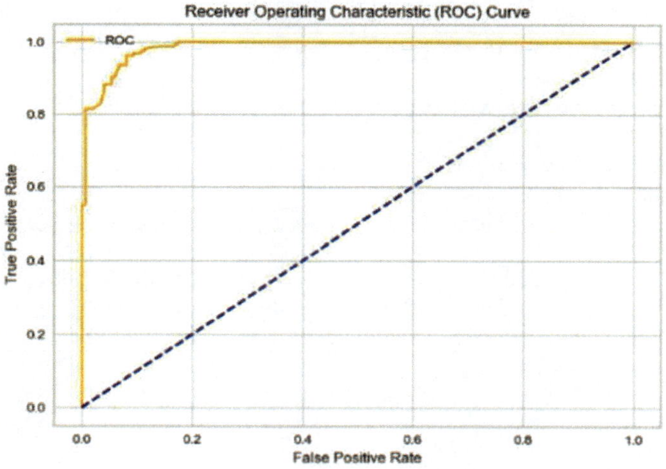

Fig. 22 AdaBoost algorithm ROC curve

a machine learning model that is capable enough to identify malicious from the entire lot.

The proposed system achieves a 95 percent overall accuracy in tests with various feature sets. The overall performance of this framework is compared to current machine learning approaches that are very simple. Having said that, the findings show that the proposed system outperforms machine learning approaches across the board. Finally, the proposed system's output is compared to a variety of other works

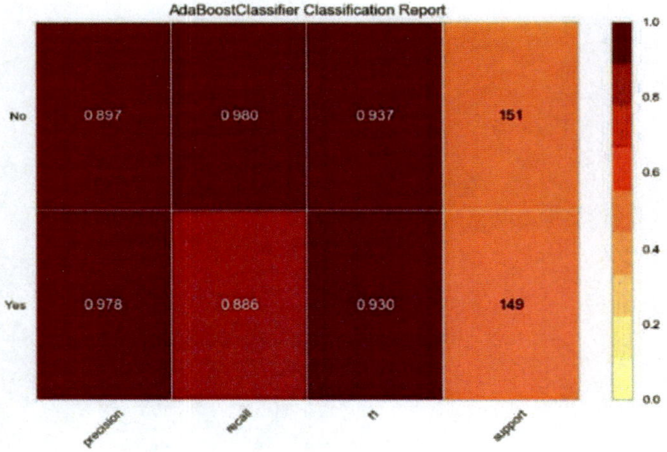

Fig. 23 AdaBoost algorithm classification report

```
GNB:Confusion Matrix:  [[242 109]
 [  1 348]]
GNB:Accuracy :  84.28571428571429
GNB:Test set
GNB:Confusion Matrix:  [[ 89  60]
 [  0 151]]
GNB:Accuracy :  80.0
AUC: 0.81
              precision    recall  f1-score   support

         Yes       1.00      0.60      0.75       149
          No       0.72      1.00      0.83       151

    accuracy                           0.80       300
   macro avg       0.86      0.80      0.79       300
weighted avg       0.86      0.80      0.79       300
```

Fig. 24 Overall Result of Naïve Bayes Classifier Algorithm Testing

in the literature, with the findings indicating that the proposed system outperforms most recent work in terms of accuracy.

As future activity, the paper proposes to create a larger number of clusters in order to support a wide variety of malware that is unknown. As a future scope, it would be of utmost significance to be able to perform analysis at installation time. In order to gain higher accuracy in classification in future, an optimal feature vector may be created by combining features such as opcode by Dalvik, Java reflection and attributes of Android Manifest.

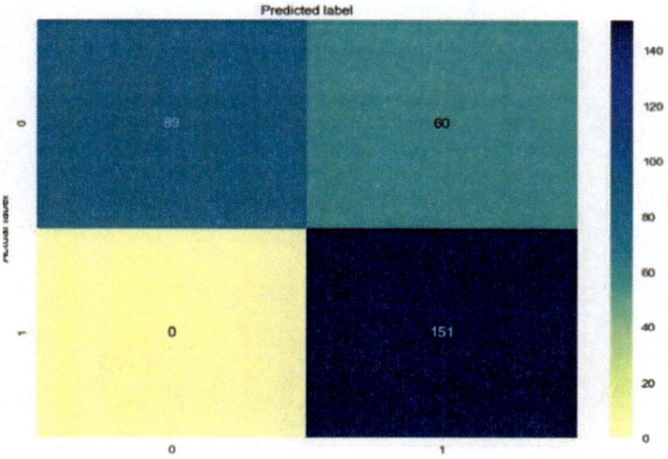

Fig. 25 Naïve Bayes Classifier Algorithm Predicted Label

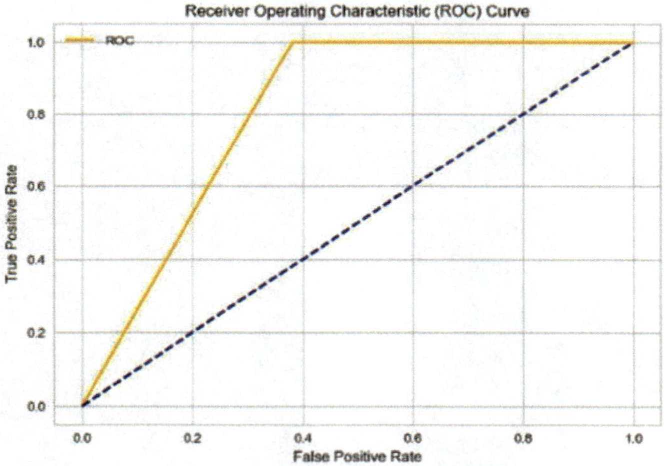

Fig. 26 Naïve Bayes Classifier Algorithm ROC Curve

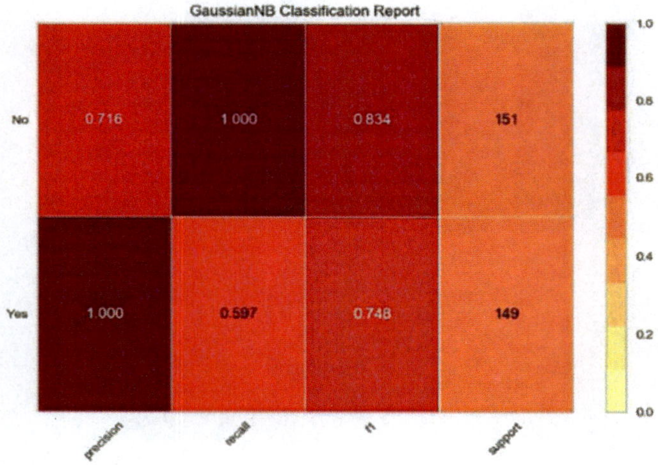

Fig. 27 Naïve Bayes classifier algorithm classification report

Fig. 28 Output for Benign Sample

Fig. 29 Output for Malign Sample

References

1. Bedford, Andrew, et al. "Andrana: Quick and accurate malware detection for android." International Symposium on Foundations and Practice of Security. Spring-er, Cham, 2016.
2. Sanz, Borja, et al. "MAMA: Manifest analysis for malware detection in Android." Cybernetics and Systems 44.6–7 (2013): 469–488.
3. A. Feizollah, N.B. Anuar, R. Salleh, G. SuarezTangil, S. Furnell, AndroDi-alysis: Analysis of Android Intent Effectiveness in MalwareDetection. Compute. Secur. **65**, 121–134 (2017)
4. Z. Yuan, Y. Lu, Z. Wang, Y. Xue, Droid-Sec: Deep Learning in Android Malware Detection. Sigcomm **2014**, 371–372 (2014)
5. Lifan Xu, Wei Wang, Marco A Alvarez, John Cavazos, and Dongping Zhang. Parallelization of shortest path graph kernels on multi-core cpus and gpus. Pro-ceedings of the Programmability Issues for Heterogeneous Multicores (MultiProg), Vienna, Austria, 2014

The Era of Deep Learning in Wireless Networks

Keren Lois Daniel and Ramesh Chandra Poonia

Abstract The rapid increase in the rise of wireless technology and wireless applications are envisioned to bring a radical change in the future through its services constituted for the exceptional demands on wireless networking infrastructure. The research focus in the future can be based on deep learning techniques with wireless communication networks. It is a subfield of Machine Learning. It works with the data-driven approaches that does not replace, the conventional design and techniques that are based on numerical-based models. Recently, a thorough stimulation is being invoked on deep learning which focuses on artificial neural networks that would be compelling for new designs and working on wireless network. The functioning of Artificial neural networks must be integrated for a better and efficient functioning of wireless in networks. The methodologies of deep learning are extensively used with training methods through all learning paradigms. It is also linked with other major development works such as reinforcement learning. The challenges that are faced are uncertain due to the dynamic environment in the network along with its dimensions that are complex in nature, and strong couplings among different wireless users with diverseness in wireless resource, air interface, and mobility. In this paper, we focus on the significance of deep learning in wireless fields of network and its prospects in the future.

Keywords Deep learning (DL) · Machine learning (ML) · Wireless networking · 5G systems · Performance optimization network management

K. L. Daniel (✉) · R. C. Poonia
Amity University, Jaipur, Rajasthan, India
e-mail: kerendaniel@stxaviersjaipur.org

A. K. Luhach et al. (eds.), *Second International Conference on Sustainable Technologies for Computational Intelligence*, Advances in Intelligent Systems and Computing 1235,
https://doi.org/10.1007/978-981-16-4641-6_28

1 Introduction

In recent times, the way we live is being controlled by the wireless communications mostly. The rapid development in wireless technologies has gained momentum recently. The wireless communication system has become the fundamental necessity that allows the user to communicate even from remote operated areas. Wireless communication works with the ease of accepting the flexibility of mobility. In the conventional wireless communication system, the transmission takes place with multiple signal processing, but with the recent technologies, the system is quite mature that can obtain global optimality. Wireless communication has relied on its own models that were accurate based on mathematical models [1]. But due to the demand for communication in complex situations where the communication channels are unknown and low-latency are needed, the model of deep learning was introduced.

Deep learning is a form where the concept of machine learning is introduced to solve complex problems when data is diverse, it is un- structured, and dependent on other source. It is a model of machine learning that inspires and creates model similar to the functioning of how the human brain works. In deep learning, the machine learning algorithms are used to perform tasks directly on complex data in the form of images, text, or sound. The best alternative for human brain is found in the abilities of DL. The algorithms are capable of accomplishing accurate outputs, and even sometimes surpassing the achievements of humans. The models are trained with learning experiences which later build up into trained model with the large set of data and neural network architectures that involve many layers of network.

Deep Learning is a prime technology that enhances network functions behind the wireless technologies that work virtually. Since the functionality of deep learning involves training the data and learning from the experiences, its procedure is termed as 'Deep'. With each passing time, this learning extends deeper and deeper into the different levels of neural network. Every data that is trained targets and centralizes its attention to intensify the performance efficiently. The efficiency of the DL Technology to smartly recognize just like a human has created a demand by data experts.

The growing heterogeneity and complexity of wireless network systems have been observed and managed by the multitude of network elements intractable. The increasing emergency need for wireless networks in different applications bring about the surge in network data traffic. In the near future, however, network services must develop smart heterogeneous designs and tools that can spawn the need that arises dynamically.

In recent trends, the system of developing a well-defined model is reflected using machine learning (ML)-based solutions to problems from the lowest network technology to the upgraded wireless technologies. These practices include the study of Linear Algebra such as Tensors, Scalars, Vectors, Matrix, Determinant, and Eigenvalues and Eigenvectors. It relates to basics of Calculus such as derivatives and Gradient Descent. Therefore, the upswing in wireless networks will be determined by the unique complexity, which make traditional approaches to network formation, design, and operations no longer adequate.

The revolution in digitization of our present culture has increased wireless network drastically. This inspired features of deep learning optimizing the conventional feature with the channel estimation and detection has enhanced the traditional communication to an impressive communication system. The growing generation of wireless communication networks have featured the innovative techniques in dense infrastructure, antenna, and use of frequency bands in the mmWave range with energy-efficient network management [2, 3], giving us a promising target of higher data rates and higher bit-per-Joule energy when compared with the existing network generations [3].

In this environment, the essence of deep learning knowledge can be disseminated efficiently and increasingly in real-time scenarios where hypothetical correlations can be obtained from the large data set, while reducing the pre-processing effort. With the system of Graphics Processing Unit (GPU), which is based on parallel computing, further enables deep learning to make decisions within nanoseconds [4]. The ultimatum for future wireless networks is the extreme heterogeneity of services to be provided with no much delay and has the flexibility to hold up to many innovative services that works with its own specifications. Therefore, new approaches toward deep learning have increased the network flexibility of attracting researchers into obstacle avoidance concept in AUVs [4].

Therefore, DL has paved a broader way by showing excellent results while dealing with many real-world scenarios. The use of DL model would help us to disseminate the complex network environment to an abstract representation as better decisions can be achieved finally for the computer network nodes to achieve improved network quality-of-service (QoS) and quality-of-experience (QoE) [5].

1.1 Advantages of Deep Learning

The advancements in computing recently has the capacity to make all possibility of executing large and complex to work much faster and efficiently. The following are the benefits of using deep learning:

(1) DL is advantageous in security and network layer functions.
(2) It helps in reducing the data to an optimal desired output. The pre-extraction of data is not required.
(3) The functions of DL help in speeding the algorithmic processing.
(4) It invokes less time in learning the techniques.
(5) Robustness to the dynamic data.
(6) Parallel computation using GPU can be applied to different volumes of data and applications.
(7) The model of deep learning is flexible.

1.2 Disadvantages of DL

1. Deep learning can do better only with large amount of data.
2. It is quite expensive as it requires complex data to work.
3. To train the complex data requires large machines and GPUs which is quite costly.
4. The use of deep learning cannot be just adopted with the knowledge of network.
5. Convolutional neural network can only help to comprehend the output.

2 Deep Learning for Wireless Networks

In Wireless networks, the primary parameters are based on the network slicing paradigm, communication signaling, channelizing the path, queue state of each node, path congestion, and resource management. Special services can be provided to customize the network specific requirements by using specialized techniques like Software Defined Networking (SDN) [6] and Network Function Virtualization (NFV) [7]. In recent research in the field of unmanned vehicles, AUVs will act as control point that can be reallocated based on heterogeneous traffic conditions to support the dynamic requests [8]. From every point of aspect, the past and present wireless communication networks are governed by mathematical models which includes some theoretical consideration.

Deep learning utilizes supervised and un-supervised learning to train the data representation by exploring the capabilities, benefits, and threats. Deep learning inherits the name from the fact that it involves itself deep into several layers of network, which also includes a hidden layer.

As the complexity of network is steadily rising, more demand for higher computational techniques is desired with faster and flexible intelligent learning algorithms that can handle the larger data set of the modern network. To meet these urgent requirements, deep learning applications in wireless networks has drawn lots of interests. DL functions on wireless network like a 'human brain': it accepts a large number parameters that are required for network, such as link signal-to-noise ratios (SNRs), the idle time in the channel, the rate of collision, the number of success links, routing delay, packet loss rate, bit error rate, etc., and performs deep analysis on the intrinsic patterns (such as congestion degree, interference alignment effect, hotspot distributions, etc.). These patterns can be used to control the protocols found in different network protocol layers. The concept might look better when we consider the alternate path while routing while the transport layer can reduce the congestion size. When compared to Machine Language, Deep Learning tends to provide better performances to abstract patterns from the data provided to make more accurate decisions.

Accomplishment of DL over wireless network can be compared with the following parameters in consideration with the human brain as the functioning of DL is similar to that of man:

Tolerance level: The ordinary human brain can tolerate disfigured images when there are some pixels missing in the image. But when data is incorrect in the wireless network, there will be distortion in channelizing the path. In Deep Learning, the methodology of deep neural network is to endure the missed or distorted input data. Such capabilities are very important to wireless networks as the possibilities to collect the node link state, node mobility, and controlling the channel failure.

Capacity of handling large amount of information: Since the human brain has the ability to absorb complex information at the same time due to the inherent quality of six senses. This gives the ability to make quick decisions. On the other hand, DL can accept large amount of data parallel from multiple set of protocol layers on the network which then determines the exact congestion in the path of a network. The potentiality of Deep learning is to analyze big data in the transmission performance through the parameters in the network layers on the huge traffic.

Making dynamic and controlled decisions: In human brains, thoughts manipulate and guide us in our behaviors. Passive learning is not much beneficial in network analysis. The ultimate goal is to use the learning results as the major output for the better functioning of the network management. DRL emerged from the use of Markov decision in DL, where the DRL works on training the data set with a reward function, and policy seeking, to have control over network optimality based on the maximum reward received. Thus, we can use DRL to achieve largescale wireless network control.

3 The Role of Deep Learning Wireless Network Layers

There are five Open System Interconnection layers of wireless network where deep learning plays different roles in each layer. From the below figure, we see that DL can be adopted in wireless network at different layers (Fig. 1).

From the above figure, we see that deep learning contributes through various applications in the wireless network. DL can be used in various layers of network. It can be used in interference alignment [9] which seems to be a promising technique for the future. Since there is a huge increase in traffic, there are possibilities of intrusion detection becoming more challenging when they pass through the filters in network. In such situation, DL is the perfect tool to perform large-scale network analysis of data. Deep learning can also be used to classify the modulations, into an efficient method of error detection in the network layer [10]. The contribution of DL in different protocol layers help us to seek an optimal routing path [11] with multi-session scheduling. Deep Learning is also quite advantageous in security and network malicious attacks. Intrusion detection over privacy protection becomes more challenging due to the heavy traffic. Dl is ideal to perform large-scale intrusion events [12].

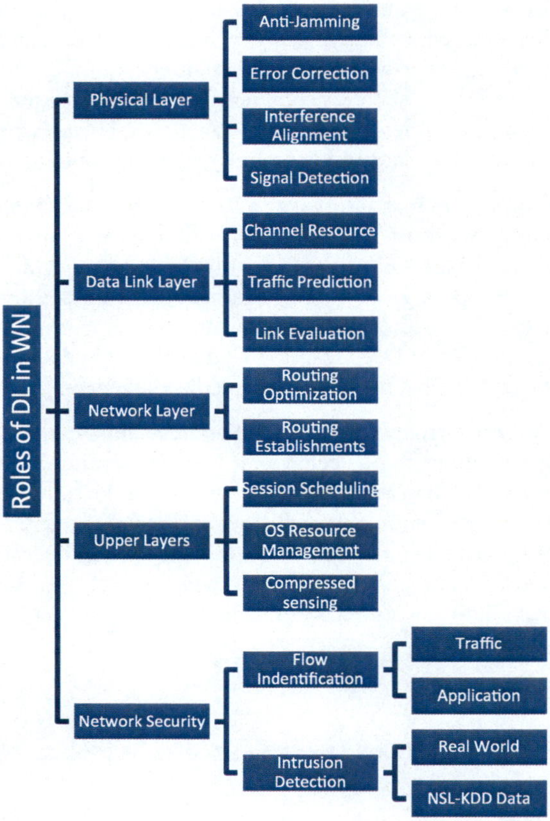

Fig. 1 Classification of DL roles in wireless network

The field of DL is still yet to be matured in many ways as there are few issues that are yet to be resolved. There are many fields that are still yet to be surveyed to enhance the wireless networks in cognitive radio, software designed networks, fog, etc.

4 Paradigms of Wireless Network Using Deep Learning

There is an excessive propagation of the use of DL in wireless network. Though the concept of DL was inherited from AI, it is quite different from the functioning of wireless communications. Basically, the functionality of wireless is based on accuracy using traditional way of using mathematical representation, such as Additive White Gaussian Noise (AWGN) channel model, which helps in designing channel estimation algorithms or channel feedback schemes [1], but Dl works in the absence of traditional models. The basic method in which DL functions is by training all parameters of DNN totally. The advantages of having classical methods of wireless

Table 1 Transmission difference between wireless and DL

	Wireless	Deep learning
Model	Works on accurate mathematical model	It does not comply according to mathematical model
Design	Each module is separately evaluated by mathematical equations	Every parameter of DNN is trained
Interpretation	Quantitative	Non-quantitative
Genral applicability	Widely used	Used according to Specification
Challenges	Unrealistic	Too many parameters

transmission are often quantitative with credible details that are widely used. On the contrary, DL are sometimes non-intuitive, which may be difficult to translate the dataset. DL uses specific applications based on requirement. For different task, we need to train DNN [18–27]. The transmission of wireless network is merited on its ideal standard of working and simplified mathematical model (Table 1).

Although the significant differences between wireless transmission and DL might seem to be disappointing, the paradigms of Dl can be used for wireless network to attain profound benefits in the future. The following few paradigms can be implemented with wireless network although there are many other paradigms.

4.1 Architecture Based on DL

The two main parameters that can be invoked for a better performance design of network transmission are:

- DL with Orthogonal Frequency Division Multiplexing (OFDM):
  To make decision in an unknown scenario is to use the OFDM in the channel to know the behaviors and also to decode the signals appropriately. In the training process, the signals are inputted into the DNN to reconstruct the output.
- DL-based point-to-point communication:
  This paradigm of DL is designed in resemblance of wireless communication systems and the autoencoder, where their output aims at equating the input message s and the output message s'. The signals pass through a transmitter in the hidden layers, an encoded signal is generated to all possibilities.

4.2 Algorithm Design

The deep learning helps in speeding the process of algorithms to improve the potentiality of performance and reduce the latency. It is convenient to implement as it relies on parallel architectures.

- **Transmission-based Algorithms**
  In wireless communication, the most expected outcome is based on reliable transmission over any network. The Channel estimation is carried out from the initial point of receiving to the deepest layer and through the other receivers till the channel is decoded.
- **Optimization Algorithm**
  The second most important parameter of a good network is to have an optimized system that realizes efficient use of radio resources that are limited in usage. Since the DL has the power to speed up the process of algorithms, it can be induced into any dense or complex network for maintaining reliable performance.

5 The Era of Deep Learning

The ERA of manipulating the complex or large data is capturing a wide interest in deep learning across different research disciplines [13, 14] and a growing number of surveys on the real scenario are rapidly emerging. The deep learning plays a vital role in multiple protocol layers. It has the ability to integrate itself with various wireless network functionalities like CRNs, SDNs, etc., for having a balanced traffic in the network. It is essential that the quantified data and the quality of the training data determine the final performance of the result which is obtained by using DL. The focus of using DL is to recognize the intrinsic patterns hidden in the large data. No other choice could be better than that of defining the model of DL for all networking functional nodes that could possibly reduce the attack on the path of transmission. According to Schmidhuber, survey on deep learning has made impressive evolutions on the methods and applications that can be used for the complex issues [15]. There are various principals of deep learning models that have emerged and are being used in various developmental applications such as Speech Processing, pattern recognition, and computer vision [16]. The applications of artificial intelligence (AI) and machine learning (ML) technologies in wireless communications have drawn compelling attention of researchers lately. The perfect tool to be implemented in various cross layer metrics of network and extracting data from the inherent network patterns for protocol optimization is deep learning. Therefore, there is a remarkable performance of deep learning with neural networks in different control problem that are created recently (e.g., video games or virtual games, etc.). As the concept of deep learning hits the stream of popularity in research many researchers are keen in using this concept. In the recent tutorials on comprehensive learning of DL, the popular applications and a prerequisite knowledge are explained covering the principles of DL functions in applications [17]. Therefore, when the two models (AI with DL) join together, it is an indispensable tool for the design and operation of future wireless communication networks, and an added advantage for better management of output.

6 Conclusion

This paper briefly brings out the knowledge of deep learning, and its role in each of the network layer is briefed with the methods of applying DL in performance enhancement of wireless management. Precisely, the concept of deep learning is very effectual for intelligent wireless network management due to its functioning as human brain in pattern recognition capability. With the present wireless hardware products, the performance of DL is easier in adoption. It plays vital roles in multiple protocol layers. The brief view of the roles of DL in the wireless network layer is summarized with its applications in different layers. Definitely, deep learning is capable of supporting all communication systems that are complex in nature and difficult in their operating scenarios. With its high training methodologies, Dl can speed-up the process of large-scale data with guaranteed output. It is believed that the implementation of DL for dynamic scenarios may be accurate in proving its great development in terms of both theory and practical innovations. DL is probably the most promising tool that can make things that seem impossible to be made possible.

Therefore, DL can be an integrated part in wireless network to gain the optimality on the centralized or distributed resource allocation and balancing the traffic functions in all the areas. The aim of this paper is to help readers to understand the basic knowledge on the era of how deep DL has made its way through the wireless networking functions in different layers with some interesting and challenging research topics to pursue in the future.

References

1. L. Dai, R. Jiao, F. Adachi, H. V. Poor, L. Hanzo, Deep Learning for Wireles Communications: An Emerging Interdisciplinary Paradigm, IEEE Wireless Communications, August 2020 1536–1284/20/$25.00 © 2020 IEEE
2. J. Andrews, S. Buzzi, W. Choi, S. Hanly, A. Lozano, A.C.K. Soong, J.C. Zhang, What will 5G be? IEEE J. Sel. Areas Commun. **32**(6), 1065–1082 (June 2014)
3. Zappone and E. Jorswieck, "Energy efficiency in wireless networks via fractional programming theory," Foundations and Trends R in Communications and Information Theory, vol. 11, no. 3–4, pp. 185– 396, 2015
4. "NGMN alliance 5G white paper," https://www.ngmn.org/5g-white- paper/5g-white-paper.html, 2015
5. Telus and Huawei, "Next generation SON for 5G," *White Paper*, 2016
6. Chaoyun Zhang, Paul Patras, and Hamed Haddadi w"Deep Learning in Mobile and Wireless Networking: A Survey "Proceedings of the IEEE
7. D. Kreutz, F. M. V. Ramos, P. Verissimo, C. E. Rothenberg, S. Azodol- molky, and S. Uhlig, "Software-defined networking: A comprehensive survey," *Proceedings of the IEEE*, vol. 103, no. 1, pp. 14–76, 2015
8. Imran, A. Zoha, and A. Abu-Dayya, "Challenges in 5G: how to empower SON with big data for enabling 5G," *IEEE Network*, vol. 28, no. 6, pp. 27–33, 2014
9. Ian Goodfellow, Yoshua Bengio, and Aaron Courville. Deep learning. MIT press, 2016.
10. Y. Hechtlinger, P. Chakravarti, and J. Qin, "A Generalization of Convolutional Neural Networks to Graph-Structured Data," eprint arXiv:1704.08165, Apr. 2017.

11. M. Defferrard, X. Bresson, and P. Vandergheynst, “Convolutional neural networks on graphs with fast localized spectral filtering,” in Proc. Conference on Advances in Neural Information Processing Systems (NIPS 2016), vol. 29, Barcelona, Spain, Dec. 2016, pp. 3837–3845.
12. P. Xie, J.-H. Cui, and L. Lao, “VBF: Vector-based forwarding protocol for underwater sensor networks,” in Proc. 5th international IFIPTC6 conference on Networking Technologies, Services, and Protocols (Networking 2006), Coimbra, Portugal, May 2006, pp. 1216–1221.
13. Qian Mao, Student Member, IEEE, Fei Hu, Member, IEEE, and Qi Hao, Member, IEEE, Deep Learning for Intelligent Wireless Networks: A Comprehensive Survey, journal of latex class files, VOL. 14, NO. 8, JANUARY 2018
14. NF Hordri, A Samar, SS Yuhaniz, and SM Shamsuddin. A systematic literature review on features of deep learning in big data analytics. International Journal of Advances in Soft Computing & Its Applications, 9(1), 2017
15. Y. LeCun, Y. Bengio, G. Hinton, Deep learning. Natu re **521**(7553), 436–444 (2015)
16. J. Schmidhuber, Deep learning in neural networks: An overview. Neural Netw. **61**, 85–117 (2015)
17. Weibo Liu, Zidong Wang, Xiaohui Liu, Nianyin Zeng, Yurong Liu, and Fuad E Alsaadi. A survey of deep neural network architectures and their applications. Neurocomputing, 234:11–26, 2017.
18. S. Buzzi, C.-L. I, T. E. Klein, H. V. Poor, C. Yang, A. Zappone, A survey of energy-efficient techniques for 5G networks and challenges ahead,” IEEE Journal on Selected Areas in Communications, vol. 34, no. 5, 2016
19. S. Abdelwahab, B. Hamdaoui, M. Guizani, T. Znati, Net- work function virtualization in 5G. IEEE Commun. Mag. **54**(4), 84–91 (2016)
20. Alzenad, A. El-Keyi, F. Lagum, and H. Yanikomeroglu, “3-D placement of an unmanned aerial vehicle base station for energy- efficient maximal coverage,” *IEEE Wireless Communications Letters*, vol. 6, no. 4, pp. 434—437, August 2017
21. S. Bi, R. Zhang, Z. Ding, S. Cui, Wireless communications in the era of big data. IEEE Commun. Mag. **53**(10), 190–199 (October 2015)
22. X. Cheng, L. Fang, L. Yang, S. Cui, Mobile big data: The fuel for data-driven wireless. IEEE Internet Things J. **4**(5), 1489–1516 (October 2017)
23. X.-W. Chen, X. Lin, Big data deep learning: challenges and perspectives. IEEE access **2**, 514–525 (2014)
24. Maryam M Najafabadi, Flavio Villanustre, Taghi M Khoshgoftaar, Naeem Seliya, Randall Wald, and Edin Muharemagic. Deep learning applications and challenges in big data analytics. Journal of Big Data, 2(1):1, 2015
25. Mehdi Gheisari, Guojun Wang, and Md Zakirul Alam Bhuiyan. A survey on deep learning in big data. In Proc. IEEE International Conference on Computational Science and Engineering (CSE) and Embedded and Ubiquitous Computing (EUC), vol. 2, pp. 173–180, 2017
26. A. Zappone, M. Di Renzo, M. Debbah, Wireless Networks Design in the Era of Deep Learning: Model-Based, AI-Based, or Both? IEEE Trans. Commun. **67**(10), 7331–7376 (Oct. 2019). https://doi.org/10.1109/TCOMM.2019.2924010
27. E. Hodo, X. Bellekens, A. Hamilton, C. Tachtatzis, and R. Atkinson, “Shallow and deep networks intrusion detection system: a taxonomy and survey,” eprint arXiv:1701.02145, Jan. 2017

A Study of Association Rule Mining for Artificial Immune System-Based Classification

S. M. Zakariya, Aftab Yaseen, and Imtiaz A. Khan

Abstract One of the most useful and quite well data mining techniques is association rule mining. The immune system's tremendous information capacities are used by the artificial immune system (AIS). To deal with a complicated search space, the artificial immune system clonal selection strategy uses the population-based search paradigm of evolutionary computing algorithms. In this paper, the accuracy of the classification method based on the artificial immune system was computed on distinct clonal variables and varied numbers of generations. The findings of these systems are presented on various benchmark datasets. A comparative analysis is performed based on the accuracy of the varying clonal factors 0.1 to 0.9 and generations 10, 20, 30, 40, 50, and 60. Six standard datasets are used to calculate the accuracy. It is noted that the method offers the highest accuracy with a clonal factor of 0.4 in each dataset for different generations.

Keywords Association rule mining · Artificial immune system · Associative Classification · Clonal selection algorithm · Accuracy rate

1 Introduction

As we all know, data mining is essential for making knowledgeable decisions. Data mining is the process of extracting relevant information from a vast number of datasets. Data classification entails extracting models to represent relevant data categories or anticipating prospective data trends. Association rule mining is used in associative Classification to gather high-quality rules that properly extend the training dataset during the rule discovery process. To develop a model (classifier) for prediction, Associative Classification combines two possibly the best data mining operations, association rule mining and classification. Classification and association

S. M. Zakariya (✉) · I. A. Khan
Electrical Engineering Section, University Polytechnic, Aligarh Muslim University, Aligarh, India

A. Yaseen
Department of Computer Science Engineering, Integral University, Lucknow, India

A. K. Luhach et al. (eds.), *Second International Conference on Sustainable Technologies for Computational Intelligence*, Advances in Intelligent Systems and Computing 1235,
https://doi.org/10.1007/978-981-16-4641-6_29

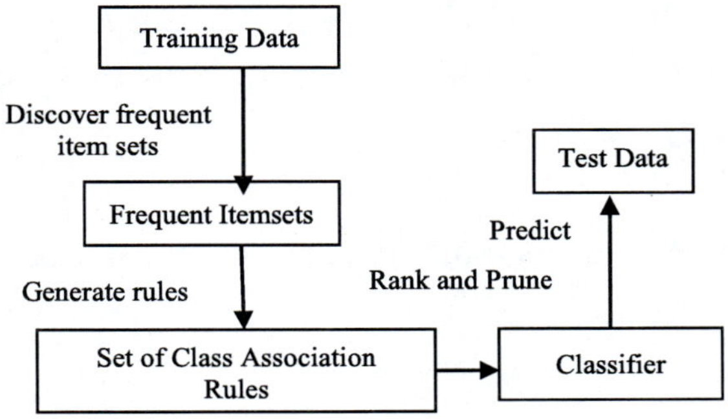

Fig. 1 Associative classification structure

rule mining are both data mining professions, with the exception that Classification's main purpose is to predict class marks. In contrast, the mining of association rules describes the relationships between objects in a transactional database. The three primary phases of associative Classification are rule discovery, rule collecting, and Classification, as shown in Fig. 1 [13].

The clonal selection method developed by AIS is comparable to mutation-based evolutionary algorithms and provides strong approximation and searching capabilities. As a consequence, the outcomes of a clonal selection algorithm-based AIS classification system are reviewed [3, 9, and 18]. The impacts of an AIS-based classification scheme are examined in this work to discover if the system performs best in terms of precision for different clonal variables and generations. All of these datasets are available in the UCI machine learning library [2]. The study aims to determine at which clonal factor and generation the highest classification precision can be achieved. The research aims to figure out which clonal component and generation yield the best classification precision. Following is the breakdown of the paper's structure. Associative categorization systems are briefly examined in Sect. 2. Section 3 discusses the tactics of artificial immune systems. The artificial immune system-based categorization is explained in Sect. 4. The results and conclusion are discussed in Sects. 5 and 6, respectively.

2 A Study of Associative Classification Schemes

Associative Classification (AC) is a subset of data processing, part of a larger scientific area. It collects high-quality rules that can accurately generalize the training dataset via association rule mining in the rule discovery process. Associative Classification generates and analyzes association rules for use in Classification. Because association rules seek powerful connections between distinct qualities, they can help overcome some of the limitations of decision-tree inference, which only evaluates one characteristic at a time. Associative Classification beats most classic classification algorithms in some trials. The three fundamental techniques being investigated are CBA [11], CMAR [7], and CPAR [15].

2.1 *Association-Based Classification (CBA)*

The CBA-main RG's goal is to locate all ruleitems with more support than minsup. The type is a ruleitem: <condset,y>, where y is the class label and condset is a collection of objects. The condset support count is the number of occurrences in D that include the condset (called condsup Count). The number of cases in D that include the condset and are tagged with Class y is the ruleitem support count (called rulesupCount). Each rule is a condset y rule that supports (rulesup Count / |D|) *100% and has a trust of (rulesupCount/condsup Count)*100%. Ruleitems that fulfil minsup are known as regular ruleitems, whereas the others are known as uncommon ruleitems. The ruleitem with the highest trust is chosen as the prospective rule (PR) for this collection of ruleitems for all ruleitems with the same condset. Pick one ruleitem at random if there are many ruleitems with the same maximum confidence. The rule is correct if the confidence is bigger than minconf. As a result, any PRs that are both regular and relevant include the set of class association rules (CARs). All of the common rules are generated by the CBA-RG algorithm [6] and [11].

2.2 *Multiple Association Rules-Based Classification (CMAR)*

CMAR uses a series of criteria to decide the class name. CMAR picks a small series of critically, strongly linked rules and examines their relationship in light of a new prediction instance. CMAR has a greater prediction accuracy than CBA, according to a comprehensive performance review.

CMAR uses a unique data structure called CR-tree to store and retrieve multiple categorization rules in a compact and efficient manner, enhancing accuracy and efficiency. The CR-tree is a robust prefix tree structure for investigating rule sharing. The CR-tree may also be used to recover rules as a rule indexing mechanism.

CMAR employs a freshly discovered FP-growth methodology variation to speed up the mining of a complete set of rules. FP-growth is considerably quicker than Apriori-like approaches employed in the preceding association-based group [7] when there are many rules, big training data sets, and extensive pattern rules.

2.3 Classification Based on Rules for Predictive Association (CPAR)

CPAR uses the following features to increase its accuracy and performance: Instead of using simply the best literal, CPAR considers all near-best literals while producing rules, ensuring that crucial rules are not neglected. CPAR generates a smaller set of laws with greater consistency and reduced redundancy than associative categorization [15].

2.4 Associative Classification Steps

The process of developing an associative classification classifier can be broken down into four steps [4].

1. The identification of all frequent ruleitems.
2. Output of all CARs that are confidential from frequent rule items extracted in Step 1 above the minconf threshold.
3. To generate the classifier, one subset of CARs from those formed in Step 2 is chosen.
4. For test data, determining the accuracy of the generated classifier.

3 Techniques of Artificial Immune Systems

A biological immune system's main job is to defend the body from foreign chemicals called antigens. Individuality, autonomy, international identification, dispersed detection, and noise tolerance are all characteristics of immune systems. Several AIS models are utilized in pattern recognition, defect detection, computer protection, and various other applications [1] and [17].

3.1 Clonal Selection-Based Algorithms

Burnet [21] developed the notion of clonal selection in 1959. The antigenic stimulus–response of the adaptive immune system is explained using this concept. Despite the apparent selection of other cells, it establishes the premise that only cells capable of sensing an antigen can grow. Several artificial immune algorithms that mirror the principle of clonal selection have been developed [8] and [19]. Figure 2 depicts the definition of the clonal selection algorithm.

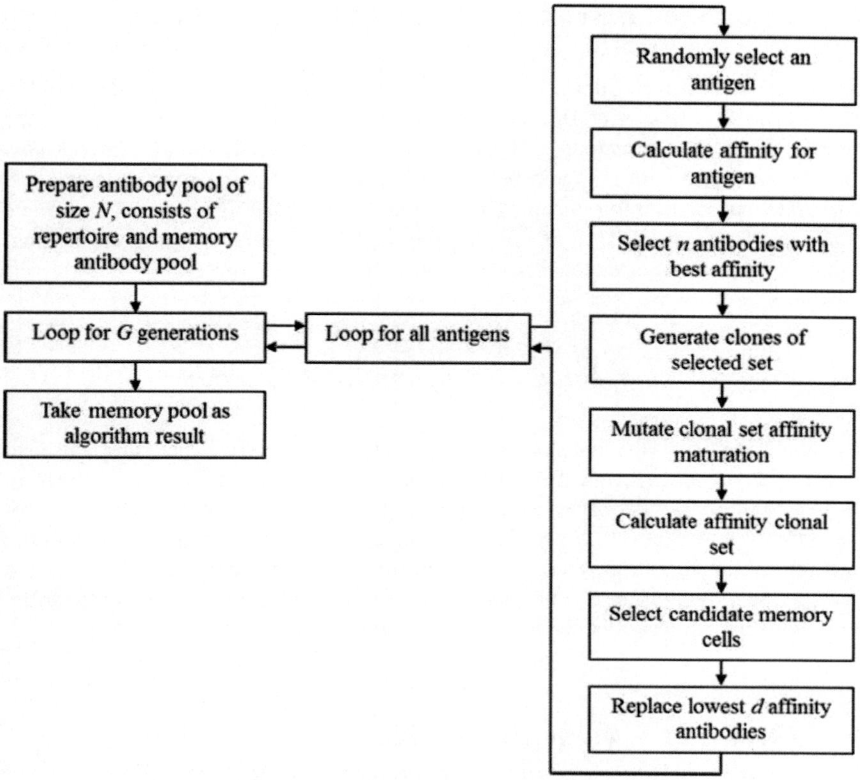

Fig. 2 Clonal selection algorithm procedure

3.2 Algorithms Based on Negative Selection

Negative selection is a natural immune system process that has influenced several modern artificial immune systems. If any self-cells are detected by a T-cell in the thymus during the immune system's T-cell maturation phase, it is eliminated before being deployed for immunological function. By removing each detector candidate that matches items from a collection of self-samples, the negative selection process generates a detector set. Negative selection-based algorithms have been employed in a variety of applications, including anomaly identification. The basic idea behind the method is to build a collection of detectors by creating candidates at random and then discarding those that identify self-data training. These detectors can then be used to identify abnormalities. Igawa and Ohashi [10] introduced the Artificial Negative Selection Classifier, a novel negative selection technique for multi-class Classification (ANSC).

4 Classification Based on an Artificial Immune System

Using association rule mining, Associative Classification examines a database of transactions for association rules. Due to the broad search room, the rule discovery process is very time-consuming. Algorithms based on artificial immune systems have robust features for optimizing problem search. Figure 3 depicts a schematic diagram of the AIS-based grouping system. The cloning protocol is designed such that a rule's clonal rate is proportional to its affinity, and the average value of each rule's clonal rate is equal to the user's clonal rate [5] and [14].

4.1 *Preprocessing, Initialization, and Rule Selection*

Preprocessing data is the first step in any data mining project. Preprocessing is used to convert the data into a format that can be quickly interpreted. Data is stored in the form of files containing transaction documents in this project. Datasets are the name for these archives. An itemset, which is a list of objects, is part of a transaction in a dataset. In a transactional dataset, there are several representation schemes for expressing itemsets. Objects present in a dataset's itemset can be interpreted as a binary string, with items present encoded as binary 1 and items missing as binary 0 [20].

4.2 *Cloning of Selected Ruleset: The Artificial Immune System-Based Classification System's Structure*

The cloning technique is set up so that a rule's clonal rate is proportional to its confidence (i.e., affinity), and the average of each rule's clonal rate equals the user's

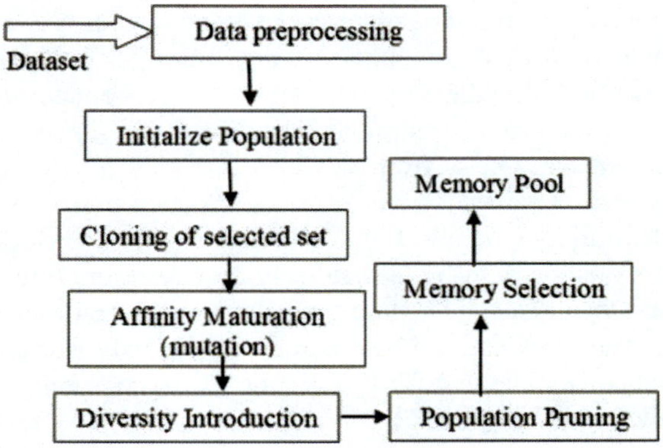

Fig. 3 The structure of the artificial immune system-based classification system

clonalRate. A rule's clonal rate is calculated in particular [12] and [22]. Let cloneRate(R) be the clonal rate of a rule R, and R1, R2, R3,..., Rn be the rules chosen at a given generation.

$$cloneRate(R_i) = A \times confidence(R_i) \tag{1}$$

Since the average value of each rule's clonal rate is equal to clonalRate, put as:

$$clonalRate = \frac{1}{n} \times \sum_{i=1}^{n} cloneRate(R_i) \tag{2}$$

Or

$$clonalRate = \frac{1}{n} \times A \times \sum_{i=1}^{n} confidence(R_i) \tag{3}$$

Thus,

$$A = \frac{n \times clonalRate}{\sum_{i=1}^{n} confidence(R_i)} \tag{4}$$

5 Result and Discussion

The results are analyzed in this paper using the WEKA (Waikato Environment for Knowledge Analysis) tool. WEKA has been made open-source, enabling the framework to be extended by introducing algorithm and tool plug-ins for the platform to academics and industry [16].

5.1 Dataset Used

The UCI machine learning repository provided the six reference datasets [2], namely Gait Classification, Codon Usage, Dry Bean, Car Evaluation, Wine, and Iris. These datasets vary in the number of classes, samples, number of items, number of attributes, and training and test datasets. The Gait Classification, Car Evaluation, Wine, and Iris datasets are tiny datasets. The Gait dataset has 48 samples only, using a training set of 34 samples and a test set of 14 samples. The Gait dataset has 4 different classes with 321 attributes and 24 items. The Car dataset has 1728 samples with 1210 samples as a training set and 518 samples as a test set. The car dataset has 4 different classes with 6 attributes and 21 items. The wine dataset consists of only 178

Table 1 The following is a list of all six datasets that were utilized

Name of dataset	Attributes in numbers	Items in numbers	Classes in numbers	Instances in numbers	Training set	Test set
Gait classification	321	24	4	48	34	14
Codon usage	69	9	6	13,028	9120	3900
Dry bean	17	32	7	13,611	9528	4083
Car evaluation	6	21	4	1728	1210	518
Wine	13	47	3	178	125	53
Iris	4	24	4	150	105	45

samples with 125 training sets and 53 test sets. The wine dataset contains 3 classes with 13 attributes and 47 items. The Iris dataset contains 150 samples which are divided into 105 training sets and 45 test sets. The Iris dataset has 4 classes with 4 attributes and 24 items. The Codon Usage and Dry Bean datasets are big. The Codon dataset has 13,028 samples with 9120 samples as a training set and 3900 samples as a test set. The Codon dataset has 6 different classes with 69 attributes and 9 items. The Dry Bean dataset has 13,611 samples with 9528 samples as a training set and 4083 samples as a test set. The Bean dataset has 7 different classes with 17 attributes and 32 items. Table 1 provides a summary of these datasets.

5.2 Evaluation Parameters

A confusion matrix containing information on the real and expected classifications created by a classifier is represented in Table 2 [6].

Accuracy is defined as the percentage of cases in which the test data collection was correct.

$$\text{Accuracy} = ((\text{P} + \text{S}))/((\text{P} + \text{Q} + \text{R} + \text{S})) \tag{5}$$

The proportion of correctly described positive cases is known as the True Positive Rate (TPR).

$$\text{TPR} = \text{S}/((\text{R} + \text{S})) \tag{6}$$

Table 2 Confusion matrix

		Predicted	
		Negative	Positive
Actual	Negative	P	Q
	Positive	R	S

The False Positive Rate (FPR) is the percentage of negative cases that are incorrectly reported as positive:

$$\text{FPR} = \text{Q}/((\text{P} + \text{Q})) \tag{7}$$

The proportion of correctly classified negative cases is known as the True-Negative Rate (TNR).

$$\text{TNR} = \text{P}/((\text{P} + \text{Q})) \tag{8}$$

The False Negative Rate (FNR) is the percentage of positive situations identified as negative incorrectly:

$$\text{FNR} = \text{R}/((\text{R} + \text{S})) \tag{9}$$

where P denotes the number of accurate predictions that an object is negative, Q denotes the number of false predictions that an object is positive, R denotes the number of wrong predictions that an object is negative. S denotes the number of positive cases.

5.3 Results with Gait, Codon, Bean, Car, Wine, and Iris Datasets

Table 3 shows the accuracy by clonal algorithm on Gait, Codon, Bean, Car, Wine, and Iris datasets with threefold cross-validation for different generations on fixed clonal factor 0.4. The clonal factor 0.4 was chosen because the method achieves optimum accuracy on all six datasets at this clonal factor. The results are tested at varying clonal factors from 0.1 to 0.9 and at different generations like 10, 20, 30, 40, 50, and 60.

Table 3 Average accuracy on six datasets with varying generation at a fixed clonal factor 0.4 by CLONALG

No. of generation	Gait	Codon	Bean	Car	Wine	Iris
10	97.566	73.846	72.848	82.736	94.365	96.678
20	99.025	73.250	72.505	82.694	93.245	98.735
30	98.823	71.876	71.845	81.720	96.496	98.029
40	98.045	73.996	72.226	83.885	95.478	97.973
50	97.755	71.075	70.985	81.965	93.782	97.165
60	98.224	72.322	71.763	82.052	94.664	97.964

The graphical representation of the consistency in all six datasets with threefold cross-validation for various generations and at a fixed clonal factor of 0.4 is seen in Fig. 4. This figure shows that generation 30 has maximum classification accuracy at fixed clonal factor 0.4 for the Gait Classification dataset.

Table 4 shows the classification accuracy on varying clonal factors at the maximum achieved accuracy on various generations for all six datasets. Figure 5 depicts the graphical representation of classification accuracy on six separate datasets, Gait, Codon, Bean, Car, Wine, and Iris, with varying clonal factors. It can be seen from this graph that at a clonal factor of 0.4, it achieves the highest classification accuracy on each dataset. Table 5 shows the maximum accuracy at fixed clonal factor 0.4 on all six datasets with threefold cross-validation. Maximum accuracy is achieved in different generations in all six different datasets. Figure 6 depicts the graphical representation of Table 5.

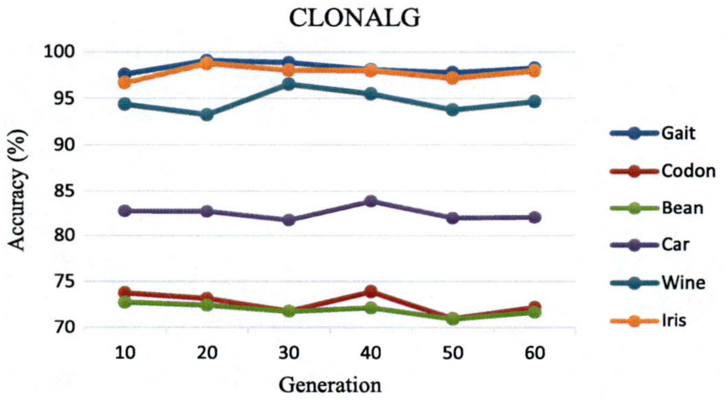

Fig. 4 Accuracy on different generation for fixed clonal factor 0.4 on Gait, Codon, Bean, Car, Wine, and Iris datasets by CLONALG

Table 4 Average accuracy on six datasets on varying clonal factor 0.1–0.9

Clonal factor	Gait	Codon	Bean	Car	Wine	Iris
0.1	96.258	70.243	70.626	80.737	92.134	95.049
0.2	96.975	71.042	71.729	81.830	95.505	97.029
0.3	96.158	73.014	70.380	80.491	92.135	95.0495
0.4	99.025	73.996	72.848	83.885	96.496	98.735
0.5	95.168	72.094	71.583	81.694	85.955	94.059
0.6	97.052	70.945	72.266	82.377	86.517	96.039
0.7	96.992	71.318	72.027	82.130	85.955	95.049
0.8	91.247	72.442	71.173	81.284	84.832	89.138
0.9	96.152	70.745	71.724	81.830	86.517	95.049

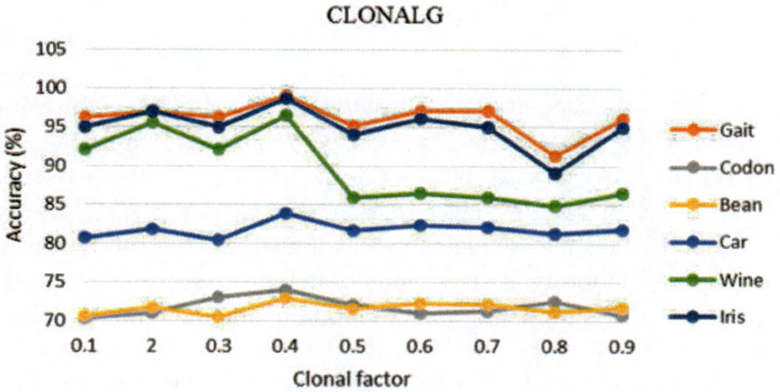

Fig. 5 Accuracy on varying Clonal Factor at fixed generations 20, 40, 10, 40, 30, and 20 for Gait, Codon, Bean, Car, wine, and Iris datasets, respectively

Table 5 Comparisons of result at maximum accuracy for all six datasets

Dataset	Training set	Test set	Generation	Accuracy (%) at 0.4 clonal factor
Gait	34	14	20	99.025
Codon	9120	3900	40	73.996
Bean	9528	4083	10	72.848
Car	1210	518	40	83.885
Wine	125	53	30	96.496
Iris	105	45	20	98.735

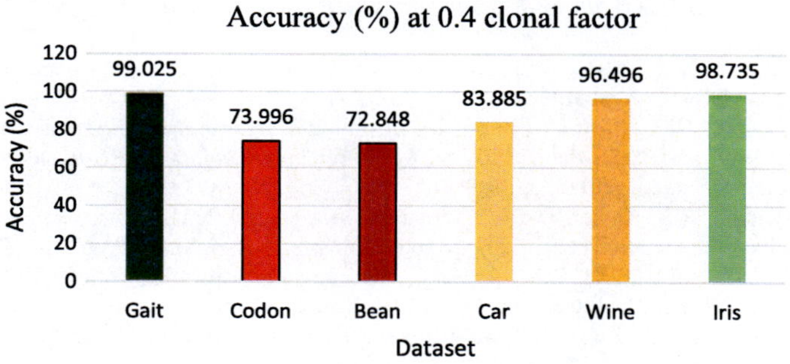

Fig. 6 Maximum accuracy at fixed clonal factor 0.4 on all six datasets with threefold cross-validation

Based on this, it was determined that the Gait dataset has the highest classification accuracy of the six datasets.

6 Conclusion

This paper evaluates the system's performance by using six benchmark datasets, namely, Gait Classification, Codon Usage, Dry Bean, Car Evaluation, Wine, and Iris. The performance is assessed in terms of accuracy as an evaluation parameter at varying clonal factors on different generations. On each dataset, the maximum accuracy is achieved at a clonal factor of 0.4. From the results, it is evident that the accuracy varies randomly with various generations for a fixed clonal factor. It is also observed that with a larger-sized dataset, the accuracy of classification decreases. Accordingly, the Gait classification dataset shows the highest accuracy of Classification owing to its small size. Iris dataset has the second-highest accuracy of Classification due to the second smallest in size.

References

1. N.E.V. Altay, B. Alatas, Performance analysis of multi-objective artifcial intelligence optimization algorithms in numerical association rule mining. J. Ambient. Intell. Humaniz. Comput. **11**, 3449–3469 (2020)
2. J. Brownlee, Clonal selection theory & CLONALG clonal selection classification algorithm (CSCA). Technical report No. 2–02, (2005)
3. F.M. Burnet, *The Clonal Selection Theory of Acquired Immunity*. (Cambridge University Press, 1959)
4. F. Campelo, F.G. Guimarães, H. Igarashi, J.A. Ramírez, A clonal selection algorithm for optimization in electromagnetic. IEEE Trans. Magn. **41**, 1736–1739 (2005)
5. L.N.D. Castro, F.J.V. Zuben, The clonal selection algorithm with engineering applications. in *Proceeding Workshop Artificial Immune System and Their Application (GECCO'00)*, (2000), pp. 36–37
6. L.N.D Castro, F.J.V. Zuben, Learning and optimization using the clonal selection principle. IEEE Trans. Evol. Comput. **6**(3), 239–251 (2002)
7. D.S.D. Cunha, L.N.D. Castro, Evolutionary and immune algorithms applied to association rule mining in static and stream data. in *2018 IEEE Congress on Evolutionary Computation (CEC)*, Rio de Janeiro, Brazil, (2018), pp. 1–8
8. D.S.D. Cunha, R.S. Xavier, D.G. Ferrari, F.G. Vilasbôas, L.N.D. Castro, Bacterial colony algorithms for association rule mining in static and stream data. Math. Prob. Eng. (2018)
9. T.D. Do, S.C. Hui, A.C.M. Fong, B. Fong, Associative classification with artificial immune system. IEEE Trans. Evol. Comput. **13**(2), 217–228 (2009)
10. H. Ian, *Witten and Eibe Frank: Data Mining: Practical Machine Learning Tools with Java Implementations* (Morgan Kaufmann, San Francisco, 2000)
11. K. Igawa, H. Ohashi, A negative selection algorithm for classification and reduction of the noise effect. Appl. Soft Comput. J. (2008)
12. W. Li, J. Han, J. Pei, CMAR: Accurate and efficient Classification based on multiple class-association rules. in *Proc. of IEEE Int. Conference on Data Mining (ICDM' 01)*, (2001), pp. 369–376
13. B. Liu, H. Hsu, Y. Ma, Integrating classification and association rule mining, in *Proc. 4th Int. Conf. Knowledge Discovery Data Mining*, (1998), pp. 80–86
14. W. Luo, X. Lin, T. Zhu, P. Xu, A clonal selection algorithm for dynamic multimodal function optimization. Swarm Evol. Comput. **50**, (2019)

15. E. Nabil, S.A.F. Sayed, H.A. Hameed, An efficient binary clonal selection algorithm with optimum path forest for feature selection. Int. J. Adv. Comput. Sci. Appl. **11**(7), 259–267 (2020)
16. D.J. Newman, S. Hettich, C. Blake, C. Merz, *UCI Repository of Machine Learning Databases, Berkeley* (Dept. Information Comput. Sci., Univ. California, CA, 1998)
17. M. Pavone, G. Narzisi, G. Nicosia, Clonal selection: an immunological algorithm for global optimization over continuous spaces. J. Global Optim. **53**, 769–808 (2012)
18. A. Sharma, D. Sharma, Clonal selection algorithm for classification. in *Artificial Immune Systems (ICARIS 2011)*, ed. by P. Liò, G. Nicosia, T. Stibor. Lecture Notes in Computer Science, vol. 6825 (Springer, Berlin, 2011)
19. G.C. Silva, D. Dasgupta, A survey of recent works in artificial immune systems. in *Handbook on Computational Intelligence: Volume 2: Evolutionary Computation, Hybrid Systems, and Applications, World Scientific*, (2016), pp. 547–586
20. Y. Wang, T. Li, Local feature selection based on artificial immune system for Classification. Appl. Soft Comput. **87**, (2020)
21. A. Yaseen, R. Ali, M. Qasim Rafiq, S.M. Zakariya, Effect of varying clonal factor and number of generation on AIS based classification. in *Proceedings of the 2011 IEEE International Conference on Computational Intelligence and Computing Research (ICCIC)*, (2011), pp. 545–548
22. X. Yin, J. Han, CPAR: Classification based on predictive association rules. in *Proceedings of the 2003 SIAM International Conference on Data Mining*, (SDM'03), (2003)

New OZCZ Using OVSF Codes for CDMA-VLC Systems

Vipin Balyan

Abstract In this paper, orthogonal variable spreading factor (OVSF) codes are used for the construction of a new optical zero correlation zone (*NOZCZ*) code set QS-CDMA used for visible light communication (VLC) systems. The orthogonal properties of OVSF codes make them suitable candidates for the construction of OZCZ codes. The BER performance is assessed for different values of code weights, code length, and zero correlation zone (ZCZ). The BER performance of the proposed code has been compared with other schemes available in the literature. The BER performance of the *NOZCZ* code is also evaluated for different sample rates in the presence of delay users for a 32-user system.

Keywords Visible light communication · QS-CDMA · Optical zero correlation zone code · LED · MAI · ROC

1 Introduction

Communication using light-emitting diodes (LEDs) termed visible light communication (VLC) is gaining attention nowadays. It has low power consumption, greater security, wider bandwidth, is less complex, and has no radio frequency interference (RFI) [1–3]. Owing to the requirements which appeared in the field of communication in recent years, the VLC is the right candidate for future generation wireless communications. The research in VLC is in the early stages and needs research to address the concerning issues before getting commercialized [4]. The main challenge in any communication network is providing access to users in large numbers [5, 6]. This is, however, limited due to interference which increases with users. CDMA is used as the most promising technology to address the increase in multiple access interference (MAI) [7–9]. The work in [10] proposes a color shift keying (CSK) modulation and the VLC system using CDMA utilizes receivers of mobile phone cameras for better practicability and capacity enhancement. The experimental work

V. Balyan (✉)
Department of Electrical, Electronics and Computer Engineering, Cape Peninsula University of Technology, Cape Town, South Africa

A. K. Luhach et al. (eds.), *Second International Conference on Sustainable Technologies for Computational Intelligence*, Advances in Intelligent Systems and Computing 1235,
https://doi.org/10.1007/978-981-16-4641-6_30

in [11] uses four femtocells and provides a 3 Mb/s rate for each user using a VLC CDMA downlink system. In [12], two resource allocation schemes are used in a multi-cell system for better data rates. A single-cell CDMA VLC system in which signals with three levels of amplitude are used has a good BER [13].

A spreading code provides better performance because it has minimal length, maximum code, maximum weight, and has good auto and cross-correlation [14]. The codes which fulfil these criteria are constructed and utilized in the VLC system. For a VLC Ethernet system [15], random optical codes (ROCs) are used which are easy to implement and increase the access to the number of users simultaneously. In [16], prime code correlation properties, with higher average light intensity, and intensity fluctuations are reduced. The work in [17] uses an optical CDMA system which is using optical orthogonal codes (OOC) with balanced incomplete block design (BIBD) codes in order to support 15 users simultaneously with a data transmission rate of 200 Mb/s in the presence of normal light.

The work available in the literature is mainly on BER performance improvement of an ideal synchronous CDMA system. The work suffers from multi-path transmission or time delay due to imperfect synchronization. Therefore, OZCZ code has been used for a quasi-synchronous CDMA-VLC (QS-CDMA-VLC) system whose correlation properties are ideal in the ZCZ [14, 18]. The numerical and mathematical analysis of which suggests that unavoidable time delay can be endured by OZCZ codes. The work in [19] proposes a new approach to generate ZCC codes that reduces MAI significantly as compared to existing codes. However, these codes suffer from the introduction of lower dimming values due to the presence of too many '0's [20–22]. The work in [23] proposes a new OZCZ code set, which includes a pair of unipolar and bipolar code sets that have good correlation properties. The work in this paper utilizes OVSF codes [24–27] for the construction of new OZCZ codes, the codes in different layers of the code tree are orthogonal to each other. The generated codes keep orthogonality even when they are used after assignment. The authors of [28] also propose an efficient channel coding for VLC systems.

2 Orthogonal Codes

2.1 OVSF Codes

The OVSF codes are generated as a binary tree [27]. The spreading factor (SF) in each layer is equal to the number of codes. A code in the OVSF code tree is denoted by C_{l,n_l}, where $1 \leq l \leq L$ and $1 \leq n_l \leq 2^{l-1}$. Also, $l = 1$ denotes layer 1 and $l = L$ denotes last layer from top to bottom. Each code in a layer generates two children.

2.2 OZCZ Codes

For two codes defined as a_i and b_j of length L, $a_i = [a_{i,0}, a_{i,1} \ldots a_{i,L-1}]$ and $b_j = [b_{j,0}, b_{j,1} \ldots b_{j,L-1}]$, the periodic cross-correlation function (PCCF) of a_i and b_j is given by

$$\theta_{a_i,b_j} = \sum_{l=0}^{L-1} a_{(i,l)}, b_{(j,l+\tau) modL}, \forall\tau \geq 0 \tag{1}$$

For the periodic auto-correlation function (PACF), $a_i = b_j$.

Let $A = \{a_i\}_{i=1}^{K}(a_{i,l} \in \{0, 1\}, 0 \leq l \leq (L-1))$ be a unipolar code set and $B = \{b_j\}_{j=1}^{K}(b_{j,l} \in \{-1, 1\}, 0 \leq l \leq (L-1))$ a bipolar code set. Both A and B are K codes of length L, and the length of zero correlation zone is denoted as Z_{cz}.

The code which satisfies the following correlation properties can be represented as OZCZ code pair

$$\theta_{a_i,b_j}(\tau) = \begin{cases} wi = j, \tau = 0 \\ 0i \neq j, \tau = 0 \\ 00 < |\tau| \leq Z_{cz} \end{cases} \tag{2}$$

w denotes the number of '1's in the code.

2.3 Proposed OVSF Code Set Pair

For quasi synchronous CDMA VLC system, a new OZCZ code pair is derived from the OVSF code tree which has parameters $(L, K, Z_{cz}) = (, K, 2)$ in the following steps:

1. Let $O_{4\times4} = \begin{bmatrix} ++++ \\ ++-- \\ +-+- \\ +--+ \end{bmatrix} = \begin{bmatrix} o_1 \\ o_2 \\ o_3 \\ o_4 \end{bmatrix}$ be an orthogonal matrix derived from the second layer of the OVSF code tree, where '+' and '−' denotes '+1' and '−1', respectively.
   A new matrix O' can be constructed using any two codes of $O_{4\times4}$, i.e., o_i and o_j where $i \neq j$

$$O^{'} = \begin{bmatrix} o_i \\ o_j \end{bmatrix} \tag{3}$$

2. In order to generate an initial matrix, children of these codes are complemented for $O_{4\times4}$, i.e.,

$$H_0 = \begin{bmatrix} o_1 - o_1 \\ o_2 - o_2 \end{bmatrix} \quad (4)$$

which can be shown as four codes

$$H_0 = \begin{bmatrix} P_0 Q_0 \\ R_0 S_0 \end{bmatrix} \quad (5)$$

where $P_0 = [p_m]_{0 \le m \le 3}, Q_0 = [q_m]_{0 \le m \le 3}$, $R_0 = [r_m]_{0 \le m \le 3}$ and $S_0 = [s_m]_{0 \le m \le 3}$ of length 4.

3. Using the iteration method a code set H_n, $\forall n \ge 1$ with $K = 2^{n+1}$, $L = 2^{n+1}$ can be constructed as

$$H_n = \begin{bmatrix} P_n Q_n \\ R_n S_n \end{bmatrix} = \begin{bmatrix} E \times P_{n-1} E \times Q_{n-1} \\ F \times R_{n-1} F \times S_{n-1} \end{bmatrix} \quad (6)$$

where $E = \begin{bmatrix} C_{2,2} \\ C_{2,1} \end{bmatrix} F = \begin{bmatrix} C_{2,1} \\ C_{2,2} \end{bmatrix}$ are the two Hadamard matrices of second order. The code set H_n consists of four codes.
$P_n = [p_{i,j}]_{0 \le i \le (2^n - 1), 0 \le i \le (2^{n+2} - 1)}, Q_n = [d_{i,j}]_{0 \le i \le (2^n - 1), 0 \le i \le (2^{n+2} - 1)}$,
$R_n = [r_{i,j}]_{0 \le i \le (2^n - 1), 0 \le i \le (2^{n+2} - 1)}$ and $S_n = [s_{i,j}]_{0 \le i \le (2^n - 1), 0 \le i \le (2^{n+2} - 1)}$.

Using the properties of the Hadamard matrix, only one code exists which has a different number of $+1$ and -1 in H_n.

A set of transmitting code T and receiving code R is obtained as follows

$$NOZCZ = < R, T >$$

$$\begin{cases} R = H'_n = \{r_k\}_{k=0}^{2^{n+1}-1} \\ T = f(R) = \left\{ t_k^{d_k} \right\}_{k=0}^{2^{n+1}-1} \end{cases} \quad (7)$$

kth user original data $d_k \in \{0, 1\}$ and $f(\cdot)$ is a mapping between R and T. Also,

$$t_k^{d_k} = \frac{1 + (-1)^{d_k} r_k}{2} \quad (8)$$

The correlation properties between r_k and $t_k^{d_k}$ is

$$\theta_{r_i, t_j^{d_j}}(\tau) = \begin{cases} (-1)^{d_j} 2^{n+2} i = j, \tau = 0 \\ 0 i \ne j, \tau = 0 \\ 0 0 < |\tau| < 2 \end{cases} \quad (9)$$

$K = 2^{n+1} - 1, 0 \leq i, j \leq K, L = 2w$, and $Z_{cz} = 2$. Using the same number of iterations, it achieves a longer zero correlation zone as compared to [23].

2.4 *Example of OVSF-Based Construction*

The parameters for OVSF-OZCZ code set pair using the parameters $(L, K, Z_{cz}) = (16, 4, 2)$ by one iteration. The steps followed are defined as follows:

1. A new matrix O' with codes o_i and o_j can be derived from $O_{4\times4}$. The O' matrix is then separated into two Hadamard matrices of second order

$$O' = \begin{bmatrix} o_1 \\ o_2 \end{bmatrix} = \left[\begin{array}{cc:cc} + & + & - & - \\ + & - & + & - \end{array}\right] \tag{10}$$

2. Generate an initial matrix and generate four codes from it

$$H_0 = \begin{bmatrix} o_1 & -o_1 \\ o_2 & -o_2 \end{bmatrix} = \left[\begin{array}{cccc:cccc} + & + & - & - & - & - & + & + \\ \hdashline + & - & + & - & - & + & - & + \end{array}\right] \tag{11}$$

$$P_0 = [++--], Q_0 = [--++], R_0 = [+-+-], S_0 = [-+-+] \tag{12}$$

Taking two second-order Hadamard matrices which are generated using OVSF codes of layer 2,

$$C_{1,1} = [11] = [++] \text{ and } C_{1,2} = [1-1] = [+-]$$

$$E = \begin{bmatrix} C_{1,2} \\ C_{1,1} \end{bmatrix} = \begin{bmatrix} +- \\ ++ \end{bmatrix}, F = \begin{bmatrix} C_{1,1} \\ C_{1,2} \end{bmatrix} = \begin{bmatrix} ++ \\ +- \end{bmatrix} \tag{13}$$

$$H_1 = \begin{bmatrix} P_1 & Q_1 \\ R_1 & S_1 \end{bmatrix} = \left[\begin{array}{cccccccc:cccccccc} + & + & - & - & - & - & + & + & - & - & + & + & + & + & - & - \\ + & + & - & - & + & + & - & - & - & - & + & + & - & - & + & + \\ \hdashline + & - & + & - & + & - & + & - & - & + & - & + & - & + & - & + \\ + & - & + & - & - & + & - & + & - & + & - & + & + & - & + & - \end{array}\right] \tag{14}$$

3. The set of transmitting code T and receiving code R is obtained as follows:

$$NOZCZ =< R, T >$$

$$R = \{h_1, h_2, h_3, h_4\} = \{r_k\}_{k=0}^{3}$$

$$T = f(R) = \left\{t_v^{d_k}\right\}_{k=0}^{3}$$

$$= \left\{ \begin{array}{c} d_v = 0 \\ \begin{bmatrix} + & + & 0 & 0 & 0 & 0 & + & + & 0 & 0 & + & + & + & + & 0 & 0 \\ + & + & 0 & 0 & + & + & 0 & 0 & 0 & 0 & + & + & 0 & 0 & + & + \\ + & 0 & + & 0 & + & 0 & + & 0 & 0 & + & 0 & + & 0 & + & 0 & + \\ + & 0 & + & 0 & 0 & + & 0 & + & 0 & + & 0 & + & + & 0 & + & 0 \end{bmatrix} \\ \begin{bmatrix} 0 & 0 & + & + & + & + & 0 & 0 & + & + & 0 & 0 & 0 & 0 & + & + \\ 0 & 0 & + & + & 0 & 0 & + & + & + & + & 0 & 0 & + & + & 0 & 0 \\ 0 & + & 0 & + & 0 & + & 0 & + & + & 0 & + & 0 & + & 0 & + & 0 \\ 0 & + & 0 & + & + & 0 & + & 0 & + & 0 & + & 0 & 0 & + & 0 & + \end{bmatrix} \end{array} \right. = \tag{15}$$

$$\begin{cases} d_v = 0 \\ d_v = 1 \end{cases}$$

3 Performance Analysis

The VLC system model is given in Fig. 1. The VLC system is LED-based which is handling multiple users. Both the encoder and decoder used are non-return to zero (NRZ); the modulator and demodulator used are an on–off keying (OOK); the signals are spreaded and despreaded with the help of *NOZCZ* codes. Also, digital

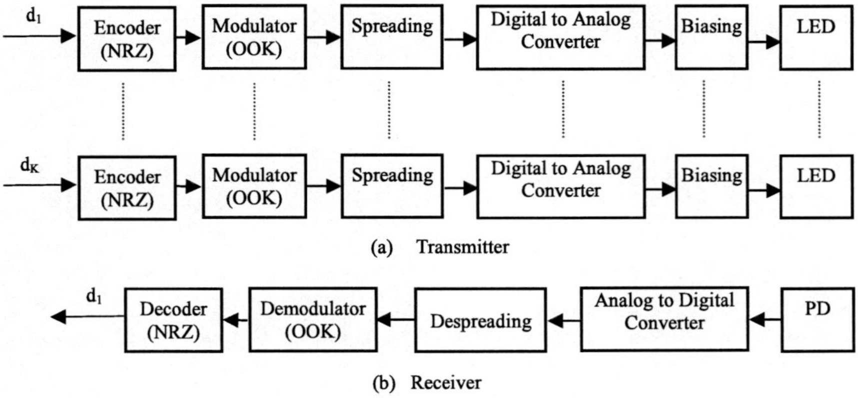

Fig. 1 VLC system model

to analog (DAC) and analog to digital (ADC) converters are used at the transmitter Fig. 1a and receiver Fig. 1b end, respectively. The transmitter also performs DC bias for generating an appropriate signal sent to LED for transmission. The line of sight (LOS) link is only considered in this paper. The transmitted optical signals pass through the optical channel received by a photodiode (PD). The PD converts them back into electrical signals. The received signal consists of both desired users, undesired user signals, and additive white Gaussian noise (AWGN). The signals from other active users produce multiple access interference (MAI). The desired user signal is despreaded using the proposed NOZCZ code and to remove the effect of MAI, followed by demodulation and decoding to retrieve the transmitted signal.

The LED broadband spectra make use of sliced wavelengths to achieve incoherent spectral coding. All active users with bit '1' occupy the wavelength for the proposed NOZCZ. The following assumptions are made for analyzing the system for K active users.

1. All active users transmit and receive with equal power.
2. The spectra of all the light sources are flat over the bandwidth $\left[f_0 - \frac{\Delta f}{2}, f_0 + \frac{\Delta f}{2}\right]$, where f_0 and Δf denote central frequency and bandwidth, respectively.
3. The first user is considered as the required user synchronized at the receiver. The remaining users are admitted with a time delay at the receiver $\tau_k = t_k T_c$, where $1 \leq k \leq K$ and $0 \leq t_k \leq Z_{cz}$. Moreover, the delay of user 1 is '0'.

At the transmitter, the data of kth user denoted as $d_k(t)$ is spreaded after encoding and modulation using proposed *NOZCZ* code. The total transmitted signals from multiple LEDs are

$$s(t) = \sum_{k=1}^{K} s_k(t) \tag{16}$$

where $s_k(t)$ denotes the transmitted signal of kth user. The symbol period $T = LT_c$, for a chip interval T_c.

Considering the spatial coherence, the root mean square value of photodetector current using Norton equivalent is given as

$$\langle i^2 \rangle = 2eI(\Delta W) + 4k_B T'(\Delta W/R_L) \tag{17}$$

where e is the charge on an electron $= 1.6 \times 10^{-19}$Coulomb.

I is the average photocurrent.

ΔW is noise in Hertz over which noise is calculated at the receiver.

k_B is Boltzmann's constant $= 1.38 \times 10^{-23}$.

T' is resistor absolute temperature in kelvins.

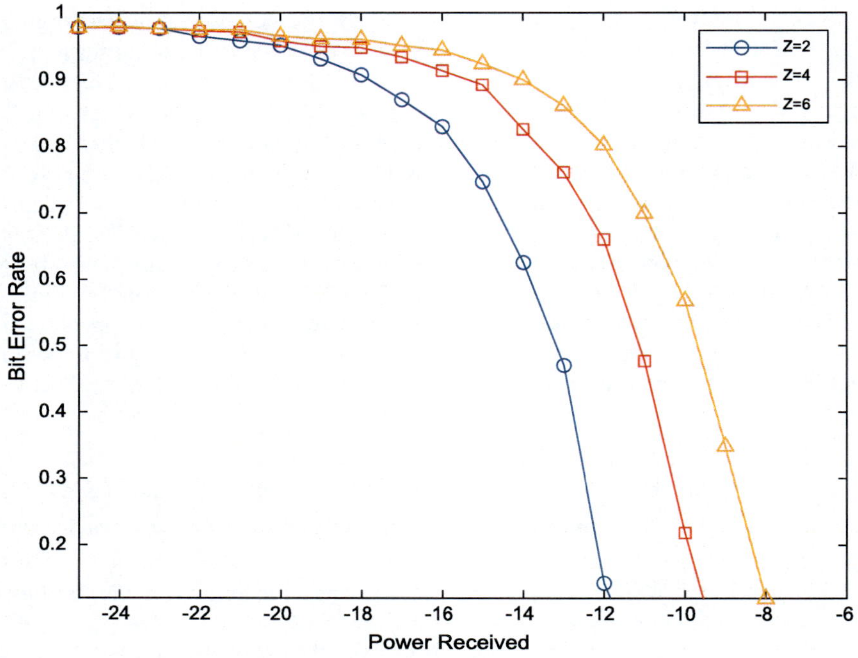

Fig. 2 System bit error rate (BER) for w = 3 and K = 4 for different zero correlation zone

R_L is the load resistor at the receiver.

The other parameters used are single-source power at the receiver of −16.77dBm, $f_0 = 480$ nm and $\Delta f = 650$ MHz.

The BERs of the system using proposed *NOZCZ* codes for $w = 3$ and $K = 4$ with different zero correlation zones $Z_{cz} = 2, 4,$ and 6 are shown in Fig. 2. The result demonstrated that increase in Z_{cz} requires higher received power for the same code weight and code length.

In Fig. 3, the BERs are compared for the proposed *NOZCZ* codes with $Z_{cz} = 2$ and $w = 3$. The increase in code length (K) allows more number of users in the system. It is evident from the plot that an increase in $K = 4, 8,$ and 12 requires more received power.

In Fig. 4, the BERs of the proposed *NOZCZ* are compared with *OZCZ* code set [14] and ZCC code set [29]. The zero correlation zone length used for comparison is 2. The ZCC code set has zero in phase cross-correlation. The *OZCZ* code set all the cross-correlation to zero. The proposed *NOZCZ* code set has all cross-correlation to zero while maintaining the orthogonality between codes after assignment to the users due to the utilization of OVSF codes which are orthogonal even with variable spreading factor.

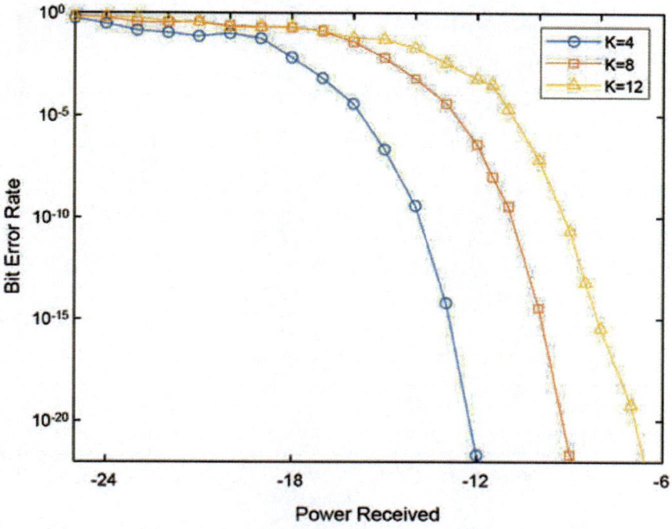

Fig. 3 System bit error rate (BER) for w = 3, $Z_{cz} = 2$ for different coding length

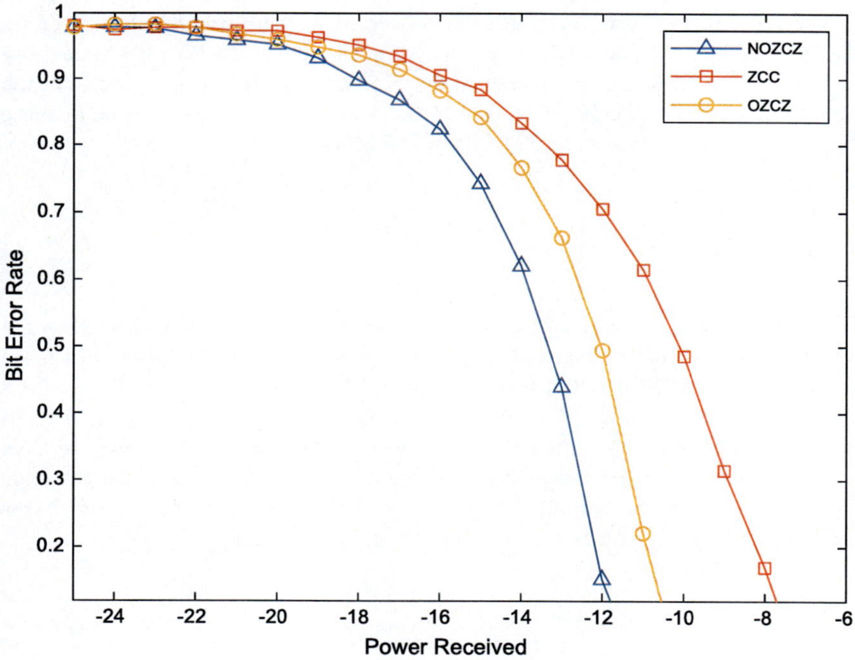

Fig. 4 Comparison of bit error rate (BER) of different schemes for w = 3, K = 4, and $Z_{cz} = 2$

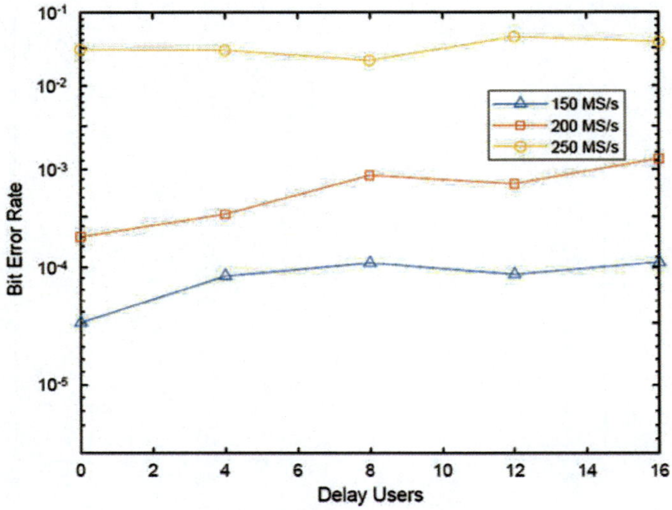

Fig. 5 Bit error rate (BER) for 32 users in the presence of delay users

The BER of the system is further investigated in the presence of 32 users in Fig. 5, taken into consideration the worst-case scenario of users having a time delay of $\tau = 1$. The BER performances for the proposed *NOZCZ* code are within the admissible range despite 16 delay users. However, BER increases with the increase in sample rate in the range of 150 MS/s to 250 MS/s.

4 Conclusion

The performance of VLC-CDMA systems adopting *OZCZ* codes is affected by the correlation properties of the used codes. In this paper, *NOZCZ* is used which are generated using OVSF codes. These are orthogonal codes that reduce multiple access interference. The proposed *NOZCZ* codes have ideal correlation properties and maintain orthogonality during transmission as they originate from OVSF codes. The BER performance is within the acceptable range for different values of code weight, code length, and ZCZ values. Variation of system rate in the presence of delay users makes BER slightly higher. However, it is still in the acceptable range.

References

1. A. Jovicic, J. Li, T. Richardson, Visible light communication: Opportunities, challenges and the path to market. IEEE Commun. Mag. (2013)
2. L. Grobe et al., High-speed visible light communication systems. IEEE Commun. Mag. (2013)
3. T. Komine, M. Nakagawa, Fundamental analysis for visible-light communication system using LED lights. IEEE Trans. Consum. Electron. (2004)
4. H. Haas, L. Yin, Y. Wang, C. Chen, What is LiFi?, J. Light. Technol. (2016)
5. H. Burchardt, N. Serafimovski, D. Tsonev, S. Videv, H. Haas, VLC: Beyond point-to-point communication, IEEE Commun. Mag. (2014)
6. H. Elgala, R. Mesleh, H. Haas, Indoor optical wireless communication: potential and state-of-the-art, IEEE Commun. Mag. (2011)
7. Y. Qiu, S. Chen, H. H. Chen, W. Meng, Visible light communications based on CDMA technology, IEEE Wirel. Commun. (2018)
8. *C. He, L.L. Yang, P. Xiao, M.A. Imran, DS-CDMA assisted visible light communications systems, in IEEE 20th International Workshop on Computer Aided Modelling and Design of Communication Links and Networks (CAMAD)* (2015)
9. H. Marshoud, P.C. Sofotasios, S. Muhaidat, G.K. Karagiannidis, Multi-user techniques in visible light communications: a survey, in *2016 International Conference on Advanced Communication Systems and Information Security, ACOSIS 2016 - Proceedings*, (2017)
10. S.H. Chen, C.W. Chow, Color-shift keying and code-division multiple-access transmission for RGB-LED visible light communications using mobile phone camera, IEEE Photonics J. (2014)
11. Z. Zheng, T. Chen, L. Liu, W. Hu, Experimental demonstration of femtocell visible light communication system employing code division multiple access, in *Conference on Optical Fiber Communication, Technical Digest Series*, (2015)
12. M. Hammouda, A.M. Vegni, J. Peissig, M. Biagi, *Resource Allocation in a Multi-Color DS-OCDMA VLC Cellular Architecture* (Express, Opt, 2018)
13. J. An, W.Y. Chung, Single cell multiple-channel VLC with 3-level amplitude-based CDMA. Opt. Commun. (2019)
14. L. Feng, J. Wang, R.Q. Hu, L. Liu, New design of optical zero correlation zone codes in quasi-synchronous VLC CDMA systems, Eurasip J. Wirel. Commun. Netw. (2015)
15. M.F. Guerra-Medina, O. González, B. Rojas-Guillama, J.A. Martín-González, F. Delgado, J. Rabadán, Ethernet-OCDMA system for multi-user visible light communications, Electron. Lett. (2012)
16. T.K. Matsushima, S. Sasaki, M. Kakuyama, S. Yamasaki, Y. Murata, A visible-light communication system using optical CDMA with inverted MPSC, in *IWSDA 2013—6th International Workshop on Signal Design and Its Applications in Communications*, (2013)
17. M. Noshad, M. Brandt-Pearce, "Application of expurgated PPM to indoor visible light communications-Part I: single-user systems. J. Light. Technol. (2014)
18. B. Fassi, A. Taleb-Ahmed, A new construction of optical zero-correlation zone codes. J. Opt. Commun. (2018)
19. M. Addad, A. Djebbari, A new code family for QS-CDMA visible light communication systems, J. Telecommun. Inf. Technol. (2018)
20. S. Rajagopal, R.D. Roberts, S.K. Lim, IEEE 802.15.7 visible light communication: Modulation schemes and dimming support. IEEE Commun. Mag. (2012)
21. F. Zafar, D. Karunatilaka, and R. Parthiban, "Dimming schemes for visible light communication: The state of research," *IEEE Wirel. Commun.*, 2015.
22. Z. Wang, W.-D. Zhong, C. Yu, J. Chen, C.P.S. Francois, W. Chen, *Performance of Dimming Control Scheme in Visible Light Communication System* (Express, Opt, 2012)
23. D. Chen, J. Wang, H. Lu, L. Feng, J. Jin, *Experimental Demonstration of Quasi-Synchronous CDMA-VLC Systems Employing a New OZCZ Code Construction* (Express, Opt, 2019)
24. V. Balyan and D. S. Saini, "Same rate and pattern recognition search of orthogonal variable spreading factor code tree for wideband code division multiple access networks," *IET Commun.*, 2014.

25. V. Balyan And D. S. Saini, “Integrating New Calls And Performance Improvement In Ovsf Based Cdma Networks,” International Journal of Computers and communications,vol. 5, no. 2, pp. 35–42, 2011.
26. D.S. Saini, V. Balyan, OVSF code slots sharing and reduction in call blocking for 3G and beyond WCDMA networks. WSEAS Trans. Commun. **11**(4), 135–146 (2012)
27. D.S. Saini, V. Balyan, An Efficient Multicode Design for Real Time QoS Support in OVSF Based CDMA Networks. Wirel. Pers. Commun. **90**(4), 1799–1810 (2016)
28. O.P. Babalola, V. Balyan, Efficient Channel Coding for Dimmable Visible Light Communications System. IEEE Access **8**, 215100–215106 (2020). https://doi.org/10.1109/ACCESS.2020.3041431
29. M. S. Anuar, S. A. Aljunid, N. M. Saad, and S. M. Hamzah, “New design of spectral amplitude coding in OCDMA with zero cross-correlation,” *Opt. Commun.*, 2009.

Statistical Study and Analysis of Polysemy Words in the Kannada Language for Various Text Processing Applications

S. B. Rajeshwari and Jagadish S. Kallimani

Abstract A language consists of *polysemy* words that have multiple meanings which have more than one sense associated with the word. Word sense disambiguation (WSD) is the process of recognizing the presence of such terms in a given input text and determining the right sense for the word in the given context. In this project, an effort is made to perform WSD for the *polysemy* words in the Kannada language. The proposed methodology employs the technique of matching the semantics of the word under consideration such as its gloss, examples, parts of speech (POS), and synsets with the target word. The semantic sense with the most number of matches with the sense of the target word would be considered as the most desirable sense of the target word. Ranks would be assigned to each of the senses. The highest rank would be assigned to such sense which would have maximum overlapping of words with the input context. The sense with the highest rank would be considered as the most appropriate sense of the polysemy word.

Keywords Kannada polysemy · Word sense disambiguation · POS tagging · Shallow parsing · IndoNet · Synsets

1 Introduction

Word sense disambiguation is the process of identifying a sense of a word in the context of its occurrence in the sentence. The word can have multiple glosses (meanings). Each gloss reflects a different sense when used along with some words. The sense of the target word (a word with multiple meanings) is dependent on its

S. B. Rajeshwari · J. S. Kallimani (✉)
Department of Computer Science and Engineering, M S Ramaiah Institute of Technology, Bangalore, India
e-mail: jagadish.k@msrit.edu

Visvesvaraya Technological University, Belagavi, Karnataka, India

S. B. Rajeshwari
e-mail: rajeshwari.sb@msrit.edu

A. K. Luhach et al. (eds.), *Second International Conference on Sustainable Technologies for Computational Intelligence*, Advances in Intelligent Systems and Computing 1235,
https://doi.org/10.1007/978-981-16-4641-6_31

surrounding words. The words around a target word are referred to as a window. The window size means the number of words around the interested word that are considered during sense disambiguation for the same.

One could similarly argue that the same technique could also be used in languages that involve abundant word set. The abundance refers to the fact that a single book or database is no good to contain information of all the words that exist or which are been used in that language. The languages consisting of a larger word base have other finer nuances which should be taken care of. Such languages might have words that have more than two senses for the same word. Generally, polysemy words on average have two senses. In the case of multiple senses that are more than two in number then other details such as parts of speech of the word, synonyms of the word, antonyms of the word, and gloss of the word should be considered to process the language.

2 Existing System

The first attempt for word sense disambiguation in the Kannada language was performed in 2011. This work inspired many to conduct experiments of NLP and construct Kannada corpora and it made use of a shallow parser and naïve Bayes classifier to disambiguate the sense of the target word [1].

The most recent work on polysemy words in the Kannada language was performed which uses a shallow parser to tag each word of the input sentence with parts of speech [2]. A database was used to store the meanings of the polysemy words. The application allowed the user to select any of the sentences with polysemy words in them to obtain the right sense of the polysemy word in the context. Using decision list, the study [3] approaches for word sense disambiguation for sentences in the Kannada language.

Here are some of the examples which are tested using Google's translation [4], an application programming interface (API) for sentences from a source language (Kannada) to the target language (English). The results seem to be quite inaccurate and unexpected.

Input sentence 1: ಎತ್ತನ್ನು ಹೊಲದ ಗುಂಡಿಯಿಂದ ಮೇಲೆ ಎತ್ತಲಾಯಿತು {Yettannu holada Gundiyinda Mele Yettalaaitu} [*An ox was lifted from a pit in a farm field*].

Google's translation output: The elevation was raised from the ground button.

Input sentence 2: ಉತ್ತರ ಕುಮಾರನು ಪರೀಕ್ಷೆಯಲ್ಲಿ ಪ್ರಶ್ನೆಗೆ ಉತ್ತರ ನೀಡಿದನು {Uttara Kumaranu Parikshеyalli Prashnege Uttara Needidanu} [*Uttara Kumara answered to a question in the exam*].

Google's translation output: North Kumar answered the question in the exam.

Input sentence 3: ಗಡಿ ಕಾಯುವ ಕಲಿಯು ತರಬೇತಿ ಶಿಬಿರದಲ್ಲಿ ರಕ್ಷಣಾ ತಂತ್ರವೊಂದನ್ನು ಕಲಿತನು {Gadi Kayuva Kaliyu Tarabeti Shibiradalli Rakshana Tantrawondannu Kalitanu}[*A soldier who looks after the border learnt a defence technique at a training camp*]

Google's translation output: Border waiting learning learned a defense technique at the training camp.

Input sentence 4: ಸಮುದ್ರಗುಪ್ತನು ದೊಡ್ಡ ರತ್ನವನ್ನು ವಿಶಾಲ ಸಮುದ್ರದಲ್ಲಿ ಹುಡುಕಲು ಹೊರಟನು {Samudraguptanu Dodda Ratnawannu Vishaala Samudradalli Hudukalu Horatanu} [*Samudragupta went in search of large gem in the vast sea*]

Google's translation output: The seafront left to find a large gem in the vast sea.

The above examples of translation by Google for source language such as Kannada involving polysemy words show that there is a lot of work that needs to be done in the field of natural language processing for regional languages.

3 Proposed System

See Fig. 1.

3.1 *The Input Module*

The input for the application would be text in the Kannada language. The application would allow providing input as a sentence. The sentence should be a valid sentence in Kannada. The validation of the input sentence in Kannada is beyond the scope of the current project. To frame the sentences in Kannada there are many online tools that would help to frame sentences in Kannada. The tools such as Baraha [5] and Virtual Kannada Keyboard [6] may help in framing the

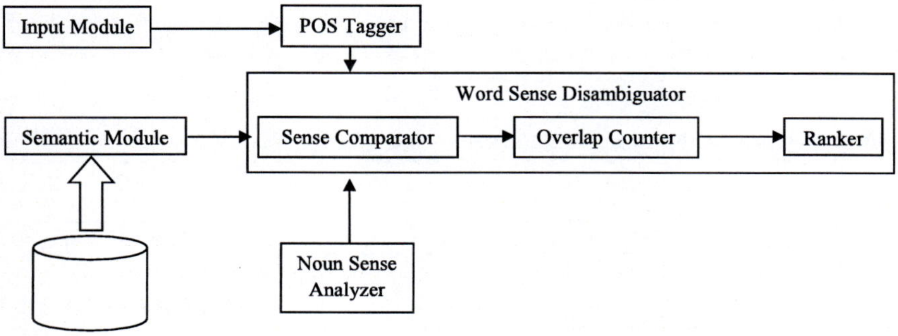

Fig. 1 The architecture for word sense disambiguation

sentences in the Kannada language. The application should be able to obtain the file and then parse the file to access the contents of the file. Consider the following example sentence containing polysemy word ಉತ್ತರ {Uttara}. ಉತ್ತರ ಕುಮಾರನು ಉತ್ತರ ದಿಕ್ಕಿನೆಡೆಗೆ ಹೊರಟನು {Uttara Kumaranu Uttara Dikkinedege Horatanu} [*Uttara Kumara went towards north direction*]. Let us consider this as an input text submitted using the input module. It means that *A person named Uttara went towards North direction.* It is a Kannada sentence that is fed as an input. The program should be able to handle sentences in the Kannada language and not English.

3.2 The POS Tagger

POS tagging is very important in NLP. It recognizes the words in the sentences and their fundamental parts. The parts refer to recognizing parts of speech for each of the words in the specified sentence. Shallow parsing [7] provides with many functionalities such as chunking, tokenizing, pruning, POS tagging, morphing and also Vibhakti computation (Table 1).

The POS tags that are tagged to each of the words for the sentence are NN and NN.

Here NN refers to the noun part of speech.

The word ಉತ್ತರ {Uttara}{*North—the direction, an answer, name of a person*} has two meanings. The meanings are as follows:

(i) A name—noun (NN).
(ii) A direction—noun (NN).

Hence it is considered a polysemy word as it has more than one meaning. Each occurrence of ಉತ್ತರ {Uttara} [*North—the direction, an answer, name of a person*] in the sentence has two different POS tags associated with the word. The first occurrence of the word ಉತ್ತರ {Uttara}[*name of a person*] refers to a name of a person which is a noun. The second instance of the word ಉತ್ತರ {Uttara}[*North—the direction*] refers to the direction, which is a noun. It could be figured that each instance has a different meaning. As the POS tagging associated each instance of the word ಉತ್ತರ {Uttara}[*North—the direction, an answer, name of a person*] occurrence it

Table 1 Kannada Words with its Parts of Speech (POS)

Word	POS
ಉತ್ತರ{Uttara}[*A person named Uttara*]	NN [*noun*]
ಕುಮಾರನು{Kumaaranu}[*The person's second name Kumaara*]	NNP [*proper noun*]
ಉತ್ತರ{Uttara}[*North*]	NN [*noun*]
ದಿಕ್ಕಿನೆಡೆಗೆ{Dikkinedege}[*towards direction*]	NN [*noun*]
ಹೊರಟನು{Horatanu}[*went*]	VM[*verb*]

is feasible to disambiguate the sense of each occurrence using the POS tagging approach.

3.3 *Identifying Polysemy Word*

The identification of a polysemy word from a given sentence involves the following steps:

(i) Tokenize each of the words in the sentence into tokens.
(ii) For each token obtain the synsets from the WordNet.
(iii) Analyze each of the semantics within the synsets for each of the tokens.
(iv) If different POS tags are associated with each of the semantics in synset, then it can be inferred that the token is a polysemy word and skip processing steps v–viii and go to step ix.
(v) If the same POS tags are associated with each of the semantics in synset then perform steps vi–viii.
(vi) Compare each of the semantics recursively within each synset.
(vii) Check for overlapping of semantics. If there is overlapping among all the semantics of synset for a token, then the token is not considered a polysemy word.
(viii) In case no overlapping of semantics is found then the corresponding semantic is considered for polysemy word.
(ix) The semantics obtained for the polysemy word identified is stored and is further compared with the contextual words to disambiguate the word sense.

The polysemy word identified would have multiple meanings. The semantics of the polysemy word would be of the same POS tags or different POS tags.

3.4 *Usage of WordNet*

A WordNet plays a critical role in natural language processing. It is very similar to a thesaurus or a dictionary for a language in the real world. It is referred to as a huge database consisting of all the required data to undertake processing on the semantics of the language. There might be multiple versions of synonyms for the same word. These versions are grouped together and are called synset [8]. The number of versions of synonyms is identified by the number of synsets for the word.

3.5 The Semantic Module

The semantic module fetches the glosses and examples for the noun-tagged instance of the polysemy from the synsets of the wordnet. Each synset contains the POS, gloss, and examples for a single instance of the word. Each row in the below table represents a synset. The semantic module fetches these details from the wordnet for the polysemy word identified in the input module. It enables to obtain all possible senses for the word ಉತ್ತರ {Uttara}[*person name, north direction, answer*] with noun-tagged POS (Table 2).

3.6 Filtering Semantics

Filtering is a process of carefully discarding irrelevant data in such a way that the essential data is retrieved and saved for further processing of the data. The filtering process also saves time for searching the required data through the huge chunks of data obtained. Filtering Talso plays a key role in determining the performance of the word sense disambiguation process. The searching time would be highly reduced, thereby increasing the turn-around time of the application.

3.7 Word Sense Disambiguator

The word sense disambiguator consists of three components:

(i) The sense comparator.
(ii) The overlap count determiner.
(iii) The ranker module.

3.7.1 The Sense Comparator

The word sense disambiguator identifies the polysemy word from the input text. Since the POS approach could not exactly identify the exact sense of the word in the context as the word which occurred twice in the sentence had the same POS tagged to the multiple occurrences of the word. Considering the word may have the same or different meanings in the sentence, the disambiguator approaches the problem using the following modules (Fig. 2).

(i) The semantic module obtains the set of words from the gloss and the examples to form a signature for a particular sense.
(ii) The input module has a set of words derived from the input text.

Table 2 Semantic module consisting of POS, Synonyms, Gloss and Example for word ಉತ್ತರ {Uttara}[*Answer, Direction*]

Sl. No.	POS	Synonyms	Gloss	Example
1	noun	ಜವಾಬು, ಮರುಮಾತು, ಮಾರ್ನುಡಿ {Javaabu, Marumaatu Maarnudi}[*Answer*]	ಯಾವುದೇ ಪ್ರಶ್ನೆ, ಸಮಸ್ಯೆಯನ್ನು ಕೇಳಿಸಿಕೊಂಡು ಅಥವಾ ಓದಿಕೊಂಡು ಅದಕ್ಕೆ ಪ್ರತಿಯಾಗಿ ಸರಿಯಾದ ಸಮಜಾಯಿಸಿ ಕೊಡುವುದು {Yaavude Prashne, Samasyeyannu Kelisikondu Athava Oodikondu Adakke Pratiyaagi Sariyaada Samajaayisi Koduvudu}[*Answering to a question, problem which is asked in form of hearing or reading*]	ನನ್ನ ಪ್ರಶ್ನೆಗೆ ನೀನು ಉತ್ತರ ಕೊಡಲಿಲ್ಲ {Nanna Prashnege Neenu Uttara Kodalilla}[*You did not answer to my question*]
2	noun	ಉತ್ತರ, ಉತ್ತರ_ದಿಕ್ಕು, ಬಡಗಣ {Uttara, Uttara Dikku, Badagana}[*north direction*]	ದಕ್ಷಿಣ ದಿಕ್ಕಿನ ಎದುರಿನ ದಿಕ್ಕು {Uttara Dikkina Yedurina Dikku}[*The direction opposite of South*]	ಭಾರತದ ಉತ್ತರ ಭಾಗದಲ್ಲಿ ಹಿಮಾಲಯ ಪರ್ವತವು ವೀರಾಜಮಾನವಾಗಿದೆ {Bhaaratada Uttara Bhaagadalli Himaalaya Parvatavu Viraajamaanavaagide }[*Himalayas is located in northen part of India*]
3	noun	ಉತ್ತರ {Uttara} [*north direction*]	ಯಾವುದೇ ಭವನ ಮುಂತಾದವುಗಳ ಮುಖ್ಯ ಪ್ರವೇಶ ದ್ವಾರ {Yaavude Bhavana Muntaadavugala Mukhya Pravesha Dwaara}[*The main entrance of a Palace*]	ಈ ಕೋಟೆಯ ಬಾಗಿಲು ಉತ್ತರ ದಿಕ್ಕಿನಲ್ಲಿ ಇದೆ {Ee Koteya Baagilu Uttara Dikkinalli Ide}[*The forts door is in the north direction*]

(continued)

Table 2 (continued)

4	noun	ಉತ್ತರ {Uttara} [*north direction*]	ಉತ್ತರ ದಿಕ್ಕಿನಲ್ಲಿ ಇರುವ ಪ್ರದೇಶ {Uttara Dikkinalli Iruva Pradesha}[*A place located in the North direction*]	ಮಹೇಶ ಭಾರತದ ಉತ್ತರದಲ್ಲಿ ಕೆಲಸ ಮಾಡುತ್ತಿದ್ದಾನೆ {Mahesha Bhaaratada Uttaradalli Kelasa Maadutiddaane}[*Mahesha is working in North India*]

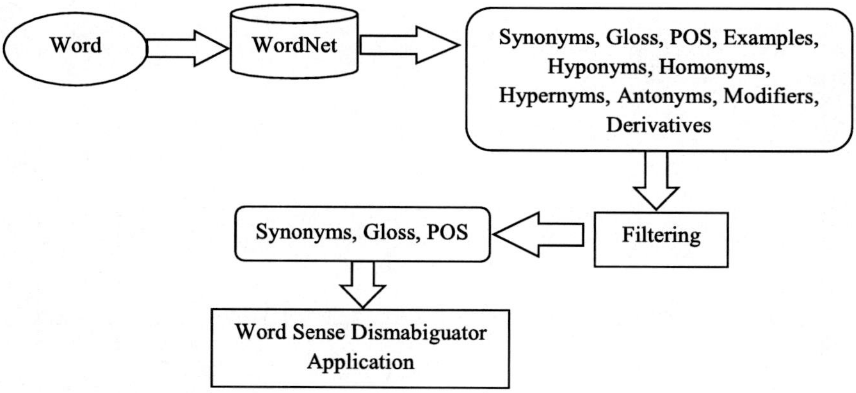

Fig. 2 Filtering process applied to discard irrelevant data from WordNet

The overview of the design of the algorithm is as follows:

```
accept input sentence s
tokenize (s) into tokens t
for each token t obtain sense g  from WordNet
          check if next sense (g+1) is available
            if g and (g+1) matches
                      continue
            end if
            else identify polysemy word as t
   end for
  for each sense g
            check if g matches with tokens in input sentence s
            if match g matches s then update count
  end for
  assign ranks in decreasing order of count obtained
  sense with highest number of matches is considered as right meaning
```

3.7.2 The Overlap Count Determiner

The disambiguator compares the words in the input module with words in each of the units from the semantic module. The words in each of the units of the semantic module comprise glosses and examples for a particular sense. The other unit if exists in the semantic module would contain the words made up of glosses and examples for other sense.

3.7.3 The Ranker Module

The ranker module comes into the frame in the scenario when there are more than one polysemy word occurring in the sentences but having different POS tagged to each instance of the target word. In this case, the overlap count determiner cannot find the overlapping count of the sense of the target word as the word sense disambiguator does not get processed at all.

3.8 Noun Sense Analyzer

The input text when consisting of nouns referring to a name of a person in regional languages such as Kannada then it would be challenging to resolve the senses of such words and determine that the sense of the word refers to a name of a person. In such scenarios, special processing of such words is required. It is because that Indian origin names for people are not named blindly or simple for the sake of doing it. The names would be named after objects such as a tree, nature, wind, water, earth, fire, mythological characters, cosmic entities such as the Sun, the Moon, a Star, or rivers, mountains, places, etc.

The following are the sub-steps for analysis:

(a) Obtain the complete word.
(b) Perform lemmatization or morphology on the word to obtain the root word.
(c) After performing morphology check if the other part associated with the root word is suffix or not. If the other constituent part is not a suffix (prefix or circumfix) then it can be concluded that the word sense does not correspond to the name of a person.
(d) If the constituent part is a suffix and conforms to the suffixes which correspond to Table 3 then the word sense refers the name of a person.

Vibhaktis are usually associated with noun forms of a word and not with verb forms of a word. For the word ಉತ್ತರ {Uttara} the multiple forms of the noun with such suffixes are in Table 3.

Table 3 denotes the suffixes (Pratyayas) associated with noun forms of a word. The noun form considered for example is ಉತ್ತರ {Uttara}[*person name, north direction, answer*].

Table 3 Noun with the suffixes for word ಉತ್ತರ {Uttara}[*person name,north direction, answer*]

Vibhakti	Vibhakti Pratyaya	Noun form
1	ಉ {u}	ಉತ್ತರನು {Uttaranu} [*Uttara*]
2	ಅನ್ನು {annu}	ಉತ್ತರನನ್ನು {Uttaranannu} [*to Uttara*]
3	ಇಂದ {inda}	ಉತ್ತರನಿಂದ {Uttaraninda} [*from Uttara*]
4	ಇಗೆ, ಕ್ಕೆ, ಗೆ {ige, kke, ge}	ಉತ್ತರನಿಗೆ {Uttaranige} [*for Uttara*]
5	ದೆಸೆಯಿಂದ {Deseyinda}	ಉತ್ತರನ ದೆಸೆಯಿಂದ {Uttarana Deseyinda} [*because of Uttara*]
6	ನ {na}	ಉತ್ತರನ {Uttarana} [*of Uttara*]
7	ಅಲ್ಲಿ {alli}	ಉತ್ತರನಲ್ಲಿ {Uttaranalli} [*in Uttara*]
8	ಏ {ae}	ಉತ್ತರನೇ {Uttaranae} [*oh Uttara!*]

4 The Implementation

The input module implemented consists of a file containing a sentence. A sentence could be any random but meaningful sentence that would be syntactically and semantically correct. The shallow parser is installed on a virtual machine on a Linux platform. The input file containing a sentence that needs to be disambiguated for the right sense is to be tagged with POS for each of the input words. To achieve this SSH (Secure Shell) network protocol [9] is used for connecting our application with the shallow parser. The input file would be remotely copied to the working directory where the shallow parser is installed using an SSH connection. The shallow parser uses a Linux environment that is able to function itself (Fig. 3).

The shallow parser also provides an efficient chunker for the Kannada language using which root words could be obtained for the tokens in the input sentence. The root word is essential in determining the polysemy word occurrence in the sentence.

The word sense disambiguation undertaken in the project has been handled for the following scenarios:

Scenario 1: The semantics for the polysemy word could be different POS tagged.

Scenario 2: The semantics for the polysemy word could be the same POS tagged.

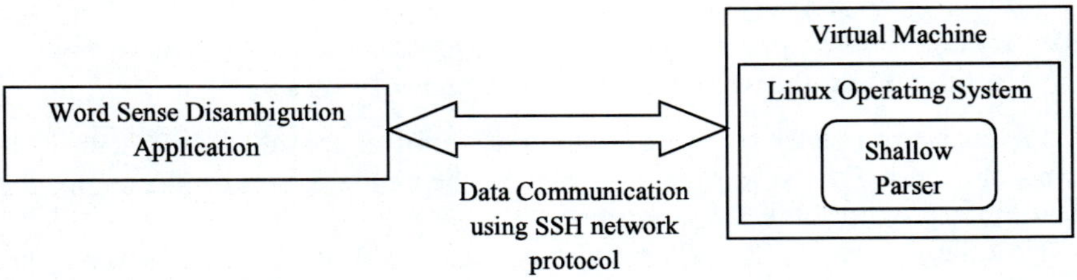

Fig. 3 Data communication using the SSH file protocol

In the case of polysemy word being different POS tagged, for instance, a noun and a verb, then the disambiguation would be straightforward as in the disambiguation semantics could be considered for noun semantics of the word or verb semantics of the word.

5 The Results

Case 1a: For the input sentence containing a single polysemy word.

The input sentence is **ಕುಮಾರನು ಪರೀಕ್ಷೆಯಲ್ಲಿ ಪ್ರಶ್ನೆಗೆ ಉತ್ತರ ನೀಡಿದನು** {Kumaaranu Pareeksheyalli Prashnege Uttara Needidanu}[*Kumara answered to a question in an examination*].

The polysemy word identified is **ಉತ್ತರ** {Uttara} [*answer, direction*] and the associated POS tag is **Noun**.

The matched word from semantics obtained which overlaps with the context of the input sentence is **ಪ್ರಶ್ನೆ** {Prashne}.

The final matched semantic associated with word ಉತ್ತರ {Uttara}[*answer, direction*] for the context of the input sentence is

ಜವಾಬು ಮಾರ್ನುಡಿ ಮರುಮಾತು ಯಾವುದೇ ಪ್ರಶ್ನೆ, ಸಮಸ್ಯೆಯನ್ನು ಕೇಳಿಸಿಕೊಂಡು ಅಥವಾ ಓದಿಕೊಂಡು ಅದಕ್ಕೆ ಪ್ರತಿಯಾಗಿ ಸರಿಯಾದ ಸಮಜಾಯಿಸಿ ಕೊಡುವುದು ನನ್ನ ಪ್ರಶ್ನೆಗೆ ನೀನು ಕೊಡಲಿಲ್ಲ {Javaabu Maarnudi Marumaatu Yaavude Prashne, Samasyeyannu Kelisikondu Athava Oodikondu Adakke Pratiyaagi Sariyaada Samajaayisi Koduvudu Nanna Prashnege Neenu Kodalilla}[*answer, reply, response, understanding or hearing to a problem and responding by providing right justification, You did not answer to my question*]

Test Case Result

The sense being matched refers to ಉತ್ತರ {Uttara}[*answer*] which is the expected outcome.

Case 1b: Input sentence containing a single occurrence of a polysemy word which is POS tagged noun.

The input sentence is **ಗಡಿ ಕಾಯುವ ಕಲಿಯು ತರಬೇತಿ ಶಿಬಿರದಲ್ಲಿ ರಕ್ಷಣಾ ತಂತ್ರವೊಂದನ್ನು ನೆರವೇರಿಸಿದನು** {Gadi Kayyuva Kaliyu Tarabeti Shibhiradalli Rakshana Tantravondannu Nereverisidanu}[*A soldier protecting the border performed a defense technique at a training camp*]

Polysemy word is ಕಲಿ {Kali}[*a soldier, to learn*].

a. **The Matched semantic polysemy word ಕಲಿ**{Kali} **is** ಸೈನಿಕ ಯೋಧ ಯೋದ್ಧಾ ರಕ್ಷಕ ರಣ ದಂಡಾಳು ಸಿಪಾಯಿ ವೀರ_ಸೈನಿಕ ಭಟ ಕಾದಾಳು ಯೋಧ ಕಂಚುಕ ಕದನವೀರ ಕಟ್ಟಾಳು ಕಾಲಾಳು ಖಡ್ಗಧಾರಿ ತಲಾರಿ ತಳವರ ತಳವಾರ ತಳಾರ ತಳ್ವಾರ್ ದಂಡಧಾರಿ ಧನುರ್ಧಾರಕ ಪರಾಕ್ರಮಿ ಪಹರೆಯವ ಪಾಯಕ ಪಾರ ಬಾಣಗಾರ ಶಿಪಾಯಿ ಸ್ಕಂದ ಹೋರಾಟಗಾರ ಹೋರು ಹೋರಾಳು ಸುಭಟ ಸಶಸ್ತ್ರಧಾರಿ ಸಿಪಾದಾರ ಸೈನ್ಯ ಅಥವಾ ರಕ್ಷಣಾ ದಳದಲ್ಲಿ ಇದ್ದು ಯುದ್ಧಮಾಡುವ ವ್ಯಕ್ತಿ ಅವನು ವೀರ ಸೈನಿಕ {Sainika Yodha Yodha Rakshaka Rana Dandaalu Sipaayi Veera-Sainika Bhata Kaadaalu Yodha Kanchuka Kadanaveera Kattaalu Kali Kalaalu Khadgadhaari Tallari Talavaara Talaara Talvaar Dandadhaari Dhanurdharaka Paraakrami Pahareyava Paayaka Paara Baanagaara Sainya Athava Rakshana Daladalli Iddu Yuddamaaduva Vyakti Avanu Veera Sainika}[*a soldier, warrior, fighter, savior, sword fighter, a person who is part of army and who fights war in the battle field*]

Test Case Result

The sense being matched refers to ಸೈನಿಕ {Sainika}[a *soldier*] which is the expected outcome.

6 Snapshots

6.1 Navigation Menu for Sense Disambiguation for Single and Double Occurrences of Polysemy Words

See Fig. 4.

6.2 Sense Disambiguation for Single Occurrence of Polysemy Word in a Sentence

See Fig. 5.

6.3 Sense Disambiguation for Double Occurrence of Polysemy Words in a Sentence

See Fig. 6.

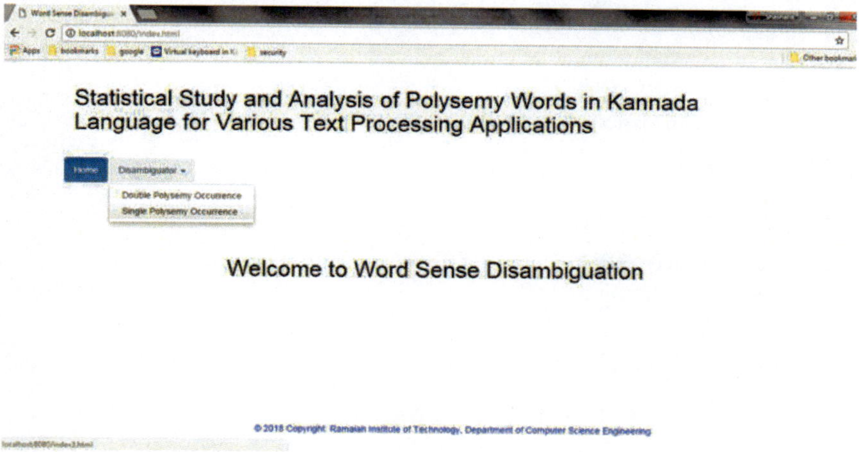

Fig. 4 Navigation menu of word sense disambiguator application

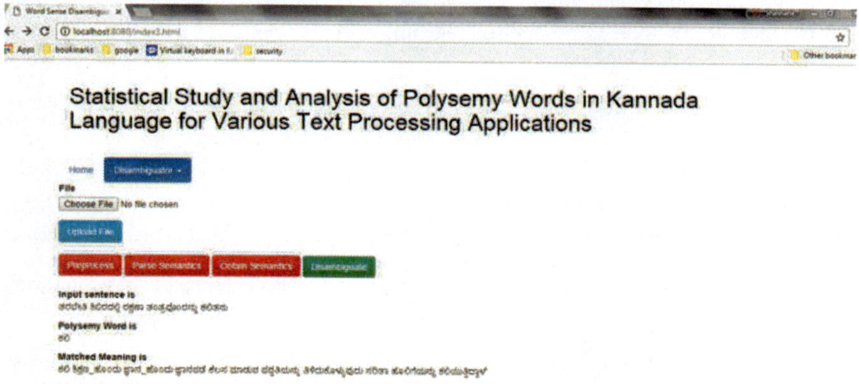

Fig. 5 The result obtained for a sentence with single polysemy word

7 Conclusion and Scope for Future Enhancement

The designed approach provides the desired accurate results. The main hindrance is that the approach is hugely dependent on the words and the gloss in the IndoNet, which is an online WordNet corpus for the Kannada language. It is necessary that the WordNet being used is huge enough to accommodate at least the most frequently used words along with its senses. This work is only the stepping stone toward word sense disambiguation for polysemy words of Kannada language in natural language processing. This is the approach that works for various text processing applications

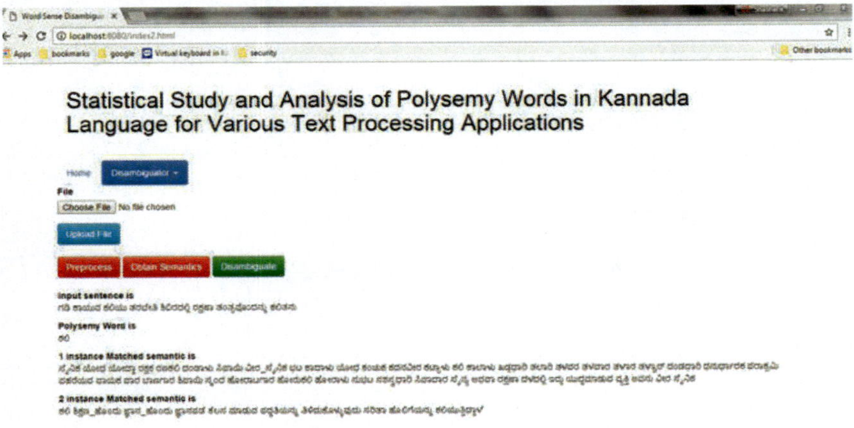

Fig. 6 The result obtained for a sentence with double polysemy words

in Kannada language. The analysis could be furthermore optimized with machine learning techniques such as supervised learning and unsupervised learning methods.

References

1. S. Parmeswarappa, V.N. Narayana, Target word sense disambiguation system for Kannada language, in *3rd International Conference on Advances in Recent Technologies in Communication and Computing (ARTCom 2011)*, (2011), pp. 269–273
2. R. Rao, J.S. Kallimani, Analysis of polysemy words in Kannada sentences based on parts of speech, in *International Conference on Advances in Computing, Communications and Informatics (ICACCI)*, (2016), pp. 500–504
3. S. Parmeswarappa, V.N. Narayana, Sense disambiguation of simple prepositions in English to Kannada machine translation, in *International Conference on Data Science & Engineering (ICDSE)*, (2012)
4. Google Translate from Kannada language to English language, https://translate.google.com
5. Baraha software for Kannada Language, www.baraha.com/index.php, 1998–2017
6. Virtual Keyboard for Kannada Language, https://gate2home.com/Kannada-Keyboard, (2018)
7. Shallow Parser Version 3.0 for Kannada language by MT-NLP Lab, Language Technologies Research Centre (LTRC), IIIT Hyderabad, (2017)
8. K. Ramakrishna, B. Padmaja Rani, D. Subrahmanyam, Information retrieval in Telugu language using synset relationships, in *15th International Conference on Advanced Computing Technologies (ICACT)*, (2013) pp. 1–6
9. SSH Communication Network Protocol, https://www.ssh.com

Author Index

A. K. Luhach et al. (eds.), *Second International Conference on Sustainable Technologies for Computational Intelligence*, Advances in Intelligent Systems and Computing 1235,
https://doi.org/10.1007/978-981-16-4641-6